THE COMPLEMENTARY NATURE

THE COMPLEMENTARY NATURE

J. A. Scott Kelso and David A. Engstrøm

A Bradford Book

The MIT Press
Cambridge, Massachusetts
London, England

© 2006 Massachusetts Institute of Technology

All rights reserved. No part of this book may be reproduced in any form by any electronic or mechanical means (including photocopying, recording, or information storage and retrieval) without permission in writing from the publisher.

MIT Press books may be purchased at special quantity discounts for business or sales promotional use. For information, please email special_sales@mitpress.mit.edu or write to Special Sales Department, The MIT Press, 55 Hayward Street, Cambridge, MA 02142.

This book was set in Stone Serif and Stone Sans on 3B2 by Asco Typesetters, Hong Kong and was printed and bound in the United States of America.

Library of Congress Cataloging-in-Publication Data

Kelso, J. A. Scott.
The complementary nature / J. A. Scott Kelso and David A. Engstrøm.
  p. cm.
"A Bradford book."
Includes bibliographical references and index.
ISBN 0-262-11291-4 (hbk. : alk. paper)
1. Brain. 2. Neuropsychology. 3. Cognitive neuroscience. I. Engstrøm, David A. II. Title.

QP376.K375 2006
153—dc22   2005052232

To our soul mates, our parents, and our children

CONTENTS

Preface	xi
Acknowledgments	xvii
PRELUDE	**1**
MOVEMENT 1 **COMPLEMENTARY PAIRS**	**17**
Contraries Are Ubiquitous	17
A Brief Trip into the History of Ideas	19
Contraries Are Complementary	35
What Is The Complementary Nature?	38
A Novel Syntax for Complementary Pairs	40
Examples of Complementary Pairs	42
The Interpretation of Complementary Pairs	49
A Philosophy of Complementary Pairs	62
Scientific Reconciliation of Complementary Pairs	63
Complementary Pairs Are Dynamical	72
Toward a Science of Complementary Pairs	75
MOVEMENT 2 **COORDINATION DYNAMICS**	**77**
Grounding the Philosophy in Science	77
A Brief Diversion into Quantum Mechanics	81

The Deep Problem of Coordination	85
Historical Roots of Coordination Dynamics	88
What Is Coordination Dynamics?	90
The Main Ideas of Coordination Dynamics	92
Foundations of Coordination Dynamics I: Self-Organizing Patterns	111
Foundations of Coordination Dynamics II: Pattern Dynamics	122
From Dynamical Systems to Coordination Law	139
How the Brain~Mind Works	142
How Brain~Minds Work Together	149
Conditions~Assumptions of Coordination Dynamics	151
A Dynamical Law of Coordination	156
Visualizing a Basic Law of Coordination	158
Mechanisms of Information Processing and Information Creation	174
A Science of Complementary Pairs and The Complementary Nature	176
Gateway to Movement 3	178

MOVEMENT 3 COMPLEMENTARY PAIR~COORDINATION DYNAMICS 183

Reconciling Philosophy and Science	183
Introduction of a Novel Philosophy~Science	187
Putting CP~CD to Work	192
Complementary Pairs of Coordination Dynamics (CP of CD)	193
Three CP of CD Vignettes	195
The CP of CD Collection	215
Coordination Dynamics of Complementary Pairs (CD of CP)	225
Three CD of CP Vignettes	227
Ways~Means of Discovering Relevant Complementary Pairs	236
The Complementary Pair Dictionary	244

A Philosophy~Science of The Complementary Nature   249
What Is The Complementary Nature?   251
End~Beginning   254

Complementary Pair Collections by Fields of Endeavor   257
The Complementary Pair Dictionary Prototype   263
Bibliography   287
Index   297

PREFACE

Scientifically speaking, nature is grounded in the laws of quantum mechanics. In quantum mechanics, a very strange situation arises. It turns out that a complete description of an atom's behavior requires both the concept of waves and that of particles. Which one is ultimately observed in empirical studies depends on the means of measurement. As the great physicist Niels Bohr taught, although waves and particles are mutually exclusive descriptions of the quantum world, they are not contradictory, but rather *complementary*. Such is the curious nature of the quantum world.

However, far removed from the vanishing dimensions of the quantum scale, our ordinary day-to-day experience of life tells us that if two descriptions of the same phenomenon are mutually exclusive, then at least one of them must be wrong. In fact, since the beginning of time, human beings have separated their life's experiences into pairs and contrived explanations for them in terms of either/or relationships. As history has proceeded, many along the way have recognized that truth seems to be less cut and dried than that, less black or white. Instead, they have held that *shades of gray* might be a more accurate description: that is, that reality lies somewhere in between opposite, polar aspects.

Be that as it may, surprisingly little language, philosophy, and science exist that adequately capture both life's myriad polar tendencies and the relationship between them. To our knowledge, there is no comprehensive, empirically based scientific theory of how the world of idealized poles and the world in between might be reconciled. But what if experiments showed that the human brain is capable of displaying two apparently contradictory, mutually exclusive behaviors at the same time? And what if the same phenomena were seen to be ubiquitous also in human behavior? What if there was a mathematically expressed scientific theory that attested directly to such complementary nature inherent in human brains and human behavior? Could this be telling us why our perception of the world appears to partition things into pairs? Could it also be telling us that there is a more enlightened way—a deeper reality that goes beyond such superficial partitioning—by which to know nature, and ourselves?

In *The Complementary Nature* the answer to these questions is a resounding yes. We discuss the history that has led to a tenable and testable scientific theory of complementary pairs and the dynamics relating them. Up to now, descriptions of complementary pairs have been predominantly metaphorical, or else they rest upon quantum mechanical interpretations of how the subatomic world behaves. A truly novel aspect of coordination dynamics is that it reconciles the usual scientific language of "states" with the novel dynamical language of "tendencies," and shows how opposing tendencies may coexist at the same time. *The Complementary Nature* offers, we think, a glimmer into the science of the in-between.

This book represents the culmination and interpretation of over 25 years of scientific research into a newer field of inquiry called coordination dynamics which is now pursued in many laboratories around the world. Coordination dynamics is a multidisciplinary approach, with theoretical and empirical facets that develop concepts of informationally based self-organizing dynamical systems chiefly in the context of the cognitive, behavioral, and brain sciences.

The success of the conceptual and methodological paradigm has been quite impressive, and has expanded into many different areas of science. More recently, coordination dynamics has even begun to interest some philosophers of mind, embodied cognition, phenomenology, neurophilosophy, and so forth. Feeling that the time was ripe for a general overview of the field, Kelso wrote *Dynamic Patterns: The Self-Organization of Brain and Behavior* (MIT Press, 1995), a slightly more technical book than the present one. And of course there have been many developments since.

The other major taproot of *The Complementary Nature* derives from a long-running dialogue between coauthors Kelso and Engstrøm on the prospects of a possible deep connection between coordination dynamics and complementary pairs. In 1991–93 and again in 1996–97 Engstrøm worked as a postdoctoral fellow of Kelso's at the Center for Complex Systems and Brain Sciences. During his first tenure, he studied dynamical transitions between reaction and anticipation in human motor behavior using the theory and methodology of coordination dynamics. During this period, the authors also discussed an apparently frequent occurrence of complementary pairs in both the theory and experiments of coordination dynamics. Some examples of the original pairs studied were reaction~anticipation, control parameter~coordination variable, individual~collective, cooperation~competition, stability~instability, qualitative~quantitative, linear~nonlinear, and perception~action. Of course, one could easily trace the importance attributed to contraries throughout the history of ideas. The fascinating prospect was that they were showing up in the center of the action here, in a tenable, testable science.

Moreover, within the context of the theory and empirical paradigm of coordination dynamics, the pairs under scrutiny happened to carry clearly defined, scientific meanings. This was a key. The authors wondered whether buried in such musings there might be some way to understand complementary pairs in general—pairs that have played such vital roles in so many philosophical traditions. The prospect was very enticing, indeed. From this original point of departure, a fascination with what seemed at first a somewhat esoteric connection between philosophy and the science of coordination began to grow. Thus was the running dialogue between the authors initiated, and the seeds of *The Complementary Nature* planted.

In the same period, as a result of determined study and computer visualization, it became apparent that there was something quite special about the *broken symmetry* version of the so-called HKB model of coordination that Kelso had developed earlier with the eminent theoretical physicist Hermann Haken, the "father of laser theory." In particular, a metastable regime of the coordination dynamics was discovered in which no stable (or unstable) equilibria were present, only transient, coexisting "tendencies"—places where the parts of the system tended to coordinate as a collective unit at the same time as they tended to function independently. In a symposium on developmental science in Stockholm in 1998, Kelso referred to this as "the Principle of the In-Between." Since then, there has been mounting evidence in the literature as well as many signs from the scientific community at large that the time has come to take coexisting tendencies and the Principle of the In-Between quite seriously. In other words, the time was ripe for the writing of *The Complementary Nature*.

We consider the work presented here pertinent to people desiring a different perspective on a world that remains almost completely dominated by a mutually exclusive either/or mentality. Five or ten minutes watching CNN or any other national news program should quickly convince anyone that some new and different approaches to reconciling diametrically opposed positions are sorely needed. This is obviously easier said than done. It is actually quite tricky, because if one attempts to repudiate either/or thinking by trying to completely invalidate it, the either/or mind-set paradoxically remains. That is, if one says, "*either* we use either/or thinking, *or* replace it with some other new, improved thinking," one hasn't escaped either/or thinking at all! Instead, what is needed is a way to put mutually exclusive either/or thinking in a more appropriate, less dominant position within a spectrum of other possible ways of thinking. *The Complementary Nature* addresses this issue directly, as its method of reconciliation includes disparate points of view rather than invalidating them, *especially* ones standing in obvious polar opposition.

The Complementary Nature not only reconciles a philosophy with a science, but makes a call for and provides ways to accomplish reconciliations among disparate fields of endeavor. It provides a way for coordination dynamics to be generally applied to advance arbitrary fields, and for arbitrary fields to make basic discoveries that could be used to advance the science of coordination dynamics. A unique feature of the material is, therefore, that it is "specifically general" and "generally specific."

The mission of this book isn't just to impart information. It isn't just written to say, "this is how it is, and isn't that interesting." It aims to engage and challenge its audience, and to provide direction for discovery and invention. It says, here is a fresh approach to life that almost any interested individual or group can study and apply to their own realms of interest and endeavor. The book provides methods, concepts, and tools to do this.

At its core, *The Complementary Nature* brings a novel scientific grounding to age-old questions that all of us ask: Which is more fundamental, nature or nurture, body or mind, whole or part, individual or collective? Here such questions are cast in a whole different light. It is contended that the way these questions are raised greatly limits their utility, because a great deal of the core essence of such dichotomized aspects seems to be located, quite literally, in what Aristotle called the "excluded middle." While a consideration of life's gray area—its ambiguities, paradoxes, complexes, mixtures, and transitions—might seem only to complicate issues, they nevertheless fill our experience and understanding of life with content, color, and flavor. It is to these wonderful difficulties that this work is dedicated. As an example, consider the following sentence found ahead:

In coordination dynamics, where apartness and togetherness coexist as a complementary pair—where a whole is a part and a part is a whole—there are no equilibria, *no fixed points at all*.

Provocative? Perhaps. But such provocations, the philosopher of science Karl Popper tells us, are necessary if science is to advance. Might this statement be of interest to fields dominated by fixed-point theorems? Might it also be of interest to brain scientists? It seems possible.

The Complementary Nature introduces a new meaning and application of the tilde, or "squiggle" character (~), as in yin~yang, body~mind, individual~collective. We think the squiggle has the potential of going the way of the @ symbol; in fact, we think the whole notion of the complementary pair has the potential of becoming a new kind of word association, like the synonym and the antonym, that may someday find its way into dictionaries. Unlike the hyphen, the squiggle does not represent a simple concatenation of words, but rather indicates the inextricable complementary relationship between them.

Thus, this novel way of writing complementary pairs could have a life of its own in contemporary usage. Actually, this practice has already begun. An example comes from the science writer and commentator John McCrone (personal communication, 2004), who has recently argued that dichotomies have great value in "opening up the mental terrain," but that we do not make best use of them if we do not model what is actually in the middle. Then, referring to the present approach, he says:

[it] breathes life back into the dichotomy by representing the opposing tendencies of autonomy and integration as a dynamic which can be tilted in either direction. The same understanding is behind familiar talk about the fruitful edge of chaos. It is not order and disorder that matter, but the stuff in-between.... A very useful notation [is suggested] to keep the dynamic nature of a dichotomy in the forefront of our minds. So they talk about determined~random, or continuous~discrete instead of the either/or slash I've been using in saying determined/random and continuous/discrete. Reality emerges within the limits set by a dichotomy rather than having to divide itself left or right to either side.

Although there is nothing special about the "edge of chaos" in the theoretical~empirical concepts contained in *The Complementary Nature*, this quote nevertheless provides a nice example of someone else who appreciates and has begun to employ the ~ character to facilitate communication of his ongoing investigations and theories concerning dichotomies.

Another more cosmetic feature of this book is that it contains little black and white stylized caricatures of historical figures, along with short quotes by each figure that tie them to the underlying themes of the book. The pictures and quotations are small and unassuming, but are intended to capture the reader's interest and enhance curiosity. A further special feature is that the book provides a nonmathematical treatment of the concepts and tools lying at the heart of coordination dynamics. This tutorial has been honed in lectures given to many audiences over the years. We hope that it will fill a crying need for those who would like to better understand these methods, and even put them to use. A more subtle feature of *The Complementary Nature* is that it is structured after its own message, and is presented in three movements: the first philosophical, the second scientific, and the third a reconciliation of the philosophy and the science.

The Complementary Nature is intended for the broadest possible audience. Complementary pairs are ubiquitous. They impinge on all aspects of the human condition. The subject matter embraces the young~old, as found in issues of development; male~female, as in issues of procreation and sexuality, chauvinism and feminism; the rich~poor, as in socioeconomic and political issues; experience~inexperience, as in educational, health, and management issues; the organism~environment, as in discussions of sustained development, deforestation, and global warming; and so on.

What seems to be needed today is a tenable, comprehensible way to reconcile polarized and conflicting mind-sets. Historically, such reconciliations have been based upon metaphors, parables, and rhetoric. *The Complementary Nature* extends such metaphorical treatments of contraries and age-old wisdom by providing a basis for them in coordination dynamics, a theory of the way human beings and human brains are coordinated.

ACKNOWLEDGMENTS

J. A. Scott Kelso wants to acknowledge the place where he comes from and everything implied by that. There is hardly a greater testament to the ubiquity of contraries than the name Londonderry: London, the capital of England, and Derry (Doire), the Celtic name for the place of the oaks. Two cultures, two religions, two sides—separate yet inextricably connected. Kelso was educated at Foyle College, named after the lovely river that flows through the town. There above all he was warned about the "the dangers in the neatness of identifications," to use that Portora boy Samuel Beckett's apt phrase. Yet when confronted with polarization Kelso turned away from it and, like many before him, left it behind. Without that rich and diverse and immensely human Scots~Irish heritage, however, Kelso would not have been drawn back to the topic of this book and the science of coordination. Whatever the drawbacks and limitations of the work presented here, he is very grateful for that. Kelso is deeply indebted to the many scientists and thinkers, old and young, dead and alive, male and female, who have contributed to the birthing and development of coordination dynamics. Their names are acknowledged specifically in the main body of the text—with sincere apologies for any omissions. Rest assured that these reflect ignorance, not intent. Kelso is also extremely grateful to his sponsors, especially to the National Institute of Mental Heath for their generous support of his research through Senior Scientist and MERIT awards.

David A. Engstrøm would like first and foremost to acknowledge his dear wife, Lene Engstrøm. Without her, this book would have been a lot less likely. In her capacity as a devoted and loving support, a sounding board, a devil's advocate, a reader, a muse, and a very, *very* patient friend and companion, she must be wholeheartedly commended. He would also like to acknowledge David Ottmar for his generous and dedicated support to the project over the years, on all fronts, physical, mental, and spiritual. In Ottmar's utter commitment to the acquisition, research into, and aesthetic appreciation of his fine artworks, he has provided a unique, almost Buddha-like example of what it means to be devoted to one's interests and endeavors in life. Last but certainly not least, Engstrøm would like to dedicate his efforts and inspiration to his friend and mentor, JASK.

Both authors are grateful to Ellen Faran, Tom Stone, Margy Avery, Sharon Deacon Warne, Matthew Abbate, Paula Woolley, Mary Reilly, and the staff at the MIT Press for their enthusiasm and dedication to this project.

THE COMPLEMENTARY NATURE

PRELUDE

POLARIZATION~RECONCILIATION

F. Scott Fitzgerald
(1896–1940)

> The test of a first rate intelligence is the ability to hold two opposed ideas in the mind at the same time and still retain the ability to function.
>
> *The Crack-Up* (1936)

The Complementary Nature (TCN) sheds new light on some ancient questions that have been pondered since the dawn of humankind: Why do we divide our world into contraries? And why do we perceive and interpret so many of life's contraries in mutually exclusive, either/or ways? Contraries are ubiquitous. They have been given many different names, such as dichotomies, duals, and polar opposites. But a rose is still a rose by any other name. Human awareness is teeming with contraries, like self~other, us~them, physical~spiritual, good~evil, friend~enemy, grief~joy, heaven~hell. Because contraries are so pervasive, it has been widely believed throughout history that understanding their nature should eventually lead to a deeper understanding of how nature works, of how *we* work. This is what TCN is about. It's about a way to understand contraries and how to reconcile them. And it's about how one might *use* such novel information productively. Why does dichotomizing seem so fundamental? The answer, we propose, lies within us, in the workings of the human brain and our new understanding of it in terms of coordination dynamics.

Although reconciliation of diametrically opposed aspects is something that seems desirable, it is much easier said than done. Consider the repeated failures to reconcile the polarized communities and cultures of Northern Ireland, of India and Pakistan, of China and Tibet, of Israel and Palestine. A question for the ages is how such contraries can be brought into harmony, and to what end. In TCN, we contend that though contraries are diametrically opposed by definition, they are nevertheless coexistent, mutually dependent and inextricable. We say that they are "complementary in nature." But how can that be? How can contraries be complementary? One might argue that it is in fact the other way around, that contraries are, well, *contrary*. Isn't that what is meant by contraries? After all, isn't it true that in some contexts, one of the two contraries, say "nature" or "nurture," is *more* basic than the other? In other circumstances, aren't they completely different things, like mind and matter, that somehow interact? In still other scenarios, aren't contraries just two aspects of a third, greater entity, in the modern lingo a kind of emergent phenomenon? Even more difficult to comprehend is how diametrically opposed aspects or ideals might somehow be brought into harmony. Trying to reconcile contraries poses many conundrums. This is what TCN is about.

We should mention here at the outset that the idea that contraries are complementary has certainly been proposed before. But in TCN, we aren't *only* going to tell you that we believe that contraries are complementary and that this complementary nature is ultimately relevant to a deeper understanding of nature itself, though we believe this to be the case. To begin with, TCN transcends metaphorically based reconciliations of "conflicting opposites" that have been frequently expressed in the history of ideas, and are still being expressed today. The reconciliation we speak of in TCN isn't purely philosophical, and it isn't based strictly in metaphorical language and rhetoric. The reconciliation we speak of in TCN is actually grounded in the principles and mathematical language of an evolving *scientific* theory called coordination dynamics. As will be shown, coordination dynamics offers a way to systematically study and understand quite general paired contraries like individual~collective, stability~instability, integration~segregation, competition~cooperation, as bona fide, experimentally testable, observable phenomena. This gives one hope that coordination dynamics may provide the scientific underpinnings of all complementary pairs in nature and that all complementary pairs possess a discoverable coordination dynamics.

Many fundamental scientific discoveries have been made via reconciliation of contraries. The reconciliation of electricity and magnetism by James Clerk Maxwell comes quickly to mind, as does Albert Einstein's reconciliation of space~time and energy~matter. Important biological things also come in pairs, like mitosis~meiosis, sperm~egg, X and Y chromosomes, and male~female. Life's genetic material, DNA, relies on complementary base pairings (adenine pairs with thymine and cytosine pairs with guanine) that form the steps of its famous double helix. In computer and informational sciences, the notion of a bit—a binary digit, the basic unit of computation—rests on a choice between the alternatives of 1 and 0. An atom can be split, but not the 1 and 0 of a bit. As these examples illustrate, TCN highlights the fact that complementary pairs play an essential role in nature and how we interpret it. As such, they form the focus of philosophical and scientific studies that seek to reconcile contraries, dichotomies, and conflicting opposites. In TCN, we draw a connection between *all* such reconciliations.

With so many religions, philosophies, and scientific discoveries critically depending upon them, one might have thought that someone must have worked out a general scientific theory for complementary pairs by now. Surprisingly enough this is not the case. A possible exception is the so-called Copenhagen interpretation of quantum mechanics, in which Danish theoretical physicist Niels Bohr and his colleagues reconciled the wave and particle theories of light. In the classical physics of everyday life, the behavior of light—as a stream of particles or undulating waves—does not depend on how one measures it. But in the vanishingly tiny world of photons and electrons, it does. Bohr called this wave/particle duality of light and matter "complementarity." By complementarity, he meant that mathematical descriptions of these two apparently contradictory aspects of light were both necessary for light's complete description. It is no secret that Bohr was fascinated by the possible connection between the concept of complementarity and Eastern philosophy, though he did not address why this might be. TCN takes Bohr's hope of establishing a more general form of complementarity and runs with it, through the emerging science of coordination dynamics and the insights it provides into the workings of the human brain and human brains working together.

As we said at the outset, the idea that contraries are essential in nature has been advanced by many great thinkers and doers. Wonderful philosophical advances and scientific discoveries have been made via the reconciliation of two polarized aspects of nature previously thought to be separate and independent of one another. So here's the mystery: Why haven't all of these many discoveries and the deep thought behind them led to a comprehensive scientific theory of complementary pairs?

In TCN, we aim to provide a way to understand the complementary nature of complementary pairs, a way to study and comprehend the entire spectrum of complementary pairs and discover how they come about, persist, and change. We do not claim that this way is the *only* way to comprehend complementary pairs. Actually, the entire conceptual and theoretical basis of TCN suggests that other ways *must* exist. And what is this way of ours? In short, it is the grounding of what we call the *philosophy of complementary pairs* in the empirically based scientific theory of coordination dynamics. Coordination dynamics embraces not only the world of idealized extremes, but also the world of reality existing in between them. TCN lays open the possibility that a reconciliation of a philosophy and a science, two unlikely bedfellows these days, will prove illuminating and useful. One potential payoff considered later is that TCN offers a fruitful and provocative scaffolding upon which to discover and invent.

Why is it so important to try to reconcile philosophy with science? Over the ages, boundaries have been created around specific fields of endeavor and philosophical outlooks that prevent people from learning from each other. Such specialization, though occasionally warranted, has in many ways impeded rather than enhanced the overall understanding of ourselves and the world we live in. In TCN, no subject is thought to be an island, complete in and of itself. One of the outcomes of the present work is that the complementary pairs found in many fields appear to share a common underlying coordination dynamics: opposing tendencies are demonstrated to shift and move, never at rest, seesawing back and forth through myriad multifunctional possibilities.

Our hope is that this book may provide an inspiration and a means for a new kind of dialogue between people in many fields of human interest and endeavor, people who share different opinions about important matters that affect us all. For example, how might the conflicting views of holists and reductionists be reconciled? How can rampant individualism be reconciled with the collective goals and needs of a caring society? How might the humanities be reconciled with the sciences? How might processes that seem diametrically opposed—like competition and cooperation—coexist? Few would argue that some kind of middle ground must exist between such contraries.

As in Aristotle's golden mean, reality must somehow fall in between such familiar and all-encompassing idealized poles as professional and amateur, expert and novice. Yet although this idea is an ancient one, surprisingly little information is available on what this middle ground consists of, and how it might feasibly work. TCN addresses this subject directly. What might an artist creating a work of art and the process of cellular replication share in common? How might the study of coordination become interesting and useful to a politician, a historian, a physicist, a pharmacologist, a molecular biologist, an artist, an activist, an anthropologist, a computer scientist, a consumer, an economist, a CEO, a sky surfer, a football player?

TCN comes in three movements. Movement 1 presents the philosophy, Movement 2 the science. Movement 3 is a reconciliation of the first two movements, a philosophy~science that is our mainspring for exploring new ideas and avenues of research. Although TCN is both philosophical and scientific in scope, it has not been written exclusively for professional scientists and philosophers. Nor has it been written exclusively for a lay audience. Like many others of the genre, it aims to provide a suitably comprehensive account in a style that engages the widest possible audience. Although TCN discusses some very recent advances in an ongoing scientific research campaign, it does so in a way that minimizes technical details and jargon as much as possible. Our hope is that reading the book will inspire people to pursue the complementary nature of their own subjects of interest and perhaps see how TCN might apply to their own lives.

MOVEMENT 1: COMPLEMENTARY PAIRS

Vincent van Gogh
(1853–1890)

So I am always between two currents of thought, first the material difficulties, turning round and round and round to make a living; and second, the study of color. I am always in hope of making a discovery there, to express the love of two lovers by a marriage of two complementary colors, their mingling and their opposition, the mysterious vibrations of kindred tones. To express the thought of a brow by the radiance of a light tone against a somber background.

letter 531, *The Complete Letters of Vincent van Gogh* (1958)

Contraries Are Ubiquitous

Movement 1 comes in three parts, one for each of three interrelated themes. In the first part, we develop the theme that contraries are ubiquitous. We are still only at the beginning of an understanding of the innermost workings of nature, and of the life that allows us the very means to wonder about it. But one thing seems clear enough. No matter where the search begins for answers to our deepest questions, all roads lead inexorably to the subject of contraries, of conflicting opposites. It is a taproot.

Not only do contraries arise in all facets of life, but it is difficult to express anything without them. Dichotomizing seems central to human cognition, one of the only ways human beings have of trying to capture reality and their own existence. Science itself operates on the methodological assumption of either versus or. Under the influence of the Viennese philosopher Karl Popper, science is said to progress through the falsifiabilty of opposing hypotheses. A good theory, according to Popper, should be bold and daring, and open itself to refutation by empirical scrutiny. Thus, to quote a popular scientific maxim, "you can't prove things in science, you can only disprove them."

Contraries have been thought about, studied, discussed, debated, implicated, worried about, forgotten, remembered again, acted on, and fought over within and around every human activity, philosophy, science, art, and craft since the beginning of human awareness. They have been expressed, albeit in different ways, by all cultures in all languages. They have played a starring role in humanity's longest-standing, toughest, and certainly hottest debates throughout history, right up to the present moment. Just think of all the fierce debates, fights, and wars centered on issues such as nature versus nurture, church versus state, order versus disorder, tradition versus progress, structure versus function, creation versus evolution, mind versus matter, stability versus instability, leader versus follower, beginning versus end, public versus private, individual versus collective, right versus wrong, us versus them, and life versus death.

Are contraries real, actual phenomena or mere mental constructs that help the human brain understand phenomena? In general, are paired aspects like genotype and phenotype, organism and environment, competition and cooperation, global and local, order and chaos, or mind and matter independent and separable, or dependent and linked? Whole religions, philosophies, and sciences are built upon very specific answers to questions such as these. TCN provides some answers too. TCN says that many answers lie in the science of coordination and its insights into the strange convergent~divergent dynamics of the human brain~mind.

Contraries Are Complementary

Our only way of avoiding the extremes of materialism and mysticism is the never ending endeavor to balance analysis and synthesis.

International Encyclopaedia of Unified Sciences (1938)

Niels Bohr
(1885–1962)

Why is it we talk about "instead of," "versus," and "rather than" most of the time? Why do we divide the world into dichotomies like genotype versus phenotype, discrete versus continuous, individual versus collective, friend versus enemy, order versus randomness, qualitative versus quantitative, internal versus external, persistence versus change, gradual versus abrupt, reductionist versus holist, and certainty versus uncertainty? The answer in each case depends on how contraries *are interpreted.* Modern science writer and commentator John McCrone (personal communication, 2004) expresses this sentiment eloquently:

We seek out what seem to be the general distinctions and then draw them out into categorical opposites. We push for the extremes of what may be the case so that we can be sure that whatever actually is the case will be caught somewhere in-between. We have the many familiar dichotomies: form/substance, random/determined, quality/quantity, continuous/discrete, vague/crisp, universal/particular, mind/matter, objective/subjective, internal/external, active/passive, stable/plastic, local/global, part/whole, nothing/everything, figure/ground, order/disorder, process/structure, etc. So the question is, what is the proper interpretation of a dichotomy?

The second theme of Movement 1 echoes the famous maxim of physicist and Nobel laureate Niels Bohr: *Contraria sunt complementa*—contraries are complementary. Contraries are not contradictory. Although Bohr himself viewed this maxim as a general epistemological position, most of his friends and colleagues saw it as a "fond hope" unlikely to be realized. Nevertheless, spurred on by the desire to reconcile apparent contraries, we make our first move: we replace all related but slightly different terms like contraries, polar opposites, duals, opposing tensions, binary oppositions, dichotomies, and the like with the all-encompassing term "complementary pairs."

From this point in the text onward, we use this term exclusively. We refer to the polarized aspects of complementary pairs as "complementary aspects." For example, "body" and "mind" are complementary aspects of the complementary pair "body~mind." We use the tilde ~ not to concatenate words or as an iconic bridge between polarized aspects, but to signify that we are discussing complementary pairs. Equally if not more important, the tilde symbolizes the *dynamic nature* of complementary pairs. As we'll show, it is not only the polar complementary aspects of complementary pairs that matter, but also all the stuff and all the action falling *in between* them.

With the tilde in hand, we provide our working definition of complementary pairs, and proceed to describe our *philosophy of complementary pairs*. "Contraries Are Complementary" concludes with a question: If contraries are indeed complementary, what does that tell us? What do we gain from this knowledge? What, in fact, is the *nature* of complementary pairs? To answer these questions, it is necessary to turn to some inevitable issues facing anyone trying to determine the nature of anything: the issues of being~becoming and of persistence~change. Where do complementary pairs come from? How do complementary aspects form, move, change, evolve, and dissolve? What do they influence and what are they influenced by? How do different complementary pairs relate and interact with each other? In short, such questions lead us to inquire into the *dynamics* of complementary pairs.

Complementary Pairs Are Dynamical

Ralph Waldo Emerson (1803–1882)

Nature is a mutable cloud, which is always and never the same.
"History" (1841)

The third and final theme of Movement 1 is that complementary pairs *move*. They persist and change, form, adapt, and dissipate—they are dynamical in nature. The dynamics of complementary pairs is quite literally *where the action is*. But where's the new message in this? Siddhartha Shakyamuni Gautama, otherwise known as the Buddha, said this 2,500 years ago. In Buddhist terminology, *skandas* come together, persist, change, and fall apart. New wine in old bottles? No. The new message has to do with how this *actually happens*. It is one thing to correctly *intuit* that nature is fundamentally dynamical, and quite another to scientifically identify the actual nature of its dynamics.

In TCN, we suggest that the limitations of many ingenious efforts to understand complementary pairs and their dynamics might be readily overcome by adopting a rather unconventional "dynamical middle-ground" approach. Plato, for instance, valorized form *over* change, and Descartes mind *over* body. One of the main messages of Movement 1 is that we cannot afford to stick to the common assumptions central to the mutually exclusive mentality that has so dominated Western science and thought, especially in the latter part of the twentieth century. How then do we propose to express this dynamic middle-ground perspective, this reconciliation of differing interpretations of complementary pairs? After taking care of a few preliminary definitions and terminological issues, *we are going to show it to you*.

We propose that a tenable middle way lies within the concepts, methods, and tools of informationally based, self-organizing dynamical systems, otherwise known as *coordination dynamics*. Scientists in this relatively new field have, over the last 25 years or so, identified key variables characterizing dynamic patterns of behavior on a number of different levels and in a number of different systems. Further, they have identified laws or regularities that underlie how these patterns evolve in time, their so-called pattern dynamics. Laws of coordination turn out to be nonlinear and context-dependent. Polarized aspects of a complementary pair appear as modes of a dynamical system that is capable of moving between boundaries even as it includes them. Coordination dynamics thus provides an opportunity to study real complementary pairs and to determine whether our interpretation of them is tenable.

Movement 1 concludes with some challenging questions. Can complementary pairs actually be explored in a nonmetaphorical, scientific manner? Can their dynamics be demonstrated in real systems? Is it something about our own brains and thought processes that makes us divide the world into polarized aspects? If so, how do we understand that? What if we were to find a way to scientifically reconcile polarized complementary aspects as well as everything in between them? What would such a science look like?

MOVEMENT 2: COORDINATION DYNAMICS

In coordination dynamics, the real-life coordination of neurons in the brain and the real-life coordinated actions of animals are cut, fundamentally, from the same dynamic cloth. Integrity is in turn preserved because it is never threatened. Psychophysical unity is undergirded at all levels by coordination dynamics.

"Preserving Integrity against Colonization" (2004)

Maxine Sheets-Johnstone (1930–)

The science that we use to ground the *philosophy of complementary pairs* is called coordination dynamics. Just as Bohr's complementarity philosophy is inextricably connected to quantum mechanics, so our philosophy of complementary pairs is inextricably tied to coordination dynamics. Coordination dynamics explains how coordination patterns form, adapt, and change through the processes of self-organization. Information that is meaningful for system function (what we call "functional information") is shown to stabilize and destabilize coordination patterns depending on context. By providing an account of how functional information is created and annihilated, coordination dynamics sheds potential light on the origins of biological agency and awareness.

In Movement 2, we elucidate the main ideas of coordination dynamics as well as how these ideas enable us to peek into, even cross the brain~mind and brain~behavior barriers. Coordination dynamics is chock full of complementary pairs both conceptual and actual: stability~instability, qualitative~quantitative, symmetry~broken symmetry, convergence~divergence, individual~collective, segregation~integration, competition~cooperation, creation~annihilation, and so forth. Coordination dynamics provides a vocabulary as well as a rich scientific basis for our *philosophy of complementary pairs*. It may also be able to explain why human beings divide their world into complementary pairs in the first place.

One reason is that the essentially nonlinear property of bistability (and in general, multistability) is a dominant, empirically observed feature of ordinary human behavior, human brains, and even the individual neurons inside brains. In terms of coordination dynamics, this means that two or more dynamically stable states of a system can coexist for exactly the same parameter values. Which one is observed depends on where the system starts from, its so-called initial conditions. By virtue of its intrinsically bi- or multistable nature, coordination dynamics thus offers a general dynamical principle through which complementary aspects, typically polar extremes, may be represented. A second reason is that in coordination dynamics transitions among stable states occur as a result of dynamic instability. Dynamic instability thus provides a universal decision-making mechanism for switching between and selection among polarized states. The third and perhaps most important reason why coordination dynamics offers insight into the complementary nature is that it provides a way to understand the gray area in between. How so?

In coordination dynamics, stable patterns or states refer to "attractors" of an underlying dynamical system. A central discovery of coordination dynamics is that not only can a system exhibit two or more possible stable coordination states, but also that two or more "tendencies" can simultaneously coexist. Such tendencies represent a crucial complementary pair: specifically, the tendency for individual coordinating elements to couple or bind together (an integrative cooperating tendency), and the tendency for them to retain and express their independence (a segregating, differentiating tendency), are present *at the same time*.

In this crucial "metastable" regime of the coordination dynamics (*meta* meaning beyond) there are no fixed points and therefore, technically speaking, no states at all, neither purely stable nor purely unstable. Let us repeat: In coordination dynamics, where apartness and togetherness coexist as a complementary pair—where a whole is a part and a part is a whole—there are no equilibria, *no fixed points at all*. Nor are the dynamics necessarily chaotic. Instead, the vast metastable world between extremes contains only *tendencies*—preferences and dispositions. In coordination dynamics, polar extremes represent ideal states of affairs while reality lies in the world in between those poles, just as Aristotle thought.

Ample evidence exists in the brain and behavioral sciences that metastability, the coexistent mixture of integrating and segregating tendencies—where individual brain regions are neither fully locked in nor fully independent—is essential to the way human brains work. In fact metastability may turn out to be *the* way brains work, and possibly many other complex organizations as well. If this is true, metastable coordination dynamics offers a new, mathematically precise principle of brain structure~function. It also means that the word *metastability* is on the rise—watch for it. Moreover, the fact that human brains can in fact work in this complementary fashion and not always in an either/or mode should be a source of inspiration to us all. This altogether new conception of the brain reconciles the two main (naturally contrary!) theories of how the brain works.

One is called the "functional segregation" or "localizationalist" theory. This theory sees the brain as a vast collection of distinct regions, each localizable and capable of performing a unique function—which happens to be partially true. The other is called the "functional integration" or "holistic" theory. It looks upon the brain not as a collection of specialized centers but as a highly integrated organ that works as a whole—which is also partially true. Metastable coordination dynamics reconciles the well-known individualistic tendencies of specialized brain regions with the tendencies of those regions to work together. Metastable coordination dynamics says apartness (functional segregation) and togetherness (functional integration) coexist in a complementary way, not as conflicting forces or theories.

In Movement 2, we hypothesize that this unique blend of integrating~segregating tendencies in the brain gives rise to, in fact *creates* new information. A key idea here is that spontaneous self-organizing coordination tendencies are the source from which the sense of biological agency springs. One might even say that coordination dynamics is offering a strong hint for how awareness of "self" could emerge from self-organization (a term that by definition paradoxically means the organization of patterns without an organizer, *without a self*).

More generally, the simultaneous ability of coordination dynamics to explain dynamic opposing tendencies (complementary aspects) as well as the creation of meaningful information together point to a scientific foundation for our *philosophy of complementary pairs*. The complementary nature is thereby seen both as the entire spectrum of coexistent dispositional tendencies (i.e., the universe of complementary pairs) and the multifunctional dynamics that guides them. Remember that this is exactly what we were looking for in Movement 1, a comprehensive scientific theory that provides an account of complementary pairs and their dynamics—a theory that captures both polarizing tendencies and all that falls in between them.

The world of coordination dynamics, a world of tendencies and dispositions, resonates strongly with our notion of complementary pairs: individual component parts coexist with the collective whole (individual~collective, part~whole). These same individual coordinative elements compete with each other to retain their autonomy while also cooperating as collectives (competition~cooperation). In the flow of the dynamics, the tendency to converge toward attractive states coexists with the tendency to diverge from them (convergence~divergence). The paths or trajectories a system follows smoothly coexist with thresholds, places where abrupt transitions occur. Qualitative changes in the form of phase transitions are produced by quantitative variation of parameters, and accompanied by quantitative consequences such as enhancement of fluctuations (qualitative~quantitative). And, of course, tendencies to integrate coexist with tendencies to segregate. Thus, coordination dynamics provides a science for some very basic complementary pairs: it explains the presence of *both* differentiating *and* integrating tendencies, and characterizes the dynamics that relates them (states~tendencies).

Contraries are all around. They are ubiquitous. Contraries are complementary, as Neils Bohr's coat-of-arms says. So far, the possible scientific basis of complementary pairs and what it means for them to be dynamical has been expressed mostly in prescientific metaphorical language or as an interpretation of the quantum. Coordination dynamics offers an interpretation of complementary pairs that is neither metaphorical nor solely quantum mechanical in origin. The complementary nature of coordination dynamics, its inherent nonlinearity that gives rise to multi- and metastability, is different but perhaps just as strange as quantum mechanics.

Coordination dynamics ties polar complementary aspects like segregation~integration to the *essentially* nonlinear property of multistability, and their mutual interplay to coexisting tendencies that arise in the metastable regime of the dynamics. It provides experimental measures of how long a system persists in a particular tendency before it changes to another, and how long it escapes for before it returns to that original tendency. Such measures are used to determine the strength of coupling among individual coordinative elements (their bonding with each other) relative to their independence (their apartness, "doing their own thing"). More generally, the presence of both convergence and divergence in the flow of the coordination dynamics attests to its truly nonstationary, transient nature, like thinking and life itself. In our view, here is where a science of the in-between puts the language of stationary states, fixed points, equilibria, as well as brain states and mental states in appropriate relief. Metastable coordination dynamics, to quote the eminent scholar and scientist Peter R. Killeen, is the "new complementarity of the twenty-first century."

MOVEMENT 3: COMPLEMENTARY PAIR~COORDINATION DYNAMICS

The mind knows exactly what is to be mixed together, what is to be kept separate, and what is to be divided off from each other.
Peri physeos (About Nature), fragment no. 12

Anaxagoras
(499–428 B.C.E.)

In the process of grounding the *philosophy of complementary pairs* in the science of coordination dynamics, we noticed that every single principle of coordination dynamics could be expressed as a complementary pair. In Movement 3 we call these the "complementary pairs of coordination dynamics." We've mentioned some of them already. Consolidating discoveries, we realized that while coordination dynamics advances our understanding of complementary pairs, complementary pairs may also advance our understanding of coordination dynamics. But what does *that* sound like? Could our *philosophy of complementary pairs* and the science of coordination dynamics be complementary aspects of a complementary pair themselves? Weird as this prospect might seem, we have no choice but to think so. Employing our syntax of complementary pairs, we write this complementary pair "complementary pair~coordination dynamics," which is the title of Movement 3. To make life a little easier, we abbreviate this term as "CP~CD."

This rather extraordinary development suggests that our philosophical reconciliation of complementary pairs, now firmly grounded in coordination dynamics, not only applies to complementary pairs under our scrutiny, but to the very process of reconciliation itself! Moreover, if coordination dynamics can successfully explain certain aspects of the dynamics of brain and behavior, then it seems entirely plausible that this "superreconciliation" of CP~CD might apply to our very means of generating theories and practices in the first place. Stop for a moment and try to imagine this: we are talking about a theory~practice that not only provides an explanation for its intended subject~object, but also explains *its ability* to explain its subject~object! As an old phrase has it, CP~CD "shines in its own light."

It is clear that in some way, human nature is nature observing itself. This involves a self-referential recursion that must somehow be drawn from the wellsprings of its own nature. CP~CD provides a way out of the infinite loop of self-reference. It is a way that is neither purely reductionist nor purely holistic, neither purely individualist nor purely collectivist, neither purely materialistic nor purely idealistic, neither purely empirical nor purely theoretical. In Movement 3, we talk about how coordination dynamics entails complementary pairs while complementary pairs entail coordination dynamics. This is the moment we reveal our theory~practice of the complementary nature, and talk about some of its novel ramifications. Movement 3 runs with the ideas that the *philosophy of complementary pairs* (CP) can be grounded in the tangible, realizable science of coordination dynamics (CD), and that coordination dynamics itself is comprised of complementary pairs that may be understood via our *philosophy of complementary pairs*. As a first step in the further development of this high-level reconciliation, we introduce two strategies to put CP~CD to work.

The "CP of CD" Strategy

The first CP~CD strategy uses the collection of complementary pairs found in coordination dynamics to advance the understanding of coordination phenomena that are not yet understood. We call this "complementary pairs of coordination dynamics," or the "CP of CD" strategy for short. For example, in the CP of CD strategy we use the complementary pairs of coordination dynamics as a context within which to understand the coordination dynamics of human learning. Research has established that how an individual human being learns depends on the relation between the learner's past experiences and prior history and the new material to be learned. This involves a subtle blend of competitive and cooperative mechanisms. In certain contexts, new information to be learned competes with the intrinsic dynamics of the learner. In others, the relationship is a cooperative one. Using the CP of CD strategy, we are able to suggest ways to improve learning in schools and to guide educational policy decisions. But the CP of CD strategy is potentially applicable in many other spheres of human activity. It is especially useful in situations where it is crucial to identify preexisting preferences and capacities, e.g., to individualize medical treatment or introduce a new form of management in an organization.

The "CD of CP" Strategy

The second strategy is called "coordination dynamics of complementary pairs," or the "CD of CP" strategy for short. The CD of CP strategy uses the concepts, methods, and tools of coordination dynamics to advance the understanding of new complementary pairs, i.e., those that are not members of the CP of CD collection. The CD of CP strategy runs on the idea that all complementary pairs entail coordination dynamics. For instance, as a starting point, one might attempt to identify a complementary pair in a field or activity outside the current sphere of coordination dynamics. Then the goal is to discover "fingerprints" of coordination dynamics in real studies of that complementary pair in that field. Such fingerprints are a strong indication that coordination dynamics can be successfully applied and used as a guide to ongoing research. In this way, the CD of CP strategy may be used to advance the understanding of complementary pairs that lie outside the range of phenomena usually treated by coordination dynamics, such as buying~selling in economics or promotor~repressor in gene regulation.

The CD of CP strategy thus offers an unconventional, entirely different way to apply coordination dynamics to different fields and levels. Of course, in the more straightforward scientific paradigm of coordination dynamics, one tries to identify not only the key pattern variables of the system of interest, but also their evolving behavior, their *pattern dynamics*. The CD of CP approach does not preclude this by now firmly established paradigm in any way. Far from it. As the coordination dynamics of new complementary pairs is revealed, the CD of CP approach strongly suggests that other, related complementary pairs will be quickly discovered. These newly discovered complementary pairs can also be elucidated by study of their related coordination dynamics. Taken to its logical end, the full landscape of coordination dynamics in a given context becomes available through a process of association, translation, and deduction.

Application of CP~CD theory, including both the CP of CD and CD of CP strategies, and of course everything theoretically lying in between them, is still in its early days. But we expect that if CP~CD is pursued in a serious and dedicated manner, it will open up new avenues of research and development, contribute to established ongoing programs of research and development, aid in the creation of new inventions, and facilitate communication between disparate fields of endeavor. Why? Because complementary pairs are ubiquitous, and their dynamics universal. They are found everywhere. They are found in all human endeavors, large~small, in all subject~objects, in all theory~experiments. With CP~CD, we have a way to understand these complementary pairs and their relation to each other. Thus, CP~CD fits the requirements for a comprehensive theory of complementary pairs, and qualifies as a candidate philosophy~science of the complementary nature.

BEGINNING~ENDING

TCN is intended for the broadest possible audience, because complementary pairs impinge on all aspects of the human condition. This is why we have tried to communicate our message in a style that is not too hard yet not too easy, not too technical yet not too superficial. We have tried to practice what we preach despite the difficulty of expressing this message in a world still almost totally dominated by either/or thinking.

We challenge our readers to set aside what they think they know or what they think is already established, and to adopt a "child mind" of openness, wonder, and curiosity, as Francis Bacon—a founder of the scientific method—would have wanted. As you journey down the path of CP~CD with us you might be surprised by the subjects and objects you encounter along the way, and by where it all ends up. But first you must begin by following us down the rabbit hole of contraries and dichotomies. Like Alice's journey into Wonderland, we guarantee that it will get *curiouser and curiouser*. As Morpheus said to Neo in the science fiction thriller *The Matrix*, "You take the blue pill, the story ends, you wake up in your bed and believe whatever you want to believe. You take the red pill, you stay in Wonderland, and I show you how deep the rabbit hole goes."

Perhaps you haven't read Lewis Carroll's *Alice in Wonderland* or seen *The Matrix*, and so aren't familiar with this "rabbit hole" metaphor. In that case, the words of Nobel Peace laureate His Holiness the XIV Dalai Lama, found in the opening remarks of his *Book of Wisdom*, can more lucidly and compassionately convey our point:

Although I speak from my own experience, I feel that no one has the right to impose his or her beliefs on another person. I will not propose to you that my way is best. The decision is up to you. If you find some point which may be suitable for you, then you can carry out experiments for yourself. If you find that it is of no use to you, then you can discard it.

MOVEMENT 1   COMPLEMENTARY PAIRS

CONTRARIES ARE UBIQUITOUS

William Blake
(1757–1827)

Without contraries no progression. Attraction and repulsion, reason and energy, love and hate are necessary to human existence.
The Marriage of Heaven and Hell (1790)

Contraries (dichotomies, binary opposites, duals, paradoxes) occur in every walk of life, at every perceivable and imaginable level. Awareness of contraries can be traced back to the very dawn of humankind. For thousands of years, human beings have been trying to comprehend nature, life, human existence, God, and countless other deep mysteries. It is in these attempts at perceiving, comprehending, interpreting, and expressing life's mysteries that contraries always seem to arise. Why contraries are so pervasive in nature is less clear. Are they purely mental phenomena, purely physical phenomena, or somehow both? Whichever, the great poet, artist, engraver William Blake hits the nail squarely on the head: contraries are necessary to human existence. The key questions for TCN are, how are we tied to contraries, how are they tied to us, and what do these ties tell us about ourselves and our world—our nature?

As a point of departure, consider the words of the eccentric and notoriously irreverent fiction writer Tom Robbins. In the following passage from *Fierce Invalids Home from Hot Climates*, Switters, one of Robbins's most paradoxical characters, ponders the many contradictions of his life and times:

To report that he was of two minds is not to imply, exactly, that he was torn by dilemma. Though hardly a stranger to contrariety, Switters had always seemed to take a both/and approach to life, as opposed to the more conventional and restrictive either/or. (To say that he took both a both/and and an either/or approach may be overstating the extent of his yin/yanginess.)

Taken together, Blake's confident epigram and Switters's unique outlook on his life choices herald our upcoming exploration of "contrariety" quite nicely. In TCN we pursue a path that is little constrained by the necessity of straight lines and neat, inflexible definitions. In fact, we actually do take *both* a both/and *and* an either/or approach to life. In this book contraries are commonplace. They are both the rule and the exception. Though hardly comprising a conventional worldview, it is the prospect of reconciling contraries and explicating the dynamics of both the both/and and the either/or (and for that matter the neither/nor) that is to be entertained in TCN.

There are, however, a few obstacles in the path of pursuing such a dramatic change to the way we think and behave. The most pervasive is the policy of basing one's actions and interpretations on a mind-set that has confounded human judgment from the most ancient days right up to the present moment. Like Switters, we refer to it as the "either/or" mind-set: the insidious and ensnaring tendency to dichotomize. As a quick illustration of its potency, from each of the following pairs of contraries, choose the one you think is more fundamental: Nature or nurture? Heart or mind? Simple or complex? Cooperation or competition? Choice or chance? Individual or society? Freedom or constraint? Life or death? Reductionism or emergentism? All questions so posed are governed by the either/or mind-set.

But perhaps you feel this was an unfair quiz. After all, you were requested to choose between alternatives, so you did. Perhaps if the choice had not been imposed in such a fashion, you might have ventured the opinion that nature and nurture are equally valid and important. Likewise, both heart and mind; both simple and complex; both cooperation and competition; both choice and chance; both individual and society, and so forth. *But would you?* Such a stance appeals to common sense, a sense of fair play, equality, and balance. But even if you are inclined toward a both/and outlook, you also realize that life isn't always like that. Somehow, despite their best intentions, people tend to behave as if one of the contraries is obviously more important and fundamental than the other.

It is often far easier to choose between ideas, valorize one policy over another, divine the better from the worse, the practical from the theoretical, than it is to see all sides of an issue. Yet, as the most common mind-set going, we are all aware of its costs and consequences. Two individuals or groups, both availing themselves of the either/or perspective, disagree about who is right, which system is good and which evil, who is the winner and who is the loser. All too often, such intransigence results in conflict and disharmony, fighting and war.

Let's be clear that we appreciate that situations and scenarios crop up all the time where it is desirable and even necessary to make a choice between alternatives. Asked if you would rather have a beer or a soda, it is perfectly appropriate to choose one or the other (or both if you happen to be thirsty!). The apparently crisp, mutually exclusive efficiency of the either/or enables us to classify objects, events, and relations in the world. Almost by definition, it dictates the way we make decisions. The problem is that if we allow it to, the urge to dichotomize can also become a troublesome and counterproductive habit, totally dominating the way we think.

TCN does not choose the both/and over the either/or any more than it chooses the either/or over the both/and. This may seem a radical and foolhardy stance, given the dominance of the either/or mind-set in everyday life, and the absolutely central role that binary oppositions have played in science and philosophy since the very beginning. But it is a position that apparently must be taken if we are to come to an understanding of how the world of idealized poles may be reconciled with the world in between, that is, the world of the complementary nature.

A BRIEF TRIP INTO THE HISTORY OF IDEAS

Heraclitus
(540–480 B.C.)

All things come into being by the conflict of opposites.
Diogenes Leartius, *Lives of the Eminent Philosophers*

That the world could not exist without the clash of opposing currents is an idea that can be traced back to Ephesus, the "birthplace of Western Philosophy," and even further back to shamanism, thought to have originated with the Tungus people of Siberia. The core belief in shamanism is that nature is divided into two coexistent worlds, the physical and the spiritual. A shaman is someone who is able to cross over from the physical to the spiritual world, acting as a coupling between them. This ancient concept of moving back and forth between a pair of contrary aspects will reappear again and again in many different guises and contexts throughout TCN. The important point here is that even ancient cultures were aware of contraries and held them in deep reverence.

The pervasive spread of shamanism had an important impact on the development of Chinese philosophy. In early China it was the emperors who were believed to mediate between the "two worlds," providing a dynamic link between heaven and earth, between spirits and humans. Although in twentieth-century China communism abolished the ancient imperial order, Chairman Mao Zedong's thinking was shaped as much by the concepts underlying ancient Chinese thought as it was by orthodox Marxist doctrine. Mao's aphoristic writing style echoed that of Confucius, and his vision of endlessly recurrent dialectical contradictions echoed the ancient conflictive harmony of yin~yang. Although the Maoist regime ruthlessly eliminated religious institutions, the ancient contraries nevertheless remained intact and are still present in contemporary Chinese thought. Indeed, Heraclitus's famous maxim remains a central ideological point for a political system that currently governs over a billion people.

Roots of Chinese Philosophy

The Tao begets one.
One begets two.
Two begets three.
And three beget the ten thousand things.
The ten thousand things carry yin and embrace yang.
They achieve harmony by combining these forces.

Tao te ching (600–400 B.C.E.)

Shamanism had another important impact on Chinese thought. It was a precursor to Taoism and Confucianism, both of which relied heavily on the principle of yin~yang. Rooted in the dynamic marriage of opposites, the principle of yin~yang is a window into the Tao, thought by Taoists to be the ineffable, ultimate source and cause of all things and events in nature. Yin is that which maintains, is immanent and enduring. Yin gravitates, contracts, moves inward. Yang is that which is generative. Yang radiates, expands, moves outward. Yin stands for togetherness. Yang stands for apartness. In both Taoism and Confucianism, all things entail the ceaseless dynamical tension of yin and yang. Important for our story, yin and yang are held to be inextricable, dynamic contraries. They are mutual, forever together. Although either the yin or the yang aspect can be individually focused upon, both are held to be equally important and fundamental.

The principle of yin~yang has become a modern worldwide icon symbolizing the fundamental conflict of opposites in nature. In his *Tao of Physics*, the theoretical physicist and mystic Fritjof Capra drew many fascinating parallels between Eastern philosophy and quantum physics:

A BRIEF TRIP INTO THE HISTORY OF IDEAS

The notion that all opposites are polar—that light and dark, winning and losing, good and evil are merely different aspects of the same phenomenon—is one of the basic principles of the eastern way of life. Since all opposites are interdependent their conflict can never result in the total victory of one side, but will always be a manifestation of the interplay between two sides.

Similarly, look at what Gary Zukav says in his very similar book called *The Dancing Wu Li Masters*:

At the subatomic level there is no longer a clear distinction between what is and what happens, between the actor and the action. At the subatomic level the dancer and dance are one.

However, although both of these authors lent provocative and insightful commentaries on the possible connections of ancient Eastern philosophy and modern physics, neither actually addressed the basic issue that concerns us, namely *why* such a connection should exist in the first place, and how it might come about.

Creator~Destroyer, Light~Darkness

Sri Bharati Krishna Tirtha (1884–1960)

In our culture there is no undue stress laid on matters purely spiritual or purely temporal. We regard life as one composite whole. And the purely spiritual and purely temporal are terms that express only parts of one unity.

Man cannot have distinctions between religion and science. As I have said, the two are inexorably intertwined. Matter is the medium through which the Spirit manifests itself.

Vedic Mathematics (1965)

The idea that all things come into being via dynamic pairs of contraries exists in all ancient world religions, philosophies, and ideologies. In Hinduism, all things spiritual and physical alike are thought to be aspects of Brahman, the Absolute. The Hindu Trimurti, which stands at the head of the pantheon of Hindu gods, is comprised of Brahma, the creator god, with Vishnu the preserver and Shiva the destroyer (whose dynamic balance Brahma embodies). Later we will show how creation~destruction arises naturally as a complementary pair in coordination dynamics.

Still another ancient example of Heraclitus's maxim was the conflict of good and evil that formed the creation myth of Zoroastrian religion, around 1000 B.C.E. This ultimate conflict of opposites—the conflict between God and Satan—was an early theme in many world religions and remains an abiding force today. For the Manicheans, a sect founded in Persia in the second century, the universe was thought to be the result of two supernatural forces: *Light*, which is the origin of all that is good, and *Darkness*, which is the origin of all that is evil. Prophets, sent to liberate the light imprisoned in corrupt matter, acted as mediators and messengers between the two worlds. Once again, we note the recurring theme: two poles and their dynamic mediation.

Dialectic~Contraries

How singular is the thing called pleasure, and how curiously related to pain, which might be thought to be the opposite of it, for they are never present to a man at the same instant, and yet he who pursues either is generally compelled to take the other; their bodies are two, but they are joined in a singular head.

from Plato, *Phaedo*

Socrates
(469–399 B.C.E.)

There is always one in every crowd: the one who compulsively brings up controversial topics; the rabble-rouser who wants to get to the bottom of the issue; the one who usually starts the argument, and never, ever abandons it. This kind of person likes to pose a question, questions the answers given to the question, and then questions the answers to that, too. The person who gave this "arguable" attribute a name was none other than the Greek philosopher Socrates. He called it "dialectic."

Socrates didn't only argue for argument's sake (by all accounts he did that too), but attempted to use dialectic as a way to explore and discover the universe, as an instrument of science. It was his way to acquire knowledge and gain wisdom. Like a marble sculptor or a cook peeling an onion, Socrates, and all dialecticians following him, tried to achieve enlightenment by chipping or peeling away falsity from truth, separating the relevant and irrelevant, the wheat from the chaff.

To Socrates, truth was truth, to be acquired through the process of question and answer. What was true was desirable and what was false must be eliminated. And if you disagreed with Socrates and his immutable truth du jour, you were headed for conflict. Not surprisingly, Socrates engaged in plenty of arguments. Indeed, he thrived on them, enough to finally get himself arrested for "infecting the minds of the young" with his controversial ideas.

Like both Jesus and Mahatma Gandhi, Socrates stood so firmly behind his principles and methods that it ultimately led to his demise. Using the very dialectic that got him in trouble with the authorities in the first place, he used his now immortalized trial as a perfect arena to practice what he preached while being tried for preaching it. Of course, this wasn't the easiest row to hoe. Sticking with the truths uncovered by means of his dialectic, Socrates elected death over banishment (so the story goes), drank the fated hemlock, and the rest is history. Game over for the controversial inquisitor!

Ironically, Socrates' concept of dialectic fared much better than its originator. Not only did dialectic survive his trial and execution, it flourished and evolved through several transformations over the next two millennia. The first such transformation was supplied by his most famous pupil, Plato. After Plato, Aristotle, Immanuel Kant, and Georg Hegel all proposed nontrivial updates and novel interpretations. The common thread among all conceptions of the dialectic is the principle of opposition, bringing us back to the bottom line: despite fundamental differences between oriental and occidental philosophies, both traditions agree about the central role of contraries. That East and West have been dichotomized into conflicting opposites only seems to vindicate their deeper connection!

Aspirant~Mentor

You cannot conceive the many without the one.
R. S. Brumbaugh, *Plato on the One* (1961)

Plato
(428–348 B.C.E.)

Unfortunately for history, Socrates didn't leave any notes of his own. What we know about his philosophy comes from the writings of his student, Plato. But Plato wasn't cut out simply to be Socrates' secretary and public voice. Plato moved forward, constructing a more comprehensive philosophical system than his mentor. Plato's thought is expressed in his dialogues, stories in which Socrates figures as the main character. This was quite a brazen maneuver, considering that Socrates was put to death for *his* ideas. Yet Plato not only publicly communicated dialectic; in the later dialogues he actually appropriated the character of Socrates as a mouthpiece for his own ideas! Plato was neither put on trial nor banished. On the contrary, he founded his Academy, which has been regarded by many as the first university and which flourished for some 900 years. Pretty impressive considering that the founder of that Academy was the star pupil of a condemned heretic.

Form~Change, Being~Becoming

Contraries are central to Platonic philosophy. Plato, however, was less interested in the process of reconciliation of contraries than in the evolution and transcendence of mind, a theme that repeats itself in the ideas of Georg Hegel and the entire Neoplatonist tradition. In his works the *Republic* and *Phaedo*, Plato postulated his theory of Forms. Forms, also called Ideas, were thought to be immutable archetypes from which the world of the senses derived its meaning. Forms were Plato's equivalent of Socrates' ultimate Truth. They were something to be aspired to, a universal perfection that could never actually be reached.

Whereas Socrates considered humanity for the most part to be living in ignorance, Plato believed that human beings inhabited a world of illusions. The seeker of truth must proceed from the "shadows" toward the "light," evolving from the imperfect imitation toward the perfect, ideal Form. This process of gaining enlightenment was suggested in Plato's famous metaphor of the cave. In the so-called uninformed state (i.e., while residing in the "cave of ignorance"), humanity is surrounded by a world of shadows or illusions. To experience the real world of Forms, we must struggle out of the dark and into the light.

Like his mentor Socrates, Plato believed dialectic was a supreme science. One proceeded by constantly questioning assumptions and by explaining a specific idea in terms of a more general one until the ultimate ground of explanation—the Form, the Idea, the universal truth—was reached. Unfortunately, the complementary nature was obscured from both Plato and Socrates by the "shadow" of the either/or habit. For Plato, being was *more fundamental* than becoming, stasis *more fundamental* than change. The immutable soul was *more fundamental* than the life he directly and daily experienced. Although Plato used contraries in his dialectic, his preference for one side over another may have kept him ultimately a prisoner in his own cave.

Thesis~Antithesis

Georg Wilhelm Friedrich Hegel (1770–1831)

To him who looks upon the world rationally, the world in its turn presents a rational aspect. The relationship is mutual.

The Philosophy of History (1837)

For Hegel, the road to truth lay in the dynamic synthesis of thesis and antithesis, a beautiful pair of contraries. Out of the conflict of thesis and antithesis comes synthesis, said Hegel (though he didn't use these particular words himself). Synthesis in turn becomes a new thesis for further progressions, ultimately culminating in an absolute or perfect idea or form. Once again, this progression, this idea of a natural evolution toward a more comprehensive and perfect idea or form, seemed to be much more important for Hegel than the particulars of the contraries themselves—dynamic contraries were a means to an end. Importantly, Hegel intended his ideas to apply not only to abstract thought but also to the material universe. For Hegel, the evolution of nature itself was based on an eternal, multidimensional progression of conflicting and synthesizing opposites!

Another essential element of Hegel's system was his belief that reality can only be grasped when examined as a whole and that any attempt to discover truth by scrutinizing a specific facet of reality was doomed to failure. By choosing generality over specificity, whole over part, even a philosopher of Hegel's caliber was not totally immune to the either/or habit.

Collective~Individual

Natural science will in time incorporate into itself the science of man, just as the science of man will incorporate into itself natural science: there will be one science.
Economic and Philosophic Manuscripts of 1844

Karl Marx
(1818–1883)

Although Marx based his philosophy of history on Hegel's dialectic, he considered his own interpretation not only different, but in direct opposition to it. In his most famous work, *The Communist Manifesto,* written with his friend and sponsor Friedrich Engels in 1848, Marx used Hegel's philosophy to formulate the notion of class struggle. Whereas Hegel thought that the real world was only the external, phenomenal form of an Idea, Marx's view was that the Idea is nothing else than the material world reflected by the human mind and translated into forms of thought—contrary positions. While commending Hegel's conception of a dynamic progression of thesis, antithesis, and synthesis, Marx criticized Hegelian dialectic as being "upside down," namely in its misplaced emphasis on the *rational* and *theoretical* over the *empirical* and *material* (note the contraries). Historical progression for Marx was to be accounted for exclusively in terms of dialectically conflicting material forces and social classes:

> The collisions between the classes of the old society further in many ways the development of the proletariat.... The proletariat of each country naturally must first settle its accounts with its own bourgeoisie ... and in the different evolutionary stages which the struggle between the proletariat and the bourgeoisie must pass through they [the communists] always represent the interests of the whole.

Like Hegel, Marx viewed history as ultimately a one-way progression, working itself toward a definite end. For Marx's "dialectical materialism," however, it was continual revolution that was eventually destined to bring about a final peaceful, classless society. Marx's interpretation of Hegelian dialectic assumed that everything is materially derived and that change takes place through the struggle of opposites:

> Thus in bourgeois society the past dominates the present; in communist society the present dominates the past. In bourgeois society capital is independent and personal, whilst the living individual is dependent and deprived of personality.

Marx held that this conflict of opposing forces leads to growth, change, and development, according to definite laws—a kind of self-movement. Dialectical materialism is deterministic. Because it was assumed that history inevitably follows certain laws, and that individuals have little or no influence on its development, the collective was taken by Marx to be *more fundamental* than the individual. The biased conclusion of the polarizing either/or mind-set strikes again!

Ubiquitous Contraries in the History of Science

In the scientific revolution of the sixteenth and seventeenth centuries, thinkers like Francis Bacon and René Descartes were among the first to favor the practice of hypothesis, direct observation, and meticulous documentation over more speculative modes of inquiry into nature. As radical a departure from the status quo of their time as these practices were, contraries remained in the center of the action, right on cue: Descartes preferred hypothesis *over* observation; for Bacon and Isaac Newton, observation took priority *over* hypothesis (well exemplified in Newton's famous dictum, *Hypotheses non fingo*). And on it went.

Most of the great debates that raged throughout the scientific revolution have been drawn along the lines of conflicting opposites: authority versus novelty, induction versus deduction, hypothesis versus observation, empiricism versus rationalism, creation versus evolution. Over and over in the history of science, contraries have continued to sit at the center of the action. For example, even though

Darwin has been dead for over a century, arguments between evolutionists and creationists linger on—more forcefully than ever, it seems.

Of course, since the time of the scientific revolution, the scientific method has been codified as the sine qua non of knowledge acquisition. Purely metaphorical speculation and description has given way to precise experimental observation, hypothesis testing, and mathematical prediction. This revolutionary change in the way knowledge is pursued was largely a product of another famous pair of conflicting opposites, rationalism and empiricism. Rationalism is the theory that *reason* is itself a source of knowledge superior to and independent of sense perceptions. Famous early rationalists were René Descartes, Baruch Spinoza, and Gottfried Leibniz. Empiricism is the theory that knowledge originates exclusively as a result of one's experience based on sensory perceptions. Famous early empiricists were John Locke, George Berkeley, and David Hume. Despite Immanuel Kant's attempted tour de force of reconciling rationalism and empiricism, the division between the two still exists, and can still be felt today. Even though it is widely understood that both theory and experiment are crucial to good science, there remains a certain uneasy tension between theorists and experimentalists in many fields, especially in those outside physics. And a good deal of turf-guarding goes on between the sciences. There is the story of the theoretical physicist who turned his hand to biology, and suggested some key experiments to a biologist colleague. The blunt reaction was that he should do them himself. *Not too neighborly!*

Physics: Relativity~Quantum Mechanics

Even if there is only one possible unified theory, it is just a set of rules and equations. What is it that breathes fire into the equations and makes a universe for them to describe?

A Brief History of Time (1988)

Stephen Hawking
(1942–)

Have the twentieth century's crowning scientific achievements of relativity and quantum mechanics finally eliminated Heraclitus's ancient weather vane and replaced it with something more tangible, seamless, and certain? Not even close. To begin with, both of the theories are themselves composed of reconciliations of contraries (energy~matter, time~space, particle~wave, position~momentum, etc.). Perhaps even more compelling is the fact that these two most successful theories yet conceived—relativity, the master theory of the macrocosm, and quantum mechanics, the master theory of the microcosm—are themselves still considered conflicting opposites to this day! Relativity is continuous and smooth, while quantum mechanics is discontinuous and jumpy. Witness the following excerpt from a summary of Stephen Hawking's *A Brief History of Time*. As you read it, keep "all things come into being by a conflict of opposites" in mind:

> After all, physical science can now be reduced to two theories that together can explain almost everything: the theory of relativity and the theory of quantum mechanics. The problem is that the two theories are incompatible with each other where they meet. Relativity accurately explains large scale physical phenomena, and quantum mechanics explains things on the tiniest atomic scale, but the two theories can't be used in tandem with any accuracy. Though Hawking hasn't yet formulated a unifying theory, he does have a strategy for how he'll do it. He has found places in the universe—black holes—*where the two incompatible theories literally collide* [italics ours].

Evolutionary Biology: Continuous~Discrete

Charles Darwin
(1809–1882)

But as my conclusions have lately been much misrepresented, and it has been stated that I attribute the modification of species exclusively to natural selection, I may be permitted to remark that in the first edition of this work, and subsequently, I placed in a most conspicuous position—namely, at the close of the Introduction—the following words: "I am convinced that natural selection has been the main but not the exclusive means of modification."

Autobiography (1876)

In 1972, Niles Eldredge and the late Stephen Jay Gould introduced the term "punctuated equilibria" into evolutionary biology. Actually, the concept that evolution displays "saltatory jumps" in which nongradual speciation alternates with longer gradual periods of relative stasis has been around for quite a while, even before the days of Charles Darwin. Darwin, of course, was a determined gradualist, viewing new species as strictly arising from gradual, steady transformations of populations. Even his friend T. H. Huxley expressed concern about that. "You have loaded yourself with an unnecessary difficulty in adopting '*Natura non facit saltum*,'" he once wrote to Darwin.

Scholars who have studied Darwin's writings, however, have actually sensed an emerging pluralism in his views. By the fourth edition of the *Origin of Species* Darwin acknowledged that it is probable that each form remains unaltered for a long time and then undergoes modification. Consequently, care needs to be exercised in the use of the term "gradual." Depending on the time scale of various observations, one scientist's "saltatory" may be another's "gradual." Aside from the scientific evidence that is still in considerable dispute, suggestions have even been made that punctuated equilibrium theory is politically motivated along the lines of Marxist views of historical change. Many others use one of the major criticisms of Darwinian evolution, the lack of evidence in the fossil record, as a scientifically verifiable plug for creationism.

Polemically vigorous responses from opponents and adherents of punctuated equilibrium rage to the present day. For us, however, polarized theories like gradualism versus punctuated equilibrium return us inevitably to a central issue concerning the nature of change: Does life *fundamentally* proceed by long, continuous, and smooth change or fast, discrete, and jumpy change? As advocates of the complementary nature, we contend that both are equally valid and fundamental. For instance, whether one calls something a structure or a function depends greatly on the scale of magnitude at which it is observed. A bone looks like a static structure, but that's only because it's changing on a slow time scale relative to who's observing it.

Molecular Biology: Genotype~Phenotype, DNA~Protein

The hypothesis we are suggesting is that the template [for genetic replication] is the pattern of bases formed by one chain of the deoxyribonucleic acid and that the gene contains a complementary pair of such templates.

Nature (May 30, 1953)

James Watson
(1928–)
and
Francis Crick
(1916–2004)

In 1962, Francis Crick and James Watson won the Nobel Prize for discovering the double helix structure of DNA, surely the icon of twentieth-century biology. The beautiful structure they revealed is a biological example par excellence of dancing contraries: two spiraling strands of DNA in which each nucleotide base in one strand is paired with its complementary partner on the opposite strand (adenine~thymine; guanine~cytosine). The molecular complementarity of the double helix is astonishing and profound, and its discovery has transformed all of biology, technology, and medicine. Among other miracles, it means that for any part of the nucleotide sequence, one strand can act as a template for the other strand.

Complementary base pairing is thus central to the fidelity of copying genetic material, as well as to other processes such as the synthesis of messenger RNA and DNA repair.

Crick and Watson are not shrinking violets. Unlike Albert Einstein, who at least in his later years valued above all his "apartness" in which to work on his unifying physical theory, Crick and Watson have been totally immersed social warriors, always appearing in the fray of science, rallying on one side or another of different scientific debates. Watson, for example, recently declared that our fate, once thought to be in the stars, can now be thought to rest in our genes. For Watson, diseases as well as traits are caused by corruption of the double helix. In the last few years, however, it has become apparent that most genetic diseases and disorders are not caused by aberration of a single gene. Instead they usually result from changes in multiple genes, and are thus mediated via dynamic alterations of gene-gene interactions. Thus, the *same* disorder can be the result of *different* combinations of genetic predispositions, conditioned or "tuned" by environmental factors, and vice versa. Such interactions are called "epigenetic," a term first coined by Conrad Waddington 50 years ago to help explain the development of a complex organism from its simpler cellular beginnings.

Waddington's epigenetic hypothesis was that although the overall complement of genes stays the same, they are switched on and off differently to make the different cells of the body. In other words, *patterns* of gene expression define each cell type, not just the genes themselves. Now, DNA in organisms with nuclei is coated with at least an equal mass of protein, creating a complex called chromatin. Both DNA and chromatin protein (not just DNA itself) are, at least in living cells, partners in the control of gene activity. Both significantly influence and determine the expression of traits and diseases. Does this sound familiar?

The late Francis Crick—the other half of the twosome—was more circumspect. In his famous little book *Of Molecules and Men*, Crick notes in a single paragraph that the organism is an "open system," meaning that it interacts with its environment, exchanging energy, matter, and information with its surroundings. This, he says, is the minimum requirement for life. But amazingly enough, that is all he says on this issue. The rest of his excellent book is devoted to the fact that organisms possess genetic material that allows them to reproduce and pass on "copies" of themselves to their descendants. Natural selection, as it were, does the rest.

Ironically, it remains unknown to this day how natural selection has actually led to the production of a gene. In fact, Darwinian selection *presupposes* the existence of self-sustaining structures like the gene. The bottom line is that for Crick and especially for Watson, life is gene-centric, a one-way central dogma. The double helix, with its deeply beautiful complementarity, should be a poster child of the complementary nature. Unfortunately, the gene has instead become a symbol of a one-sided, polarizing, and technologically centered dogma in biology today. Have we gone too far in saying this? We don't think so. Let us give the last word on this topic to the eminent Berkeley biologist emeritus Richard C. Strohman:

> The Watson-Crick era, which began as a narrowly defined and proper theory and paradigm of the gene, has mistakenly evolved into a revived and thoroughly molecular form of genetic determinism.... Molecular geneticists are finding genetic databases insufficient to explain function either in development or in evolution.... At the same time, human medical geneticists are working on the disease side to show that genetic predisposition, to the extent that it exists, is dependent on (all?) other genes AND on the natural life history of the individual. We just don't understand how it all comes together.

These wise words anticipate coordination dynamics, a theory that seeks to understand how things come together and split apart in living things—not only at the genomic level, but at *all* relevant levels.

Computer Science: Logical~Physical

Science is a differential equation. Religion is a boundary condition.
from J. D. Barrow, *Theories of Everything* (1991)

Alan Turing
(1912–1954)

Few figures have so changed the course of twentieth-century history and yet been so little known by the public as Alan Turing. Way back in 1936 Turing conceived of the most general kind of machine that could deal with symbols, the so-called universal Turing machine. This was the forerunner of the all-important personal computer that is probably sitting on your desk right now. The universal Turing machine was a "machine" that, through the simple operations of reading, writing, erasing, and shifting through a sequence of ones and zeroes on a tape, could compute any number normally encountered in mathematics. Turing's inspiration was the human mind, each "state" of mind of the human computer being represented by a configuration of the corresponding machine.

The Turing machine connected rather than reconciled abstract symbols and the physical world. For Turing, only logical patterns of states mattered. The conceptual universal Turing machine was thus independent of the laws and constraints of physics. Turing mused that whatever a human brain does—all that we think of as human—it does so because of its logical structure, not because of what generates that logical structure inside the head. Turing reckoned that this logical structure could just as easily be carried out on some other piece of physical machinery. All that mattered was the "program," a set of logical instructions provided to the machine to compute whatever a person wanted it to compute. This remains a provocative yet necessarily limited theory. Why is that?

Because in the physical universe, logical structure does not hover in a vacuum. Even in the very simplest sense, every logical operation a computer accomplishes costs energy. Moreover, with respect to computers and brains, a key point is that just because an algorithm can be written to simulate what an organism or a brain does, it doesn't necessarily follow that biology uses such algorithms. In fact nature's self-organizing systems don't necessarily need explicit instructions (programs) to carry out their pattern-forming functions.

Ironically, Turing himself had something very important to say about this alternative to programs written by programmers that prescribe desired outcomes. In 1952, which happened to be the same year that Watson and Crick made their discovery of the double helix, Turing made a profound inquiry into how biological matter manages to self-assemble patterns and structures, like the fivefold symmetry of the starfish, the pattern of stripes on a tiger, and the dappling patterns on the coat of a deer. How could such recognizable, repeatable patterns arise and take shape across literally millions of cells? This is the problem of embryonic form, of "morphogenesis."

Turing showed that a mixture of chemical substances, merely by reacting and diffusing with each other, could settle into recognizable macroscopic patterns (reaction~diffusion). The chemical balance between reaction and diffusion becomes unstable due to the presence of a catalyst or the effect of temperature, and "reaction~diffusion" patterns arise spontaneously. This is a beautiful early example of simple self-organization, namely spontaneous pattern formation achieved without any specific instructions, algorithms, or programs. Such patterns seem to put themselves together. Later work by scientists such as Lewis Wolpert and Brian Goodwin showed that dynamic concentrations of various chemicals called morphogens can actually switch genes on and off to effect cell differentiation and to provide cells with positional information, like whether they are going to participate in forming a head or a tail, etc.

So Turing's brain held keys that unlocked the secrets of two vital yet apparently opposite ways that nature might process and use information in order to do work. On the one hand, the tortured genius came up with the idea of a universal programmable computing machine based purely on logic; on the other, he made fundamental contributions to the physico-chemical basis of self-organized biological morphogenesis. Following the themes of TCN, it might occur to you that this would put Turing in the perfect position to make a rather profound reconciliation. But were the "two sides" of Turing actually reconciled? Unfortunately not. As his biographer Andrew Hodges cogently comments in *Alan Turing: The Enigma* (1983), Turing himself could never reconcile the logical and the physical: "Thinking and doing; the logical and the physical; it was the problem of his theory, and the problem of his life."

Turing's creative work seems almost to cry out for the concept of complementarity as espoused, for example, by the eminent physical biologist Howard Pattee:

Complementarity ... is a sharpening of the paradox. Both modes of description [here the *logical, symbolic* and the *rate-dependent, dynamic*] though formally incompatible, must be a part of the theory, and the truth is discovered by studying the interplay of opposites.

What might Turing have accomplished had he been able to reconcile his two stunning scientific achievements? Perhaps if society hadn't imposed such a blatantly judgmental either/or attitude toward his personal tastes and habits, we might have found out: shortly after his famous paper on morphogenesis appeared, he was arrested and tried for "gross homosexual indecency" and then subjected to "scientifically based" treatments (chemical castration) designed to cure his "problem." Two years later, at the age of only 41, Turing was discovered dead with a cyanide-laced apple by his bedside.

Ubiquitous Are Contraries

We contend as many others have before us that the entire history of ideas may be usefully comprehended in terms of the many ingenious attempts to understand specific pairs of opposites and to illuminate their nature. The more pressing point, though, is that contraries are influencing everything and everybody *right now*, including we the authors and you the reader. Contraries are ubiquitous. They are everywhere. Contraries impinge on all aspects of our lives, scientific and otherwise.

A century has passed since Emma Goldman's famous lecture in Philadelphia in 1904, but her words still ring true:

The problem that confronts us today, and which the nearest future is to solve, is how to be one's self and yet in oneness with others, to feel deeply with all human beings and still retain one's own characteristic qualities. This seems to me the basis upon which the mass and the individual, the true democrat and the true individuality, man and woman, can meet without antagonism and opposition.

Efforts toward understanding contraries as a subject in themselves, which have been pursued at least by the philosophers for centuries, have been largely overshadowed. These days, performance is often more important than ideas, that is, *doing* is deemed more important than *thinking*. The explosion of the information age, with its barrage of multimedia, rapid evolution of computers, Internet, and telecommunications, color-screened digital camera MP3-playing mobile phones, as well as the futuristic advances of medicine and molecular biology and so many other technological areas, continues to distract us from this ancient and abiding theme. And yet, what all of this technological advance hasn't done is eliminate the importance of contraries, the ubiquitous predisposition of both nature itself and human nature to perceive and express dynamic pairs of opposites.

Contraries have been important in the past, are no less important now, and we expect them to continue to be important in the future. It is vital to understand them better. Despite their unmistakable ubiquity and their dramatic, essential role in all facets of nature and our explanations of nature, human beings still lack a comprehensive understanding of how and why nature deals in contraries, how "things come into being by conflicting opposites." Should we, for instance, really describe our current age as a "digital age"? Although the means might be digital, numerical, precise, speedy, and efficient, the desired ends are all too analog: we use our insidiously flexible digital devices predominantly to process and produce analog, wavy, colorful, imperfect, *sensual* output.

As a final demonstration of how contraries are so central to humanity's notions about the workings of the world and life itself, we present a small historical survey in the form of a table (following pages). In this table, famous personalities from the history of ideas are listed chronologically, along with their fields of expertise and an example of complementary pairs crucial to their work. Of course, this table is very far from being comprehensive. That would constitute a reference book of its own! From the names and fields presented, though, the theme that contraries pervade the history of ideas is clear enough. Finding more examples to add to this list waits only for time and inclination.

Contraries, as the table illustrates, are not exclusive to mental or physical arenas, nor confined to the level of atomic structure or any other specific level. Indeed, they are present and pertinent at every scale of magnitude or level of consideration. No matter what we experience or think about, contraries are present, whether one wishes to account for them or not. By virtue of their ubiquity, paired contraries are *everyone's* business. And how one interprets them becomes paramount to the future successes~failures of humanity.

CONTRARIES ARE COMPLEMENTARY

Contraria sunt complementa
Maxim on Bohr's coat of arms, 1947

Niels Bohr
(1885–1962)

The Frederiksborg Castle in Hillerød, Denmark, is old and beautiful. There is plenty to see inside. One of its inner attractions is the castle's chapel, where about six centuries' worth of historical shields with coats of arms painted on them line the walls. As you peruse the history of Denmark shield by shield, you eventually come upon one that seems a little out of place. The prominent feature of this coat of arms is a blue and red yin~yang symbol. "Hmm, that's odd," you think, "what is a blue and red chinese yin~yang symbol doing on a Danish shield hanging in the middle of a Christian chapel?"

Upon closer examination, you would find that the shield is in honor of the world-famous theoretical physicist Niels Henrik David Bohr. Bohr was knighted into the prestigious Order of the Elephant in 1947 for "outstanding achievements in science and important contributions to Danish cultural life." The Order of the Elephant is usually bestowed upon members of the Danish royal family or foreign heads of state. Bohr was one of only three Danish commoners in the twentieth century to receive this honor. This aside, you might still wonder: What is a blue and red yin~yang doing on the coat of arms of an atomic physicist? Wouldn't a cartoon of, say, Bohr's famous model of the atom be a more appropriate adornment?

It gets stranger still when you read the coat of arms' motto: *Contraria sunt complementa*—contraries are complementary. It's a riddle, you surmise, an ironic statement perhaps? Perhaps it is an inside joke? No, it is none of the above. It is actually a straight-shooting motto behind Bohr's basic interpretation of light and matter called the "Copenhagen interpretation of quantum mechanics." Together with some of his illustrious colleagues such as Erwin Schrödinger, Wolfgang Pauli, and Werner Heisenberg, Bohr proposed an explanation of the dual nature of light called complementarity. Complementarity is defined as the "complementary relationship of electromagnetic wave and electromagnetic particle theories in explaining the dual character of light and other quantized radiation."

Table 1 Complementary pairs found in diverse fields of endeavor and the people who explored and elucidated them. This table can be extended to any field of interest and endeavor.

Dates	Thinker	Field	Contraries
540–480 B.C.E.	Heraclitus	philosophy	permanence~change
470–399 B.C.E.	Socrates	philosophy	question~answer
428–348 B.C.E.	Plato	philosophy	form~ideal; being~becoming
400–320 B.C.E.	Lao Tzu	philosophy	yin~yang
384–322 B.C.E	Aristotle	philosophy	cause~effect
350–260 B.C.E	Sakyamuni Buddha	religion	happiness~suffering
6 B.C.E.–30	Jesus	religion	friend~enemy; love~hate
1225–1274	Aquinas, Thomas	theology	faith~reason
1564–1642	Galilei, Galileo	cosmology	science~religion
1564–1616	Shakespeare, William	literature	comedy~tragedy
1596–1650	Descartes, René	philosophy	*res cogitans*~*res extensa*
1642–1727	Newton, Isaac	physics	terrestrial~celestial
1724–1804	Kant, Immanuel	philosophy	rationalism~empiricism
1770–1831	Hegel, Georg	philosophy	thesis~antithesis
1809–1882	Darwin, Charles	biology	natural selection~random mutation; variation~invariance
1815–1872	Lovelace, Ada Byron	computer science	hardware~software
1818–1883	Marx, Karl	philosophy	proletariat~bourgeoisie
1831–1879	Maxwell, James	physics	electricity~magnetism
1857–1952	Sherrington, Charles	physiology	reflex~synergy
1869–1948	Gandhi, Mohandas	law	discipline~disobedience
1875–1961	Jung, Carl	psychiatry	conscious~unconscious
1875–1968	Dale, Henry	pharmacology	transmitter~receptor
1879–1955	Einstein, Albert	physics	time~space; energy~matter
1883–1946	Keynes, John	economics	central controls~laissez-faire
1885–1962	Bohr, Niels	physics	wave~particle; discrete~continuous
1896–1980	Piaget, Jean	psychology	accommodation~assimilation
1901–1976	Heisenberg, Werner	physics	certainty~uncertainty
1901–1978	Mead, Margaret	anthropology	primitive~civilized
1902–1984	Dirac, Paul	physics	electron~positron
1904–1989	Dalí, Salvador	art	foreground~background
1904–1979	Gibson, James	psychology	organism~environment; affordance~effectivity

Table 1 (continued)

Dates	Thinker	Field	Contraries
1905–1980	Sartre, Jean-Paul	philosophy	essence~existence; *pour-soi~en-soi*
1908–1986	Beauvoir, Simone de	literature	immanence~transcendence
1910–1976	Monod, Jacques	biology	chance~necessity (le hasard~la nécessité)
1911–	Wheeler, John	cosmology	gravity~radiation
1911–2003	Katz, Bernard	physiology	graded potential~action potential
1912–1954	Turing, Alan	computation	logic~dynamics; reaction~diffusion
1916–2004	Crick, Francis	molecular biology	metabolism~replication
1917–2003	Prigogine, Ilya	chemistry	equilibrium~nonequilibrium; being~becoming
1918–1988	Feynman, Richard	physics	particle~antiparticle
1926–	Haken, Hermann	physics	order parameter~control parameter; slow~fast variables
1926–	Pattee, Howard	biology	symbol~dynamics
1928–	Chomsky, Noam	linguistics	competence~performance
1929–	Gell-Mann, Murray	physics	simplicity~complexity
1929–	Edelman, Gerald	biology	selection~instruction
1941–2002	Gould, Stephen	biology	gradualism~saltationism

In an attempt to express the essence of this mathematically rigorous, physical theory of the subatomic world, a great physicist and deep thinker chose an ancient Chinese religious icon symbolizing the dynamic, universal marriage of opposites. The appearance of the yin~yang symbol as part of a Danish coat of arms is quite extraordinary. Obviously, Bohr was compelled by some profound relationship he saw between the ancient principle of yin~yang and his modern principle of complementarity.

A possibly deep relationship between Taoism and quantum mechanics was raised in the 1970s in such popular science works as Fritjof Capra's *The Tao of Physics* and Gary Zukav's *The Dancing Wu Li Masters*. These works were originally intended to spark an interest (controversy?) within the scientific community on the intriguing possibility that a nontrivial relationship existed between modern physics and ancient Chinese philosophy. Though the two sets of principles were separated by over two thousand years, both seem to be attempts to discover the secrets of nature, and, in some mysterious way, both were actually reaching similar conclusions.

The Tao of Physics and *The Dancing Wu Li Masters* remain very popular, well-written, and informative works, and have sold millions of copies. However, it seems fair to say that this unconventional reconciliation of ancient Chinese philosophy and modern quantum mechanics did not strike its intended chord among the greater professional scientific community. A curiosity for the serious scientist or philosopher, perhaps, and a fun read, but not much more. And yet this exact relationship is implied on the shield symbolizing the work of one of the greatest scientists of the twentieth century, the man who invented complementarity. Could there be more behind Bohr's choice of icon than some intriguing but loose metaphorical connection?

Although complementarity *by definition* is applicable only in the vanishingly small dimensions of quantized particle~waves, the principle of yin~yang is tacitly assumed to be level-independent. It is assumed to be applicable everywhere, all around, at all levels and dimensions—ubiquitous. Note that Bohr's shield doesn't say "really, really tiny contraries are complementary." It says "contraries are complementary." Could there be another, as yet undiscovered scientific basis for complementary contraries in nature that is mathematically rigorous, scientifically sound, and experimentally testable, but that is not confined to the realm of the very microscopic? This is what we are after. How and where should we look for it? Bohr's maxim that "contraries are complementary" and Heraclitus's maxim that "all things come into being via the conflict of opposites" give us two great clues. Maybe they are telling us that we might do well to focus on trying to develop a scale-independent science of contraries and their dynamics.

WHAT IS THE COMPLEMENTARY NATURE?

Kahlil Gibran
(1883–1931)

Some of you say, "joy is greater than sorrow," and others say, "Nay, sorrow is the greater." But I say unto you, they are inseparable. Together they come, and when one sits alone with you at your board, remember the other is asleep upon your bed.

The Prophet (1923)

WHAT IS THE COMPLEMENTARY NATURE?

If nature is defined as the collected set of principles responsible for the genesis, existence, and evolution of the universe as we know it, then in the simplest sense "the complementary nature" is meant to signify that these same principles are complementary in scope:

complementary   1: of, relating to, or suggestive of complementing, completing, or perfecting; 2: mutually dependent, supplementing and being supplemented in return.

Weaving these definitions of "complementary" and "nature" together gives us a definition of "the complementary nature" that we can use both as a reference and as a point of departure:

the complementary nature   a set of mutually dependent principles responsible for the genesis, existence, and evolution of the universe relating to or suggestive of complementing, completing, or perfecting relationships and being complemented in return.

This initial definition is intended to express the idea that nature is complementary in scope at all levels and at all scales, in all of its uncountable forms and forces, large and small, actual and virtual, living and dead. We may then describe our quest as the search for a comprehensive scientific theory of the complementary nature. When setting out on such a quest, history teaches us that one of the main prerequisites is the (at least temporary) setting aside of some basic assumptions. For this reason as well as others that will become clearer as we go, from now on we use the generic term "complementary pairs" in place of all the diverse yet related terms such as contraries, dichotomies, binary oppositions, dyads, opposing tensions, and the like.

We do this to remind ourselves that contraries are complementary, and to refrain from the habit of assuming that "to be contrary is to be divided." Further, we refer to the contraries themselves, like body and mind, sensory and motor, time and space, public and private, friend and enemy, as "complementary aspects." By themselves, each complementary aspect is insufficient to capture the complementary nature of that particular complementary pair. Rather, both complementary aspects are needed for an adequate, abiding description. Unlike Bohr's complementarity, though, we desire our theory not to be bounded by a particular scale. And unlike the connection drawn by Capra and Bohr between complementarity and the Tao, we desire our theory of the complementary nature to transcend metaphor. For that new science is needed.

We use complementary pairs as a window into the complementary nature, as a means to study and understand it: the complementary nature is most likely to be revealed via exposure to, experience with, contemplation and expression of complementary pairs. Many, possibly most complementary pairs are so familiar that they have become virtually invisible to our everyday perceptions and actions, like breathing (inhalation~exhalation, inspiration~expiration) and walking (standing~falling, stance~transfer). In fact, when we look into the roots~branches of any human activity, complementary pairs are always discovered there. One of the main aims of this book is to awaken a diverse, cross-discipline, cross-agency, cross-cultural interest in complementary pairs and their dynamics, and thereby to foster a newfound respect for and interest in the complementary nature.

A NOVEL SYNTAX FOR COMPLEMENTARY PAIRS

It behooves us to find some way to keep the inextricable aspects and dynamic nature of complementary pairs in the forefront of our minds. As a beginning, let us introduce a way to write complementary pairs that helps to capture their essence and facilitate their interpretation. We use the tilde, or "squiggle" character (~), as part of a novel syntax for writing complementary pairs. For example, we write the complementary aspects "wave" and "particle" as the complementary pair "wave~particle"; the complementary aspects "mind" and "body" are written as the complementary pair "mind~body"; the complementary aspects "supply" and "demand" are written as the complementary pair "supply~demand," and so on. We use this notation whenever we are speaking of and writing about complementary pairs, whenever we are trying to understand the complementary nature. Although it constitutes only a tiny addition to contemporary parlance, the squiggle may make all the difference when one is thinking, talking, and writing about complementary pairs.

Now, why might this small graphical device be important? That is, why are we so particular about how we write down complementary pairs? It is because a major difficulty in perceiving and appreciating the complementary nature is its familiar invisibility. Normal, everyday language usually carries with it an embedded and potentially damaging bias toward mutually exclusive thinking. As words themselves, complementary aspects (e.g., body, mind) are easily separable. The squiggle, however, helps one not to forget that complementary aspects are inextricably linked, even though each aspect retains its singular character. It is easier to separate the *words* "body" and "mind," "sensory" and "motor," than it is to actually separate them in a human being.

Luckily, it hasn't been necessary to change the language that much. We have simply assigned a novel meaning to the ~ character. But can a single novel meaning attributed to a text symbol really make that much difference in our lives? Well, just for fun, can you think of any other previously used keyboard symbol that has been added only recently into contemporary parlance, and that represents a radical change in the way we communicate? Can you find it in the following string: "lifechangingsymbols@ubiquitous.com"?

It is critical to point out that the ~ character is not meant to have an independent existence. That is, it does not represent a bridge or bridging concept, a third thing via which one can cross from one complementary aspect to another. Let's repeat that: The ~ character is NOT a bridge. It doesn't stand for a separate piece, link, or glue "holding" complementary aspects together or mediating between them. Nothing like that at all. It is simply a way to write and think about complementary aspects in a way that does not let us forget that they are really inextricable aspects of a complementary pair. Remember the problem: What is the basis for dividing the world into pairs in the first place?

Notice also that complementary pairs may be written in two directions, depending upon which complementary aspect is written before and after the ~. Why do philosophers and scientists usually refer to the "mind~body problem" rather than the "body~mind problem"? On the other hand, many Buddhists and alternative thinkers use the word "bodymind" and not "mindbody." This brings up an interesting point. Is there, or should there be, a difference in meaning based upon which complementary aspect is written first? Of course, the order in which a pair of complementary aspects is written may be employed with a special meaning in mind, but let's leave this issue alone for the moment. We will return to it in Movement 3.

At any rate, it seems possible that complementary pairs with their ~ symbol might eventually be codified as a category of word association like the antonym and the synonym. That, of course, will depend on whether the use of the ~ becomes widespread and people begin to use it (and complementary pairs) to advantage. In this new parlance, two complementary aspects like "body and mind" comprise the complementary pair "body~mind." We call them complementary "aspects" instead of, say, complementary "things," because complementary pairs aren't always things. For instance, complementary aspects can be actions. They can even be ideas, like reductionism and emergentism, deduction and induction.

By our definition, complementary aspects of a complementary pair are fundamentally mutual. They are distinguishable yet coexistent. They are coemergent, codefining, and coimplicative. Being mutually coupled, complementary aspects are also dynamic: they flow in and out of each other in subtle and seemingly mysterious ways. What one perceives affects what one does and what one does affects what one perceives. What we want influences what we think about and what we think about influences what we want. Complementary pairs are like this by nature. Each of them is a manifestation of, and "windowless window" into, the complementary nature.

EXAMPLES OF COMPLEMENTARY PAIRS

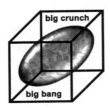

The universe starts with a Big Bang, expands to a maximum dimension, then contracts and collapses. No more awe-inspiring prediction was ever made.

John A. Wheeler (1911–), in C. W. Misner, K. S. Thorne, and J. A. Wheeler, *Gravitation* (1973)

(Stare at the Necker cube figure. Which words ["big bang" or "big crunch"] are located in the "front" square? Apply this same question to the many Necker cube figures in the text to follow.)

Big Bang~Big Crunch

If you think about it, even cosmological theories of the universe entail a complementary pair right at their heart. The currently dominant cosmological paradigm states that ever since the Big Bang, the universe has been steadily proceeding down a one-way street toward its eventual demise. Due to entropy, or the amount of disorder in a closed system, the ultimate state of the universe is going to be heat death—the degradation of all matter and energy in the universe. Put another way, as the universe expands, its battery is running down and will eventually run out.

As a result of this one-way march toward total cosmic disorder, all life and living must eventually end. So if the universe actually *is* a closed system, then, like hot coffee in a tightly closed thermos, it will slowly and finally grow cold. How depressing. All our achievements, inventions, knowledge, wisdom, great loves, passions, and inspirations will eventually be reduced to an ultimate disorder, an ill-fated, universal nothingness.

Of course, we know that the universe isn't all about degradation and disorder. After all, there would be nothing to speak of and no one to speak of it if it were. Our physical universe is our physical universe because things have also come together—they aggregate and order themselves, organize and coordinate. Living things are nothing if not functional dynamical aggregations. Consider, for example, the outlook of James Lovelock, a founder of the "Gaia hypothesis" that the earth's biosphere itself is a superorganism:

Through thinking about *Gaia* I see our universe as self-organizing and driven by the energy of its running down. The decline of the universe, like sand trickling through an hourglass, is essential for maintaining the flux of free energy that makes life possible. Indeed, it permits the most amazing self-organization to take place, with us as one of its intricate examples. To be gloomy about the second law [of thermodynamics] is as foolish as expecting to use a flashlight to see in the dark, and have the battery last forever.

On the other hand, we don't want too much aggregation: the prospects of a "Big Crunch," an ultimately recontracting boomerang scenario of the universe's future, are as unsavory as universal heat death in an ever-expanding one. If everything in the universe were eventually to be crammed into an infinitely dense singularity, there would be little chance for living systems to survive. So however it got here, life seems to depend upon both order and disorder. Order and disorder are distinct processes, yet you can't have one without the other. Order~disorder is a complementary pair. It sounds right, doesn't it? We need things to come together as well as separate, but not too much or too little. It doesn't work to have one without the other: too much order or disorder and it's game over—no chance for life!

This ordering~disordering of everything in the universe is happening right now as we speak, anywhere we want to look as well as everywhere we don't. No level of observation is immune. Neither the vast macrocosm of galaxies, nebulae, stars, and planets, nor the invisibly small microcosm of atoms, photons, and quarks, nor all that stuff falling in between, like cells, plants, and ourselves. On any level we choose to study, we find an endless number of different aspects and qualities existing there, some in the process of aggregating and ordering, some in the process of dispersing and disordering. To make it even harder to think about, it's *in motion*—it moves and changes, ever-transforming, at every level. The mind boggles at how all the different elements and patterns at all the different levels of nature are somehow stitched together—the grand, medium, small, and minuscule, each in untold numbers, all intertwined and interdependent.

This dynamic tapestry is being woven not only "out there" in the galaxies, stars, and celestial bodies and "in there" within the nuclei of atoms but also *here*, on our own terrestrial level, the human level. Of course, one can have no self-image at all without one's nervous system: a grand ordering and disordering of a billion, billion cells acting in an opera of absolutely staggering complexity, that somehow provides us awareness of ourselves and our environment, our world, our universe. It is truly a grand coordination, the complementary nature. But grand to whom? Grand to us human beings, whose awareness allows us the possibility to wonder and to appreciate.

Conscious~Unconscious

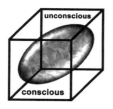

The psychological rule says that when an inner situation is not made conscious, it happens outside as fate. That is to say, when the individual remains undivided and does not become conscious of his inner opposite, the world must perforce act out the conflict and be torn into opposing halves.

Carl Gustav Jung (1875–1961), *Collected Works* (1959)

Awareness. Without our brain's capacity for awareness, there would be nothing to write about, no one to write about it, and no one to read what was written. The complementary pair conscious~unconscious is intended to remind us that conscious and unconscious are really forever together. They are intermingled, complementary. However, even though we can accept in principle that they are both equally valid and fundamental, it is easy to ignore this and act as if consciousness is the most important. Consciousness, for instance, is the difference between being asleep and being awake, and that is a *big* difference.

It was the psychiatrist Carl Gustav Jung who reminded us that when we report our dreams, which are supposed to come from the unconscious, the so-called conscious mind is affected by the unconscious and vice versa. Jung called the contents of consciousness "unconscious and conscious at the same time." That is to say, consciousness means "conscious under one aspect and unconscious under another." This remarkable statement may be worth remembering when we come to describe the coordination dynamics of the human brain later on. In the context of recent brain research Jung's intuition offers hope for a novel scientific reconciliation of the psychical and physiological.

Much of Jung's writing is thoroughly entrenched in complementary pairs. For instance, he wrote, "We must accustom ourselves to the thought that the conscious and the unconscious have no clear demarcation, the one beginning where the other leaves off. It is rather the case that the psyche is a conscious-unconscious whole." Clearly, for Jung as for us, no independent process "bridges" the conscious and the unconscious. Analogous to the alchemist's *lapis*, Jung's philosopher's stone was the *self*, a spiritual union of opposites that science itself could not bridge.

Jung's great crusade was for humanity to free itself from the antagonism of opposites that underlies all of Western thought. In our words, the crusade would be to rid ourselves of the addictive domination of either/or thinking. It is enticing to wonder how a scientific account of complementary pairs and their dynamics might have (and still might) shed light on Jung's work.

Consider his famous division of psychological types into extroverts (active, outgoing) and introverts (passive, self-absorbed) and his mandala-like functions of sensing~intuiting and feeling~thinking. Unfortunately, what was almost certainly an effort by Jung to emphasize a set of dynamical archetypal complementary pairs has largely gone the same way as the principle of yin~yang. Under the judgmental and limiting auspices of either/or thinking, these complementary aspects of behavior have been used instead to typecast and categorize. Thus people are typed as either introverts or extroverts. They are labeled as either sensual or intuitional, feelers or thinkers, perceptive or judgmental. We do not believe that Jung's dedicated attempt to comprehend the psyche was intended to be used as a binary categorization scheme. As his words reveal, Jung was clearly not a dualist, even if some later critics of his conceptual framework have concluded otherwise.

Unfortunately, without a clearer picture of what complementary pairs might *actually* be, it is too easy to cast them into binary distinctions. The erroneous labeling that results almost always leads to conflict and misunderstanding. One of Jung's most stunning insights about the human psyche was that it may well contain a mixture of both individual and collective aspects, and that these aspects are coordinated in some mysterious but obviously dynamic way.

Individual~Collective

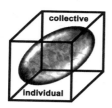

In individuals, insanity is rare, but in groups, parties, nations and epochs, it is the rule.

Friedrich Nietzsche (1844–1900), *Beyond Good and Evil* (1886)

The individual is always mistaken.

Ralph Waldo Emerson (1803–1822), *Essays: Second Series* (1844)

Consider for a moment arguably the most notorious of all contraries, at least where human beings are concerned, namely "individual~collective." Collectives imply populations. According to international demographers, the planet earth now carries over 6 billion human beings. The number looks like this: 6,000,000,000 individuals. That's a *lot* of individuals. But why is the subject of individual~collective so notorious? Because it is closely related to another complementary pair that greatly concerns all human beings: freedom~ constraint.

In any system, an individual is an individual if it is free to exhibit its individuality, if it can exercise some degree of autonomy, if it has "degrees of freedom." A collective is a collective if an aggregation persists, and always requires constraints that bind its individuals together. Now think of the complementary pair individual~collective: if individual and collective are inextricable complementary aspects, then individuals must entail collectives, and collectives must entail individuals. Most importantly, as complementary aspects within a given context, neither individual "parts" nor their collective "whole" should be considered *more fundamental*. Whatever constitutes a part and whatever constitutes a whole, they are mutually fundamental. A part is a whole and a whole is a part. They are together~apart.

Now arguments, fights, and even wars have raged through the ages over this matter of individual freedom versus collective constraint. Though one might accept as self-evident the general idea that both individual and collective should be equally valid and fundamental complementary aspects, again and again history shows that people and policy dictate otherwise. One finds a long list of conflicts between champions of the individual and champions of the collective. For example, to solve the world's economic problems, should we allow individual self-interest to determine outcomes, or use central controls to ensure the good of the collective? Perhaps more enlightened policies would emerge if the practice of pitting these two complementary aspects against one another was replaced by a more flexible practice of reconciling them as a complementary pair. One of the key discoveries to be described later on is that the individual~collective complementary pair emerges naturally from the science of coordination dynamics: tendencies for individual expression coexist simultaneously with tendencies for individuals to coordinate themselves as a collective.

Organism~Environment

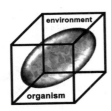

So closely coupled is the evolution of living organisms with the evolution of their environment that together they constitute a single evolutionary process in which life has, quite literally, fashioned the environment to suit itself.

James Lovelock (1919–), *Gaia: A New Look at Life on Earth* (1979)

Here in the first years of the twenty-first century, the subject of organism and environment has reached important, even epic proportions. We speak of global warming, of acid rain and deforestation as possible threats to our very existence. We speak of "sustainable development," a catchphrase whose meaning and significance occupy center stage at world summits. Sustainable development refers to the economic and social challenges that must be met in order to accommodate the present needs of the world's population without compromising the ability of future generations to meet their needs. Ignore the basic needs of human beings in the impoverished, developing nations and millions will suffer and be in peril. Ignore the environmental ecosystem, and perhaps *everyone* in the world will suffer and be in peril. Organisms and their environments can only be artificially separated. They are differentiable but mutual. Organism~environment, like individual~collective, integration~segregation, cooperation~competition, etc., is a complementary pair that exists *at all levels*.

The complementary pair organism~environment applies not only at the commonsense level of everyday experience, for example as concerns animals living in their niches, but on all levels within the organism too, stretching down to the gene and beyond. From the perspective of TCN, a body can be conceived as an environment of its constituent organs. Likewise, that same organ can be conceived as an environment of its constituent cells. At the level of the gene, the overwhelming bias toward the identification and hypothesized significance of primary DNA sequences is giving way to a renewed appreciation of and research into the vital role proteins and protein networks play in the *control* of DNA expression.

As we all know, DNA is replicated via a set of proteins, which are themselves translation products of the DNA. In fact, DNA and proteins depend upon this complementary relationship for their (and as a consequence *our*) very existence. Enter the complementary pair DNA~protein. It is quite interesting that more and more attention seems to be focused on the role of epigenetic factors such as DNA methylation in the regulation of gene activity, particularly in understanding pathological events originating from the cell nucleus. Unlike DNA sequences, which remain stable throughout an organism's life, epigenetic modifications can occur during meiosis and result in significant phenotypic differences in the offspring. On a broader scale, epigenetics reflects the essential, irreducible relationship between organisms and their environments.

This point was hammered home in Kelso's *Dynamic Patterns*. There, evidence and theory were presented demonstrating that sensory and motor, perception and action, are reconciled in the form of coupled, coordinated dynamical systems. Until recently, however, few have taken the complementary nature of organism~environment as a serious topic of scientific investigation. A notable exception was the late James Gibson, a psychologist at Cornell University whose life work and the ecological school he founded were and are committed to understanding the mutuality of organism and environment. Gibson introduced the term "affordance" to communicate his main thesis:

The verb "to afford" is found in the dictionary, but the noun "affordance" is not. I have made it up. I mean by it something that refers to both the environment and the animal in a way that no existing term does. It implies the complementarity of the organism and the environment.

Indeed. Thinking about organism~environment as a complementary pair reminds us that to understand organisms more deeply, we must also come to understand their environments, and vice versa. Further, as with all complementary pairs, organism~environment is both more than and different from the sum of its complementary aspects considered in isolation. This suggests that we stand to gain new insights and make new discoveries by studying the complementary nature of organism~environment, rather than isolating organisms from their environments, or treating environments without regard to the organisms that dwell in them.

Anabolism~Catabolism

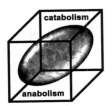

anabolism *n* constructive metabolism, opposed to catabolism.
catabolism *n* destructive metabolism involving release of free energy ... opposed to anabolism.
Merriam-Webster's Unabridged Dictionary (2000)

It is useful to remember that the process of DNA replication and protein synthesis, albeit a crucial one, is but one aspect of the complex living process. Consider all the many conditions that must be met in order for us to stay alive. All of the uncountable bits and pieces of our bodies must somehow coordinate with each other in such a way that we don't just, well, fall apart. The water in our blood cannot boil or freeze. Our blood acidity must remain within narrow ranges. Our hearts must pump that blood (systole~diastole). Our lungs must keep breathing (inhalation~exhalation), carrying oxygen from the cellular environment to the blood, and so on. Our myriad cells and organs must sustain themselves. We take in energy and raw materials that our bodies need to sustain themselves by eating and drinking. The food we eat is broken down into the molecular raw materials—carbohydrates, fatty acids, amino acids, vitamins and minerals, etc. This breaking-down process is called catabolism, and the products broken down are called catabolites.

In the reverse process, called anabolism, cellular constituents are woven from raw materials provided for it by catabolism. The entire system including anabolism (building up) and catabolism (tearing down) is referred to as metabolism. *Meta*bolism is an important and familiar word and concept, as it signifies the whole opera of potential and action in both the building up and breaking down directions. Since we don't want to forget that anabolism and catabolism are inextricable aspects of metabolism, though, let's write anabolism~catabolism as a complementary pair. Life as we currently understand it is not possible without anabolism~catabolism. Energy is stored and then expended, pulsing through the body~mind tapestry, even as molecules and patterns of molecules are formed, altered, and eventually eliminated. Among other important activities, anabolism~catabolism provides the cellular energy needed for genetic replication and protein synthesis, and metabolic control and signaling pathways play a key role in gene expression.

Of course, for anabolism~catabolism to be effective, it is necessary for the organism to acquire and expel material from its environment. For one thing, there must be air available in the atmosphere to breathe. We need oxygen to survive and prosper. And where does the oxygen in the air come from? From plants and photosynthesizing algae. We inhale the oxygen, allowing our bodies to continue the process of oxidative metabolism that keeps us alive. In the very same breath, we expel carbon dioxide, eliminating it from our bodies back into the atmosphere. Plants depend upon carbon dioxide to sustain themselves. Plants and animals depend upon each other for their very survival, for this complementary exchange of oxygen and carbon dioxide. Together, plants and animals help each other stay alive. This complementary relationship brings us back full circle to organism~environment and beyond.

THE INTERPRETATION OF COMPLEMENTARY PAIRS

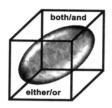

A central premise of TCN is that, for better and for worse, the way one interprets contraries dramatically influences how one interprets literally everything else in one's life. As we have taken pains to point out, there is more than one way to interpret contraries. We have already mentioned an unusual interpretation of contraries, the so-called either/or~both/and perspective, which goes hand in hand with our definition of a complementary pair.

Recall that the either/or~both/and reconciliation suggests that contraries are "complementary aspects": opposing, conflicting, and antithetical on the one hand, and complementary, mutual, and inextricable on the other. Note also that this either/or~both/and interpretation has been cast as a complementary pair in its own right. Venturing toward the poetic, it is not only like a lamp shining in its own light, but a lamp *made* of its own light. A good part of what follows explores the nuances and scientific grounding of this interpretation.

There are at least four basic interpretations of a given complementary pair of the form ca1~ca2 (where ca stands for *c*omplementary *a*spect). Though these interpretations aren't difficult to comprehend, they are perhaps easier to visualize, and so we provide figure 1.1. The four basic possibilities consist of two mutually exclusive either/or interpretations, and two types of both/and interpretations, as follows.

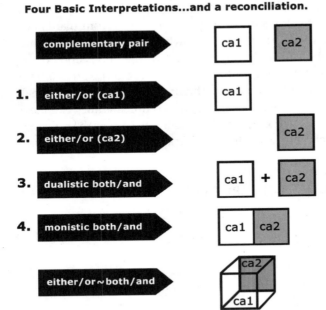

Figure 1.1 Four basic interpretations of complementary pairs ca1~ca2 and their reconciliation.

The Mutually Exclusive Either/Or Interpretations

In this interpretation, one of the two complementary aspects is valorized as the predominant or most basic aspect. Figure 1.1 shows that for any given complementary pair, one can say that *either* ca1 *or* ca2 is the more fundamental aspect. The two either/or perspectives are mirror images of one another—they are really the same kind of interpretation, but of opposite sign. Nevertheless, they lead to two possible polar monisms. One valorizing ca1 (a "ca1-ism") and the other valorizing ca2 (a "ca2-ism") as the fundamental essence, substance, aspect, concept, ideal, and so on. For example, using mind~body as a representative complementary pair, we can readily note the contrary positions of idealism (mind is more fundamental) versus physicalism (body is more fundamental).

The Dualistic Both/And Interpretation

In this interpretation, ca1 and ca2 are both thought to be fundamental. In dualism, ca1 and ca2 are extricable—they can be separated. As irreducible substances, they must somehow interact. Classic examples of dualism are the light/darkness dualism of the ancient Manicheans, the mind/body dualism of René Descartes, and the phenomenon/noumenon dualism of Immanuel Kant.

The Mutually Inclusive Both/And Interpretation

In this interpretation, $ca1$ and $ca2$ are thought to be aspects of a third, *more fundamental* "oneness." Examples are the neutral monism of Baruch Spinoza and the double-aspect theory of Aristotle.

The Reconciliation: Either/Or~Both/And Interpretation

The so-called either/or~both/and reconciliation harmonizes all four basic interpretations, which are then treated as equally valid dynamical modes, tendencies, or dispositions of the complementary pair $ca1$~$ca2$. We will develop this idea further just ahead.

Although many elaborate explanations and perspectives are manifest in as many described philosophies throughout history, when applied to specific complementary pairs, these four basic interpretations summarize a massive amount of philosophy surprisingly well. Of course this summary can easily be criticized by professional philosophers as overly simplified. We accept that. Nevertheless, in our defense, our own philosophical research verifies that there have been very real historical difficulties in the successful elimination of *any* of these four basic interpretations, whether they be general or specific, not to mention deciding between them. The checkered, emotional history of the more than three-hundred-year-old mind~body debate is a poignant case in point. Now let's briefly look at some of the problems that emerge when one of these four basic interpretations is adopted as the one "final" pure, true, and abiding interpretation of complementary pairs.

Demoting the Mutually Exclusive Either/Or Interpretations

There are truths on this side of the Pyrenees which are falsehoods on the other.
Pensées (1660)

Blaise Pascal
(1623–1662)

When differences between complementary aspects are overemphasized, as they so often are, the ability to perceive the complementary nature of complementary aspects is impeded. This impediment takes the form of the mutually exclusive either/or interpretation of complementary pairs. When tied to a political, ethical, or religious agenda, the outcome is often disastrous. It is safe to say that the mutually exclusive either/or as a modus operandi has caused and continues to cause a great majority of strife and suffering in the world today.

It would certainly be a lot harder for conflicting parties to remain in conflict if both sides accepted that their mutually exclusive causes might actually compose a complementary pair. Were that seen to be the case, the policy of championing one of the sides while excluding or annihilating the other would be an ill-informed policy *in principle*, no matter how compelling it might be to follow. This exact point has been made many times throughout history, by some of humanity's most famous pundits. Practicing complementarity, however, in the face of actual conflict and its concomitant amplification of self-interest is another issue entirely.

The problem, it seems to us, stems from trying to impose an *exclusively* mutually exclusive either/or interpretational framework upon complementary pairs. Just because something is perceived to have sharply contrasting aspects (e.g., individual versus collective) doesn't necessarily imply that those aspects are mutually exclusive or self-contradictory. On the other hand, it is much easier for human beings to decide between two clearly opposite choices than to reconcile them or to handle their possible intrinsic ambiguity. For the most part, people find ambiguity unsettling and difficult to accept.

In a leader, for instance, ambiguity is often construed as a sign of weakness, even though it may be critical for that leader to see both sides of an issue and recognize the merits of each. A leader who is cocksure of what is "right" and "wrong" is comforting to the masses, and thus perceived as strong and worthy. For instance, Adolf Hitler was convinced of what he believed to be right and wrong. For all of his faults, he was a strong, charismatic, effective leader. Hitler is a sobering reminder of what can happen when a leader or regime is predominantly run on a rigid, limited either/or mentality. And, unfortunately, contemporary examples of such thinking are woefully easy to find.

By and large, human beings want things to be either determined *or* random. We prefer things to be either totally constrained and controlled (as do emperors, company bosses, and school masters) *or* totally unconstrained and free (as subjects of the empire, company employees, and most students). This emotional need for neat, two-valued identifications seems to be a core element of human nature. Despite its transparent simple-mindedness, this habit is so pervasive that very few people are immune to it.

Reduction of complementary pairs to their idealized complementary aspects as an immutable modus operandi must be rejected for the simplest of reasons: The complementary nature simply doesn't work that way. You don't have to take our word on this. You can very easily verify for yourself that, ultimately, it's not nature over nurture or vice versa; central versus peripheral or vice versa; body over mind or vice versa; individual over collective or vice versa; man over woman or vice versa, young over old, rich over poor, and so on.

Carefully note again that what we are rejecting here is *not* the natural polarization of complementary aspects per se. Far from it. What we are rejecting is mutually exclusive either/or interpretations of complementary pairs in which one complementary aspect is valorized, reified, and finally deemed more fundamental than the other. Despite its satisfying crispness, this worldview is just too simplified and is found wanting. As we will show, it is quite possible, even reasonable, for a dynamic system to be stabilized in a mode that makes it appear *as if* one aspect is more fundamental. Indeed, in that context, such an aspect is in fact temporarily dominant. Ironically, if this were not possible, mutually exclusive either/or thinking itself wouldn't be possible! But this should not conceal the deeper fact that life's dynamic systems are inherently information-seeking, context-dependent, and multifunctional, and are able by their very nature to explore multiple valid interpretations of their own complementary nature, of which the mutually exclusive either/or is only one of many.

The Dualistic Both/And Interpretation

If someone calls you a dualist these days, the term is almost certainly meant to be derogatory and should be adopted with caution. Nevertheless, from a religious point of view, the great majority of people alive today are dualists. Most Christians are dualists, as are shamanists, Hindus, Jews, and Muslims. What's curious (and a bit insidious) about dualism is that even though it means two irreducible and fundamental substances, concepts, processes, etc., one of the two is almost always held to be superior: spiritual superior to physical, God superior to human, heaven superior to hell, light superior to darkness, right superior to wrong, genes superior to environment.

If this sounds a bit like the mutually exclusive either/or perspective, it is. The big difference is that in dualism, both aspects are irreducible and coexist by definition, or the pair isn't a dualism. You can't get rid of either one of them no matter how much you desire to. But since the aspects are *extricable*, that is, they can in fact be separated from one another, we are once again at liberty to valorize one of the aspects. The end result is that we often end up behaving in exactly the same way as we might were we following the mutually exclusive either/or interpretation. In TCN, we reject dualism as a self-erected policy or modus operandi. At the same time, in TCN, complementary pairs in certain scenarios, systems, and contexts can be perceived in such a way that they appear to be dualistic. Two of the best examples are Descartes's division of the universe into the physical and mental, and Immanuel Kant's dualism of phenomena (things as they appear to us) and noumena (things as they actually are).

Cartesian Dualism

... the two operations of our understanding, intuition and deduction, on which alone we have said we must rely in the acquisition of knowledge.

Rules for the Direction of the Mind (1626)

René Descartes
(1596–1650)

One cannot speak of dualism without mentioning the most notorious dualist of all, René Descartes. With his method of "radical doubt," Descartes desired to rid himself of the fetters of classical thought espoused by the reigning doctrine of Scholasticism, the philosophy of Aristotle as interpreted in the Middle Ages by Thomas Aquinas and others. Descartes wanted to produce a philosophy to match the new and budding method of science, yet *at the same time* retain the church's sovereign domain over the human spirit. And what was the basis of Descartes's provocative and controversial philosophical stance? It was a troublesome interpretation of the complementary pair mind~matter called "dualism," epitomized in Descartes's famous maxim, *Cogito ergo sum*—"I think, therefore I am."

Setting up a complementary pair (mind~matter) as a scaffolding for his philosophical platform was not in itself a radical philosophical departure. After all, it was just another juxtaposition of contraries, the so-called metaphysical exigency. Descartes's radical departure was the emphasis he placed upon the *separability* of the complementary aspects mind (*res cogitans*) and matter (*res extensa*). This separability of mind and matter is sometimes referred to as the "Cartesian partition," a cut made, from the very beginning, with a double-edged sword.

On the one hand, partitioning things this way allowed the new scientists (or natural philosophers, as they were known at the time) to treat everything besides mind as the "mere" motion of bodies. As a result, they were able to carve up the "extended matter" as they saw fit, thereby expanding the possibilities to explore, discover, and invent. This also worked well for the church, as it retained the mind (soul) within its stern parochial jurisdiction. So everyone was happy, right?

Well, not really. Cartesian dualism came at a heavy price, one that we are still paying for today. For starters, it opened up the possibility that *other* complementary pairs besides mind~matter could or should be interpreted identically, namely as fundamental duals. Further, the notion of the irreducibility of mind and matter led to the so-called problem of interaction: that is, how thoughts of the mind set matter in motion, and how matter in motion impinges upon thoughts of the mind. In other words, how can two utterly dissimilar substances like mind and matter be causally connected? How does that work?

Mind/matter dualism made a mess of philosophy. And every attempt since to transcend the mind/matter dichotomy and the problem of their supposed interaction has fallen foul of the same set of deep conundrums. Meanwhile, freed from the fetters of the liturgical prerogative, the unlimited potential for scientific penetration of "extended matter" was swift, intoxicating, and terribly effective. It gave rise to the "mechanists," who treated the universe as a cosmic clockwork—a gigantic universal mechanism. The *mechanism* was Descartes's root metaphor, and remains the predominant mind-set in science to this day. But this success has always come at the expense of a dubious assumption made by the self-proclaimed master of doubt, namely, that complementary aspects could be permanently divided into irreducible duals. As the eminent philosopher Alan Whitehead explains, problems have been inherent in the mind~matter conundrum since its beginnings:

> Thereby, modern philosophy has been ruined. It has oscillated in a complex manner between three extremes. There are the *dualists*, who accept matter and mind as on an equal basis, and the *two* varieties of *monists*, those who put mind inside matter, and those who put matter inside mind. But this juggling with abstractions can never overcome the inherent confusion introduced by the ascription of *misplaced concreteness* to the scientific scheme of the seventeenth century.

"Ascription of misplaced concreteness," one might add, of the *universal mechanism* metaphor. Following Descartes, this same paradigm of divide-and-conquer reductionism could be applied to any complementary pair. Schisms led to isms, and the conflicting camps began new battles for the dominant root metaphors of their day, a conflict that continues today.

Kant's Reconciliation of Rationalism~Empiricism

Immanuel Kant
(1724–1804)

> The inscrutable wisdom through which we exist is not less worthy of veneration in respect to what it denies us than in respect to what it has granted.
>
> *Critique of Pure Reason* (1787)

The philosopher of the eighteenth century who ranks with Aristotle and Plato of ancient times is Immanuel Kant. Kant spawned a chain of unprecedented ideas that humanity has continued to ponder ever since. The rationalists of Kant's world contended that universal truth is ultimately revealed by the mind ("I think, therefore I am"), while the empiricists held that universal truth is to be discovered only by the senses ("I'll believe it when I see it"). Kant's attempt to reconcile rationalism and empiricism (rationalism~empiricism) and analysis and synthesis (analysis~synthesis) lives on in contemporary efforts by people like E. O. Wilson and the late Stephen Jay Gould to mend the rift between the sciences and the humanities.

In his *Critique of Pure Reason*, Kant tried to set up a contrast between things that exist in the outside world and actions of the human mind. The former, he contended, are real, but the human mind is needed to give them order and form, and to see the relationships between them. For Kant, pure reason without reference to the outside world was impossible, an idea he borrowed from the Scottish empiricist David Hume. What one knows, according to the empiricists, is the result of what one has gathered up with one's senses.

The rational mind, however, is needed to comprehend the empirical world of sensations. In addition to the ideas of time and space, which he called "pure forms of intuition," Kant saw the mind as necessarily containing a number of archetypal conceptual mental categories that can then be applied to raw perceptions in order to gain knowledge. The catch in Kant's philosophy is that human beings only have access to the phenomenal world. They can have no knowledge of the noumenal world, the true nature of "things-in-themselves." Thus, to Kant, our conceptual knowledge can only be of the world as it is for we humans who experience it. Phenomenon and noumenon constitute a dualism, two fundamental, irreducible forms.

Considering that Kant's aim was to reconcile two contrary philosophies, each of which was a candidate theory for how human beings acquire knowledge of the world, this was a curious turn. By proposing conceptual categories of mind and pure forms of intuition, Kant treated rationalism~empiricism as a complementary pair, and did quite a lot with that insight. Yet in the end, he left us with the dualism of phenomena/noumena. In this, we see just how difficult it can be, even for a genius like Kant, to rid oneself of the mighty tug of dualism.

The Mutually Inclusive Both/And Interpretation

The final basic interpretation of complementary pairs brings us much nearer to TCN. The mutually inclusive both/and perspective sees the complementary pair as part of something bigger, some multifaceted activity that expresses itself in ways that identify with both complementary aspects. It departs from dualism in that complementary aspects cannot be thought of as irreducible, extricable substances, even in principle. This returns us to the wisdom of Heraclitus, where it is abundantly clear that conflicting opposites are not only necessarily complementary but *inextricable* expressions of a many faceted nature:

God is day~night, winter~summer, war~peace, satiety~hunger; he undergoes alteration in the way that fire, when it is mixed with spices, is named according to the scent of each of them.

Heraclitus's epigrams are entirely resonant with the ideas of his contemporaries in the East. In fact, one can hardly distinguish them from Taoist poetry and the paradoxical sounds of the principle of yin~yang:

The path up and down is one and the same;

Things taken together are wholes and not wholes, something which is being brought together and brought apart, which is in tune and out of tune; out of all things there comes a unity, and out of a unity all things;

Would that strife might be destroyed from among gods and men: for there would be no musical scale unless high and low existed, nor living creatures without female and male, which are opposites;

They do not understand how, being separated, it is united with itself. There is a backward-stretching tension, as between the bow and the lyre.

The mutually inclusive both/and perspective comes in different flavors. Although none of them quite captures the complementary nature of complementary pairs, they certainly warrant serious mention. Here we consider briefly some of the chief exponents: Aristotle, Thomas Aquinas, and Baruch Spinoza, each in his own day and ways a reconciler par excellence.

Aristotle's "Golden Mean" and Double Aspect Theory

Aristotle
(384–322 B.C.E.)

Now it is a mean between two vices, that which depends on excess and that which depends on defect; and again it is a mean because the vices respectively fall short of or exceed what is right in both passions and actions, while virtue both finds and chooses that which is intermediate.

Nicomachean Ethics, Book II, Part vi

For Aristotle, in contrast to both Socrates and Plato, dialectic was neither a science nor a method of science. That's because Aristotle thought the conclusions of dialectical reasoning to be only probable, based on suppositions of self-evident truths—root metaphors. This view was not intended to be a fatal criticism of dialectic, just a different interpretation. Intermediate between science and rhetoric, dialectic could serve both. Aristotle considered dialectic useful in philosophy because it developed one's skill in criticizing definitions and in the process of asking and answering questions. As he said, "The ability to raise searching difficulties on both sides of a subject will make us detect more easily the truth and error about the several points that arise."

For Aristotle, the dialectical process was expected to end neither in a synthesis of incomplete opposites (as in Hegelian dialectic) nor in a rejection of perceived contraries as illusory (as in Platonic dialectic), but rather in the realization of a more tenable or probable view in between them: If someone saw reality as black and another saw it white, Aristotle thought it was far more likely to be a shade of gray. Aristotle called his metaphorical reconciliation of complementary pairs the "golden mean." In it, virtue informed by reason is found in a middle path between the extremes of excess and deficiency. Courage, for example, was the golden mean between the extremes of recklessness and timidity. Generosity was thought to be a golden mean between wastefulness and stinginess, and justice, a proper balance between sacrifice and malice.

In an early example of his double-aspect theory, Aristotle proposed that form, with the exception of a "Prime Mover," had no separate existence, but rather was immanent in matter. Thus, in Aristotle's hylomorphism, both form and matter together constituted concrete individual realities. This proposition was a definite departure from the polarized Platonic paradigm of form over matter.

A Reconciliation of Faith~Reason

Thomas Aquinas
(1225–1274)

> Consequently, those who are more adapted to the active life can prepare themselves for the contemplative by the practice of the active life, while none the less those who are more adapted to the contemplative life can take upon themselves the works of the active life so as to become yet more apt for contemplation.
>
> *Summa Theologica* (1266–1273, ed. 1947)

Aristotle's mutually inclusive both/and interpretation of complementary pairs figured centrally in the philosophical work of St. Thomas Aquinas. Medieval rediscovery of Aristotle had challenged the reigning Augustinian philosophy that true knowledge is possible *only* through faith. Early in the thirteenth century, Islamic scholars concluded that truth could be discovered by reason as well as faith, none other than an application of Aristotle's double-truth doctrine. As one might imagine, the double-truth doctrine alarmed orthodox Catholic thinkers and challenged church doctrine, provoking a theological crisis.

Aquinas sought a reconciliation of faith and reason because he believed they were complementary (faith~reason) rather than antagonistic. He insisted that some truths, such as the mystery of the incarnation, may be known only through revelation, whereas others, such as the composition of material things, may be known only through experience. Still others, such as the existence of God, are known through both. Aquinas held that knowledge originates in sensation, but sense data can be made intelligible only by the action of the intellect (a stance nearly identical to the one later adopted by Kant, a small miracle in its own right).

As for religious matters, the path to revelation of the highest truths was achieved via faith. On the other hand, evoking Aristotelian arguments, Aquinas held there were limits to God's omnipotence: God could not create a circular square nor thwart the laws of Aristotelian logic. This was obviously not the *safest* contention in those days! But rather than calling it a golden mean between two polar ideals, Aquinas described the complementary pair as a sort of contextual dynamic: Knowledge is gained sometimes by approximating pure faith, sometimes by approximating pure reason, and in other situations through a mixture of both. That Aquinas achieved this rather modern, pluralistic reconciliation in the thirteenth century is quite stunning, and attests to his reputation as one of the world's greatest philosopher~theologians.

Aquinas's application of reason to faith, rejected though it was by Catholic officialdom, contributed to the rise of the secular attitudes of the Renaissance, with its new physics, astronomy, and humanism, and revolutionized philosophical thought. Aquinas practiced a principle of tolerance toward ideas that were held by most to be conflicting, antithetical, and mutually exclusive. As an Aristotelian, he believed that body and soul were inseparable (body~soul), in direct contrast to the prevailing Neoplatonic notion of the soul as an eternal entity in a temporary bodily existence (body versus soul). In his unprecedented scholarship and endless desire to reconcile faith and reason, Thomas Aquinas showed an unmistakable appreciation of the complementary nature, and an understanding of the importance of reconciliation of complementary aspects as complementary pairs.

Spinoza's "Mystic Monistic" Reconciliation

There is no hope unmingled with fear, and no fear unmingled with hope.
Ethics (1677)

Baruch Spinoza
(1632–1677)

As a final contribution to the mutually inclusive both/and perspective, we come to the ideas of another genius, Baruch Spinoza. In his attempt to solve the mind/body problem, Spinoza adopted Descartes's geometrically oriented rationalism but rejected his dualism. His replacement was a pure monism, in which God and all that is are one substance, a belief still followed today by pantheists. This unitary infinite substance, *Deus sive natura*, was thought by Spinoza to entail an infinity of aspects, of which we humans are able to perceive only two, mind and matter—the double aspect of universal substance. Spinoza believed that *all* things, whether physical objects, animate creatures, or immaterial ideas, are transient modifications or *modes* of a *single* universal substance. Actually, this idea of a single complex system able to express itself in multiple forms or modes is strikingly modern. We now know that systems with uncountably large numbers of degrees of freedom can self-organize, that is, spontaneously form orderly structures and patterns of behavior.

Allied closely to the ideas of Spinoza were those of William James, the father of American psychology, who developed the doctrine of "neutral monism." For

James, the substance of the world is pure experience, and consciousness is a function of the brain. Neutral monism was also advocated by such neorealists as Bertrand Russell and A. N. Whitehead. In Russell's *A History of Western Philosophy*, we find the following insightful words (notice the complementary language):

> While physics has been making matter less material, psychology has been making mind less mental.... Thus from both ends physics and psychology have been approaching each other, and making more possible the doctrine of "neutral monism" suggested by William James' criticism of "consciousness." The distinction of mind and matter came into philosophy from religion, although, for a long time, it seemed to have valid grounds. I think that both mind and matter are merely convenient ways of grouping events. Some single events, I should admit, belong only to material groups, but others belong to both kinds of groups, and are therefore at once mental and material. This doctrine effects a great simplification in our picture of the structure of the world....
>
> There remains, however, a vast field, traditionally included in philosophy, where scientific methods are inadequate. This field includes ultimate questions of value; science alone, for example, cannot prove that it is bad to enjoy the infliction of cruelty. Whatever can be known, can be known by means of science; but things which are legitimately matters of feeling lie outside its province.

Thus, it appears that Russell gave his blessing to neutral monism for the sake of its aesthetic appeal, but seemed unwilling to do so without the concept of value judgments—a concept so basic that it rivals physical existence. Somehow, what physical laws allow us to do must be supplemented by the addition of moral laws or values. The complementary nature strikes again, in another of the twentieth century's deepest thinkers!

The Four Basic Interpretations of the Complementary Pair Fall Short

For any philosophically significant complementary pair, the four basic interpretations eventually arise, coming in stronger and weaker versions, each with its champions and advocates. In our view, none of the basic interpretations is sufficient *in its own right* or even in summary combination to provide an adequate and abiding comprehension of complementary pairs. As such, we do not feel that they are sufficient to capture the complementary nature. If this bold claim is correct, what do we propose instead? Well, to begin with, let's return to Frederiksborg Castle and meditate once again on Bohr's coat of arms. Remember what it says: "*Contraries are complementary.*" Bohr was alluding to his complementarity via Eastern philosophy. So what was Bohr's summary statement on complementarity again?

A PHILOSOPHY OF COMPLEMENTARY PAIRS

Niels Bohr
(1885–1962)

The quantum postulate forces us to adopt a new mode of description designated as complementary in the sense that any given application of classical concepts precludes the simultaneous use of other classical concepts, which in a different connection are equally necessary for the elucidation of the phenomena.

"The Quantum Postulate and the Recent Development of Atomic Theory" (1927)

In ordinary, everyday language, we encounter Bohr's concept of complementarity when faced with a choice between two contradictory alternatives each of which can be justified by itself but both of which are necessary for understanding.

We have nearly come full circle. Recall Bohr's epistemological position we presented earlier, that complementarity not only provides a consistent and harmonious interpretation of atomic behavior, but also requires a thoroughgoing revision of our attitude toward physical reality. This is exactly what we are after in TCN—a thoroughgoing revision, a novel and contemporary worldview that advances the theme that was so dear to Bohr but was not realized by him in his own lifetime. Bohr's coat-of-arms with its quizzical motto hangs silently as thousands walk by it day after day, year after year, the yin~yang beckoning like an arrow of wisdom drawn on a taut bow-string pulled 25 centuries into the past, just waiting for someone to come along and let it fly it into the future. That time may be at hand.

Starting with Bohr's maxim that "contraries are complementary," we have defined the complementary nature, complementary pairs, complementary aspects, and the syntactical squiggle (~). Having looked at the four basic ways complementary pairs have been interpreted, we would now like to reconcile them. Using Bohr's principle of complementarity as one scientifically relevant example of reconciliation—in his case, reconciliation of seemingly incommensurate theories of radiation—we now proceed to our philosophy of complementary pairs. In the interests of brevity, we offer a simple summary. The philosophy of complementary pairs:

- *recognizes the ubiquity and historical significance of complementary pairs;*
- *acknowledges the complementary pair as a valid subject~object of contemporary research;*
- *provides a novel nomenclature for working with arbitrary complementary pairs;*

- *seeks to reconcile the four basic interpretations of complementary pairs on scientific grounds.*

Thus far, we have covered the first three points of this little manifesto. We have recognized the proliferation of complementary pairs throughout the history of thought and noted their utility in several fields—*ubiquity and historical significance*. We have also come to appreciate the complementary pair as a valid subject~object of research. We have noted that it is through the "windows" of complementary pairs that the complementary nature may come to be known. We have provided a nomenclature, a novel syntax for working with arbitrary complementary pairs. Next, we propose that some sort of reconciliation of the four basic interpretations is in order. It is desirable that this reconciliation—our philosophy of complementary pairs—should go beyond metaphor and be grounded in science. In other words, we are going to try to achieve our reconciliation of the different interpretations of complementary pairs scientifically.

So let's use the remainder of Movement 1 to set the stage for what comes next, namely the scientific grounding of our philosophy of complementary pairs. Let's begin by taking a quick look at some of the most dramatic reconciliations of complementary pairs found in science. The vignettes that follow are designed to demonstrate that Bohr's maxim, "Contraries are complementary," is as relevant to the history of science as it is to the history of philosophical thought. Finally, we will discuss the third and final basic feature of complementary pairs, namely that complementary pairs are *dynamical*. Context-dependent, multifunctional dynamics of complementary pairs signals the transition from philosophical metaphor to scientific foundation. It is the path we follow in order to achieve our reconciliation.

SCIENTIFIC RECONCILIATION OF COMPLEMENTARY PAIRS

Reconciliation means a harmonizing, a bringing together of that which has previously been considered incommensurate. The great scientific reconciliations have been quite different from their illustrious philosophical counterparts. In science, elegant argumentation must be backed up with observations and predictions. It is a lot harder to *argue* that electricity and magnetism are fundamentally two different substances when Maxwell's equations showed that they are one and the same. And it is even more compelling when repeatable physical phenomena bear out the predictions made by those equations and when those equations lead to inventions.

Such successful reconciliation of theory and experiment could be called the scientific prerogative, or the prime objective of science. Science asks questions, tests predictions, explains, and then moves on. Implicit in any scientific reconciliation is that neither aspect of a complementary pair can be considered more primary or fundamental than the other. Whereas both elements have individual, differentiating characteristics, they are always coexistent, coherent, and coextensive. Neither of the aspects can survive without the other. You can't have electricity without magnetism and vice versa, and that's that. From a scientific point of view, of course, this contention is one that leaves itself open to test.

Terrestrial~Celestial Mechanics

Isaac Newton
(1642–1727)

It is true that we may consider one body as attracting, another as attracted; but this distinction is more mathematical than natural. The attraction resides really in each body towards the other, and is therefore of the same kind in both.

Principia Mathematica (trans. A. Motte, 1729)

Isaac Newton's reconciliation of terrestrial and celestial mechanics stands out as the crowning development of the mechanical worldview that almost completely dominated Western thought from the seventeenth century through the beginning of the twentieth century, and remains influential today. His most famous discoveries were the three laws of motion and the law of universal gravitation. Just as a reminder, Newton's three laws are the following:

1. Objects at rest or in motion remain in that state, the state of "inertia," unless acted upon by an outside force (force~motion).
2. An object's acceleration, the rate of change of its velocity, is directly proportional to the force applied to it and inversely proportional to its mass ($F = ma$).
3. For every action there is an equal and opposite reaction (action~reaction).

With these laws taken alongside Kepler's laws of planetary motion, Newton derived his law of universal gravitation. The attraction between two bodies, such as the sun and earth, is directly proportional to the product of their masses and inversely proportional to the square of the distance between them. The law of gravitation contained a revolutionary and vastly influential insight and reconciliation: An apple falling from a tree and the moon orbiting the earth are governed by the same laws.

While Newton's *Principia* furnished a highly successful mathematical account of force and motion, it did so without respect to the nature of the force per se. He told us what gravity does without telling us how it actually does it. Descartes was concerned with trying to explain *how* gravity does what it does, but Newton proved that Descartes's mechanism, his hypothetical vortices, were inconsistent with both Kepler's laws and the inverse-square law, thereby undermining Descartes's mechanistic philosophy. Newton was interested in the nature of the causes operating in the world, but forever transformed science by postponing the issue of causality (considered crucial since the time of Aristotle) in order to investigate the observable properties of the phenomenal world. Newton was a contradiction himself—a man who wrote a treatise called *Hypothesis of Light* yet adamantly refused to entertain hypotheses—yet his work stands as one of the first great scientific reconciliations.

Geometry~Dynamics

By his method of calculating tangents on curves, Newton was able to calculate rates of change. In this stroke of genius, a geometrical task was reconciled with a kinematic task (geometry~dynamics). That is, to measure curvature on a graph was to determine a rate of change. Rate of change was itself an abstraction of an abstraction; what velocity was to position, acceleration was to velocity. This process was called "differentiation" in the later language of calculus. Likewise, he calculated areas by infinitely partitioning curves and infinitely adding the partitions, the mathematical process called integration. Differentiation~integration is a complementary pair. What one aspect does, the other undoes.

In creating this mathematics, Newton embraced a paradox. He believed in a discrete universe of indivisible atoms, yet he built a mathematical framework that was not discrete but continuous, based on a geometry of lines and curves. Almost like alchemy (of which Newton was an avid practitioner), the complementary pair discrete~continuous was forged. Henceforth, space would have dimension and measure, and motion would be subject to geometry.

Newton was faced with Zeno's paradox in his attempts to describe motion: Whereas matter might consist of indivisible atoms, motion needs mathematical points: To go from point *a* to point *b*, a body must pass through all points in between. Only by embracing the infinite and the infinitesimal together was it possible to handle Zeno's paradox. Newton struggled with this, as did Galileo: "We are among infinities and indivisibles," said Galileo, "the former incomprehensible to our understanding by reason of their largeness, and the latter by their smallness" (infinite~indivisible). Keep in mind that these were days when technologies for measuring time and speed as well as the concepts needed to quantify motion were totally lacking. Amazing.

Electromagnetism (Electricity~Magnetism)

We can scarcely avoid the conclusion that light consists in the transverse undulations of the same medium which is the cause of electricity and magnetism.

"On Physical Lines of Force" (1861)

James Clerk Maxwell
(1831–1879)

James Clerk Maxwell, the great Scottish physicist, is widely acknowledged as the nineteenth-century scientist whose work had the greatest influence on twentieth-century physics. His work harmonizing theories of electricity and magnetism constitutes a scientific reconciliation of the highest order. Yes, electric currents can deflect a compass needle, and yes, a moving magnet can induce a current in a loop of wire, but how are the two related? The ingenious English experimenter Michael Faraday, with his crude drawings of invisible field lines extending out like iron filings around a magnet, intuited the presence of a mysterious electromagnetic field.

While others mocked the elderly Faraday behind his back, Maxwell took this idea seriously. As one of the greatest mathematicians of his time, Maxwell was able to show that beneath Faraday's childish sketches lay a beautiful mathematical structure. Maxwell's famous equations showed the relationship between changing electric and magnetic fields in free space. A changing electric field (say from an antenna) induces a changing magnetic field which causes a changing electric field that propagates away from the antenna with great speed. And what was this speed? No less than the speed of light.

To the ordinary man walking down the streets of Edinburgh, nothing could be more different than magnetism, electricity, and light. Yet Maxwell showed that these phenomena were simply different manifestations of the same fundamental laws. And lo and behold, light emerged as just another kind of wave like the "mutual embrace" of electricity~magnetism. Twentieth-century communication technology stems largely from Maxwell's work. Radio, television, radar, and satellite communication all have their origins in his electromagnetic theory. As in the case of Newton, Maxwell's work is an excellent example of how powerful an impact scientific reconciliation of a complementary pair can have on day-to-day life.

Ironically enough, Maxwell's electromagnetic field equations were designed to explain the aether, a material medium that was thought to fill space and transmit vibrations through it (as in the quote above). But instead they paved the way for

Einstein's special theory of relativity, which established the equivalence of mass and energy (another complementary pair) and led to the demise of the aether concept. The biographer Abraham Pais reports that Einstein in his later years wrote of Maxwell's novel way of dealing with forces as follows:

> Since Maxwell's time, Physical Reality [Einstein's capitals] has been thought of as represented by continuous fields ... not capable of any mechanical interpretation. This change in the concept of Reality is the most profound and the most fruitful that physics has experienced since the time of Newton.

Maxwell's electromagnetic theory linking electromagnetism with light and later with radio waves was a great contribution toward the reconciliation of the then extant theoretical structure of physics. For our story it is important to keep in mind that electricity~magnetism represents a "real" physical complementary pair—not just a metaphor.

Einstein's Revolution

Science without religion is lame; religion without science is blind.
Out of My later Years (1950)

Albert Einstein
(1879–1955)

Albert Einstein's life story is the stuff of modern legend. The amazing success of Einstein's methods abolished the prevailing notion since Newton and Hume that hypotheses may be derived only from observation—a dubious proposition to say the least. For Einstein, creative ideas leading to deductions (hypothesis~deduction) were just as important, if not more so, to generating theory as inductive generalizations from sensory experience (experience~induction):

> We now know that science cannot grow out of empiricism alone, that in the constructions of science we need to use free invention.... The more primitive the status of science is, the more readily can the scientist live under the illusion that he is a pure empiricist.

Yet there is evidence that Einstein veered too far toward the formal, especially in his later years. His biographer, Abraham Pais, remarks that in a lecture at Oxford, "On the Method of Theoretical Physics" (1933), Einstein says he is convinced that "pure mathematical construction enables us to discover the concepts and the laws connecting them, which gives us the key to the understanding of the phenomena of Nature." Pais comments that Einstein was inclined to overestimate the capabilities of the human mind, perhaps even his own. Einstein, remember, is the man who said that, had his theory of gravitation somehow been empirically disproved, for instance if that puzzling motion of the planet Mercury were not as he had predicted, it would have been "too bad for God"!

Einstein's theory of relativity led to entirely new ways of thinking about time~space, matter~energy, and gravity~radiation, as well as to many scientific and technological advances. From our point of view, Einstein's multiple successes rested, in part, on his uncanny ability to reconcile scientifically relevant complementary aspects as complementary pairs.

Wave~Particle

In 1905, Einstein used Planck's quantum theory to explain how charged particles such as electrons are emitted when electromagnetic radiation strikes certain materials, the "photoelectric effect." Einstein also proposed that electrons, besides emitting electromagnetic radiation in quanta, also absorb it in quanta (emit~absorb). Einstein's work demonstrated that electromagnetic radiation has the characteristics of both a wave and a particle. It acts like a wave because the fields of which it is composed rise and fall in strength, and like a particle because its energy is contained in separate "packets."

Einstein proposed that light is also composed of separate packets of energy, called "quanta" or "photons," which have some of the properties of particles and some of the properties of waves. This idea redefined the theory of light, much of which had come from Newton's treatise on optics. It also explained the photoelectric effect, the emission of electrons from some solids when they are struck by light. Television and other inventions are practical applications of Einstein's discoveries. The notion of wave~particle also led to a controversy possibly as important as the mind~body debate: Is an electron *actually* a wave that acts like a particle, or is it *actually* a particle that can act like a wave? Sound familiar? The answer, which Einstein never liked, is that *both* are necessary for a full understanding of light's properties. Wave and particle behavior furnish contrary pictures of light and matter. And that was the dropped ball picked up by Niels Bohr: *Contraria sunt complementa*.

Time~Space

In 1908, H. Minkowski, a Russian mathematician, couldn't have heralded the complementary nature better when he remarked:

Henceforth space by itself, and time by itself, are doomed to fade away into mere shadows, and only a kind of union of the two will preserve an independent reality.

From its beginnings in an essay written at age 16, Einstein's special theory of relativity showed that time and motion are relative to the observer, assuming the speed of light is constant and natural laws are the same everywhere in the universe. This paper introduced an entirely new concept to the world. Although Newton was by no means comfortable with the concept of absolute space, absolute rest, and "action at a distance," in his mechanics, time and space were assumed to be separate and independent.

Subsequent to the special theory of relativity, the intuitive notion that light must be carried by a medium, the aether, was permanently laid to rest. Light was divested, as Abraham Pais has noted, of its chief mechanical property, absolute rest. Likewise, in relativity there is no absolute time, but rather a connection between time and signal velocity. From Einstein onward, the separation of time and space was replaced with time~space, obviously a most important complementary pair. With the arrival of special relativity, classical mechanics gave way to a new kinematics that embraced the complementary nature.

Energy~Matter

In the nineteenth century, the conservation of energy and the conservation of mass were considered two separate and distinct natural laws. Not so after the changes wrought by Einstein's special theory of relativity in 1905. Relativity says that mass is a form of energy. Ask anyone off the top of their head to name the most famous mathematical formula in the world, and they are likely to say "$E = mc^2$." This is, of course Einstein's most famous equation, known as the "energy-mass" relation. It says that the energy (E) inherent in a mass (m) equals that mass multiplied by the velocity of light in a vacuum squared (c^2).

The formula implies, among other miracles, that energy and matter can each be converted to the other. One of the startling implications of this equation is that there exists a shocking amount of potential energy in a very small bit of matter. Of course, the energy~mass relation doesn't apply when you weigh yourself on your bathroom scale. Only if someone orders up an atomic bomb or wants to generate nuclear energy does this application of Einstein's famous formula come into play. Nevertheless, ever since $E = mc^2$, energy~matter has been considered a complementary pair. And the world has certainly never been the same since.

Gravity~Radiation

Sometime around 1907, Einstein had the "happiest thought of his life." There he was, ruminating in the patent office at Bern, when all of a sudden a thought occurred to him: "If a person falls freely, he will not feel his own weight." Also, "if a person while in free fall drops some bodies, then these remain relative to him in a state of rest or uniform motion, *independent of their particular or chemical nature*." Einstein was startled. This simple thought made a deep impression on him. The matter-independence of the acceleration of fall impelled him toward a theory of gravitation, requiring him to extend his special theory.

The great intellectual breakthrough that Einstein made in his general theory of relativity was that he renounced the flatness of space. The consideration of all possible kinds of relative motion requires that space is curved, the amount of curvature dictated by the density of matter. Matter, by its gravitational action, "shapes" space. Gravity then is not a force, but a curved field in the space~time continuum created by the presence of mass.

Here again, this hasn't much to do with the proverbial apple falling from the tree, whose motion is adequately captured by Newton's laws. Einstein's general theory reproduces all of Newton's results in the limit of very weak gravitational effects. Worldwide fame came to him in 1919 when the Royal Society of London announced that the predictions of the general theory had been confirmed. Not only was time~space a complementary pair, it was *mutable*—both time and space could be bent!

Einstein's Blind Spot: Complementarity

All of Einstein's reconciliations may be said to have involved complementary pairs. In one lifetime tour de force, he reconciled the complementary pairs of space~time and energy~matter, aspects whose unification lie at the very core of understanding the physical universe. Of course, this is a kind of interpretation, but it is the way great science works. Things that are disparate "come together" in the hands of geniuses like Maxwell, Newton, and Einstein. However, as we noted above, there was one complementary pair that Einstein was not able to reconcile, at least in his own mind. Despite his great discovery of the photoelectric effect, Einstein could not accept the complementarity interpretation of wave and particle.

Paradoxically, as early as 1909, according to Pais's magnificent biography *Subtle Is the Lord*, Einstein had already written that "the next phase in the development of theoretical physics will bring us a theory of light that can be interpreted as a kind of fusion of the wave and the particle theory." He himself had demonstrated that when an atom is excited it is not possible to predict either the time or the direction in which the photon is emitted. This disturbed him greatly because it violated his classically grounded notions of prediction and causality.

Einstein believed that if one knows the initial state of an isolated system, it should be possible to predict its behavior at some later time. But this was not necessarily so, as he himself had shown. Later on, when not only light but also matter were shown to exhibit particle~wave behavior, Einstein could not accept their complementarity despite the concerted efforts of Niels Bohr to convince him so. He simply could not reconcile in his own mind that particle behavior and wave behavior—though seemingly mutually exclusive—are both necessary, both fundamental to understanding. We all have blind spots, and this was Einstein's. As a result, he set himself apart from the rest of the physics community. It is said of him that his fame was so well established by 1925 that he could have spent the last 30 years of his life fishing. In a sense that's what Einstein did in his own unique and solitary fashion.

Discovery~Implication

It is abundantly clear that some of the most important scientific discoveries have been the result of theoretical~empirical reconciliation of two polarized aspects that were formerly assumed to be separate and independent phenomena or ideas. In all such cases, what had previously been partitioned off as separate and independent forces or ideas were eventually replaced with complementary pairs. The great success story of the complementary pair in the history of philosophy and science is exemplified in the reconciliations of such incredible thinkers as Aristotle, Lao Tzu, Thomas Aquinas, Isaac Newton, Immanuel Kant, James Maxwell, Carl Jung, Albert Einstein, and Niels Bohr.

Of course, there are many, many more. Later on, in Movements 2 and 3, we will provide some contemporary examples of complementary pairs and their dynamics, in the brain, behavioral, and social sciences. Many crucial complementary pairs still remain to be reconciled, a challenge that confronts all who choose to pursue the paradigm of the complementary nature. The great danger in science and technology is that one aspect can seem so attractive as an absolute (e.g., genetic determinism) that its complementary aspects are ignored. As a result, the nature of complementary pairs so important for understanding the development and treatment of disease has yet to be comprehensively addressed. Similar situations arise in the social, cognitive, and brain sciences.

Given the stunning collective success of these reconciliations in physics, might they be indicative of a broader principle, possibly a deeper reality to be had? Might they be telling us that nature is complementary through and through? That the complementarity that emerged from the dual properties of waves and particles is merely one of many complementary pairs that make up the complementary nature? If so, might this also indicate that the reason we human beings can't seem to escape our fascination with complementary pairs is that human sentience is also an instantiation of the complementary nature? Perhaps along with the dogged pursuit of the details of the great twentieth-century reconciliations and their consequences, twenty-first-century science might also be greatly advanced by *mastering reconciliation*?

Now, some might suggest that if this could be done it already would have been done. But recall the mesmerizing impact of the Cartesian partition and its habitual interpretation of contraries in terms of the dualistic and mutually exclusive either/or valorization of complementary aspects. These powerful influences still dominate today. Science still proceeds largely on the basis of the "reductionist" prerogative alone, along with its addictive pursuit of almighty mechanism. Yet slowly, this mind-set seems to be losing ground. A revolution seems to be brewing in a number of quarters.

Will this revolution in science involve complementary pairs? It's a good bet: The greatest scientific discoveries of the twentieth century were based on reconciliations. Might humanity be hovering at the brink of a new understanding of complementary pairs? Might we be on the brink of a scale-independent scientific principle of reconciliation that underlies our philosophy of complementary pairs? Let's give a tentative yes to both of those questions. Clearly, satisfying answers to the deepest questions of our time are unlikely to lie in pure reductionism or pure holism ("emergentism" as it is now called). They may, however, be hidden in the theory of coordination dynamics which recognizes the complementary nature of both.

COMPLEMENTARY PAIRS ARE DYNAMICAL

For nature is a perpetual circulatory worker, generating fluids out of solids, and solids out of fluids, fixed things out of volatile and volatile out of fixed, subtle out of gross and gross out of subtle, some things to ascend and make terrestrial juices, Rivers and Atmosphere; and by consequence others to descend ...

"An Hypothesis Explaining the Properties of Light" (1675)

Isaac Newton
(1642–1727)

From Dynamical Metaphors to Dynamical Science?

Many years of dedicated research were required to establish the complementarity viewpoint of quantum mechanics. Although complementarity is still not universally accepted, to use the words of the eminent physicist John Archibald Wheeler, "there is no going back on it." In classical science, if something can be described in two mutually exclusive ways, one of them must be wrong. Such a climate hardly favors acceptance of the both/and *and* either/or perspective! The more comprehensive view of the complementary nature that we seek is still in an embryonic stage of development. In our opinion, it will require at least two kinds of reconciliation and a great deal more research.

First is a scientifically testable reconciliation of the four main interpretations of complementary pairs. Second is a reconciliation of the philosophy of complementary pairs with a science that, in order to fully capture the richness of complementary pairs, will have to be dynamical in some way. So, in closing Movement 1, let's address both.

How might one reconcile all interpretations of complementary pairs? We envision a scenario in which everyone who has an interpretation of complementary pairs is partially right. Yet being partially right is insufficient in itself to capture the nature of the complementary pair. How can all four positions be partially right yet be insufficient on their own? Let's put this question another way: How could advocates of one particular view come to the conclusion that that *their* view is the only correct and abiding one? Two possibilities come to mind:

1. *The complementary pair in question conforms to multiple interpretations.* In other words, it is able to behave in multiple ways, at least one of which can be interpreted in the way one interprets it to behave. Like the Necker cubes that adorn these pages, and also like quantum mechanics, *how* you look at a complementary pair influences what you eventually see. As preposterous as it might seem as a scientific principle, this is actually the case in the wave~particle complementarity interpretation of quantum mechanics.
2. *Complementary pairs are not crisp, but vague and uncertain.* Such ambiguity leaves the possibility open for diametrically opposing or contradictory premises to coexist. As a matter of fact, quantum mechanics also entails such uncertainty. As expressed by Heisenberg's Uncertainty Principle, it is impossible to measure simultaneously the position and momentum of a particle with an accuracy beyond that set by Planck's constant.

But notice, we aren't just talking quantum mechanics here; it only sounds as if we are. In terms of our prescientific metaphorical description of the philosophy of complementary pairs, what this means is that somehow:

1. *Complementary pairs are multimodal and dynamical.* In such a dynamical account, each complementary aspect considered in isolation represents a polar mode that exists only as an idealization. Complementary pairs are capable of change, transforming in space~time depending on circumstances.
2. *Complementary aspects can behave both as "tendencies" or "dispositions" and as well-defined states.* This raises the possibility that complementary pairs are open to multiple interpretations.

According to the foregoing metaphorical analysis, each of the complementary aspects can persist and change in a unique and distinguishable fashion, as if a defining boundary existed between them. Yet, at the same time, each influences the other's persistence and change, as if no defining boundary existed between them. In other words, each affects the other's behavior and capacity to behave. Moreover, the dynamics of complementary aspects are susceptible to moment-to-moment alteration. One moment, complementary aspects can seem to be in opposition, discernible and distinct, exhibiting individualistic tendencies. The next, they seem to merge, no longer existing as separate and distinct aspects but rather as the tendencies of some kind of as yet unidentified "multifunctional dynamic." As mutually coupled dynamic tendencies, complementary aspects are inextricable, free of duality and dichotomy.

By emphasizing that complementary pairs are dynamic and dispositional, we mean that it is not just the apparently static character of each complementary aspect that is important. As tendencies in complementary pairs, complementary aspects evolve moment by moment. Emphasis can change from one complementary aspect to the other and their dominance can switch. In addition, these polar tendencies can actually coexist! Defining boundaries between categories can disappear; at one moment complementary aspects may merge into one another and the next moment segregate and coexist simultaneously. This sounds a lot like Heraclitus again:

Upon those that step into the same rivers different waters flow.... They scatter and ... gather ... come together and flow away ... approach and depart ... ;

And the same thing there exists in us living and dead and the waking and the sleeping and young and old; for these things having changed round are those, and those having changed round are these.

Or Fritjof Capra:

The dynamic unity of opposites can be illustrated with the simple example of a circular motion and its projection (onto a line).... The circular movement will appear as an oscillation between two opposite points, but in the movement itself the opposites are unified and transcended.

These timeless metaphorical images of Heraclitus and Capra are beautiful ones, and resonate very well with our philosophy of complementary pairs. But dynamical explanations of complementary pairs based purely on philosophical metaphor (including ours so far) are insufficient to satisfy our current aim of producing a comprehensive scientific theory of the complementary nature. For complementary pairs to be taken as objects of science in their own right, they must be in some way tangible, demonstrable, testable, and repeatable.

Coming up shortly is a scientific account of complementary pairs that is based neither purely in root metaphor nor purely in quantum mechanics. It ties polar complementary aspects to *nonlinearity* in coordination dynamics, and their mutual interplay to coexisting tendencies that arise in the metastable regime of the coordination dynamics. Such coexisting tendencies appear to be one of, or perhaps *the* signature feature of how human brain~minds work, both singly as individual entities coupled to their worlds and together as human collectives (couples, groups, societies, etc.).

In a sense, the dynamics of coexisting tendencies frees us from contrarieties and the antagonism of opposites, from the habit of interpreting complementary pairs by using limiting either/or thinking. Yet it is these same contraries that constitute our ordinary macroscopic experience. Thus, the science that is needed must be able to describe this complementary world of ordinary experience that includes both polarization and reconciliation. It must, we think, be a science in which complementary pairs are not only conceptual descriptions of the complementary nature as Kant would have it, but must also constitute its actual phenomena.

TOWARD A SCIENCE OF COMPLEMENTARY PAIRS

Henri Poincaré
(1854–1912)

It is the possibility of parrying the same blow which makes the unity of these different parries, just as it is the possibility of being parried in the same way which makes the unity of the blows of such different kinds that can threaten us from the same point in space. It is this double unity that makes the individuality of each point in space, and in the notion of such a point there is nothing else but this.

Science and Method (1909)

So what is it that we are saying underlies the complementary nature of complementary pairs? Anticipating the transition from Movement 1 to Movement 2, let's list them for you now:

• We are saying that the nature of complementary pairs and, by extension, the nature of the either/or~both/and perspective can be grounded in multifunctional, metastable coordination dynamics (this is a mouthful, but bear with us).

• We are saying that coordination dynamics, to the extent that it explains how the human brain~mind works, lies at the core of why the world seems to come in pairs, and why we perceive, reflect, and express our experiences of these pairs in our nature and in our lives.

- We are saying that coordination dynamics gives us a means to understand complementary pairs—what they represent, how they work, and why they are important—and hence a means to understand ourselves.

- We are saying that complementary pairs and hence the complementary nature are fundamentally dynamical. There is no attraction without repulsion, no stability without instability, no coupling without components to couple, no segregation without integration, no cooperation without competition, no information without dynamics, no persistence without change. All of this has been said before, but we now wish to take the next step, which is to show how these and other complementary pairs arise directly from the science of coordination dynamics.

- We are saying that just as complementary pairs are intrinsic to coordination dynamics, so coordination dynamics reveals some startling things about complementary pairs—how they come about, persist, and change.

- We are saying that coordination dynamics offers a means to ground our philosophy of complementary pairs in a comprehensive and ubiquitous science of life, of everyday experience. Coordination dynamics, by virtue of its ability to exhibit coexisting tendencies, frees us from the antagonism of opposites, the either/or mind-set that has dominated Western thinking throughout the ages.

MOVEMENT 2   COORDINATION DYNAMICS

GROUNDING THE PHILOSOPHY IN SCIENCE

Samuel Beckett
(1906–1989)

The search for the means to put an end to things, an end to speech, is what enables discourse to continue.

The danger is in the neatness of identifications.

"Dante...Bruno.Vico..Joyce" (1929)

It is time to "put an end to speeches" about complementary pairs. If we are going to get to the heart of the complementary nature—to *touch it, see it, hear it, smell it,* and *taste it,* to *really* understand it—we must attempt to ground the philosophy of complementary pairs in science. This is the ending~beginning we seek that will, in Samuel Beckett's astute words, enable our discourse to continue. Being able to identify and describe complementary pairs and complementary aspects, though necessary first steps, is not enough. To proceed beyond words, beyond metaphor, we must attempt to determine what complementary pairs *actually* are, how they *actually* form~change, what they can *actually* do. This is just what science, through an organized process of theory~experiment, synthesis~analysis, is supposed to do: to discover aspects of nature and connections between them that weren't previously known or thought to exist. Unfortunately, this is often much easier said than done. As we have taken pains to point out in Movement 1, the dominant strategy used by almost everyone still remains to polarize in order to understand.

Overcoming Either/Or Thinking in Science

Contemporary science, for instance, relies heavily upon either/or hypothesis testing approaches. Indeed, the need to distinguish between alternative, competing explanations of phenomena has been drilled into most practicing scientists as the sine qua non of science. According to this line of reasoning, scientific progress is only really made if one can successfully choose between two—to falsify or reject one of two or more alternatives. By determining what is not, science hopes to slowly and steadily converge on what is.

Now there is little doubt that hypothesis testing and statistics are necessary and useful tools in many fields of scientific research. And certainly a case can be made that the more we have learned, the better our hypotheses have become. Nevertheless, many serious thinkers have voiced the nagging suspicion that the ease with which scientists (and people in general) propose tidy, either/or explanations of things may cloud a deeper, underlying reality. Perhaps the natural desire to understand nature has become waylaid by a preoccupation with the neatness of our identifications.

But what if the more frequently pursued either/or hypothesis testing methods could be successfully reconciled with both/and concepts and methods? A whole world of new knowledge might become available. Note that reconciliation doesn't require us to throw out contrary methods. This is an important point. We aren't saying that science should stop using techniques and methods that it has labored long to develop. What we are suggesting is that it might be productive to take either/or hypothesis testing methods down a notch, placing them in a more reasonable position as a complementary aspect with both/and methods. We are suggesting that either/or~both/and thinking constitutes a potentially useful paradigm shift, a novel and organized set of ways to think about and study life.

Now, are we implying that to successfully ground the philosophy of complementary pairs in the scientific method, the scientific method *itself* might need to be updated? This suggestion is less heretical than it might sound. In fact, there are many contemporary calls for a fundamental paradigm shift in the sciences, for a so-called Kuhnian revolution, made by scientists working on a wide range of fields and levels. For example, Huda Akil, herself a distinguished scientist and a former president of the Society of Neuroscience (a massive scientific organization composed of some 30,000 individuals), has called for "discovery-based" approaches to complement hypothesis-driven research in order to come to grips with the complexity of the brain and brain dysfunction. In her words,

We have long been aware that neural function requires the orchestration of numerous signals, across multiple cells and in well-defined circuits, but we have never had the tools to attack this level of complexity in a coordinated manner.

"Discovery science" is conceived to be a novel approach to knowledge acquisition in the "post-genomic" era, and will require novel tools and strategies to handle torrents of data generated by new science~technology.

But calling for a paradigm shift is not the same as making one. Although there are plenty of candidate paradigm shifts on the scientific floor today, those that rely heavily upon either/or thinking will most likely be problematic. In fact, either/or thinking is so insidious, it not only affects the development of the candidate paradigms themselves, but even influences how we go about assessing the merits of different candidate paradigm shifts once they have been proposed! Already, for example, a good deal of debate has arisen about whether discovery-based research should *replace* hypothesis-driven approaches, as if the two are somehow in direct competition. Such limited either/or thinking can be nipped in the bud by adopting a more flexible either/or~both/and stance.

Reconciling Reductionism~Holism in Science

Along with its love of hypothesis testing, contemporary science thrives on breaking things down into elementary units or building blocks—on explaining the larger through the smaller. This is referred to as the "reductionist program" of science. Thus, added to the practice of deciding what is by deciding what is not, reductionistic science aims to unify by splitting things apart! Holists, on the other hand, contend that by isolating the parts from their contexts we miss something essential about how things work together in nature.

In terms of Thomas Kuhn's thesis in his famous book *The Structure of Scientific Revolutions*, one could propose that the analytical/reductionist paradigm which boasts tremendous successes in fields like molecular biology has by and large *displaced* the older, synthetic/holistic paradigm. But what if, in keeping with the philosophy of complementary pairs, neither reductionist nor holistic paradigms was treated as the more fundamental in providing a comprehensive understanding of nature, life, and mind? Instead of trying to displace one of these paradigms for the other à la Kuhn, it may be much more productive to recognize the complementary nature of reductionism~holism, and trying to discover what this insight implies for the way we acquire knowledge (analysis~synthesis).

Sometimes it is a good thing to take things apart and break them down into simpler components or processes in order to understand them. This strategy, for example, has allowed us to design and build machines. It has allowed us to probe microcosms and discover the bizarre world of quantum mechanics. It has led to the discovery of DNA, and so on. On the other hand, it is evident to all but the most ardent Cartesians that nature is more than a machine of perfectly determined and logically ordered components.

Nature is full of uncertainty, noise, and error, and in many ways is not very machine-like at all. Furthermore, in nature, new properties emerge at every level of organization, which are not predictable from lower levels *in principle*. Such novel "emergent properties" are not at all unique to living things like plants, animals, and humans. As exemplified in Kelso's *Dynamic Patterns*, they are a general feature of nature's tendency to self-organize, that is, to form new structure~functions under certain critical conditions. The point to be made here, which follows the philosophy of complementary pairs, is that naked reductionism is inadequate *in its own right* to handle such emergent phenomena.

On the other hand, it makes no sense at all to *replace* or supplant reductionism with emergentism, as the Nobel laureate physicist Robert B. Laughlin has argued in his recent book *A Different Universe* (2005):

I think science has now moved from an Age of Reductionism to an Age of Emergentism, a time when the search for ultimate causes of things shifts from the behavior of the parts to the behavior of the collective....

What we are seeing is a transformation of worldview in which the objective of understanding nature by breaking it down into ever smaller parts is supplanted by the objective of understanding how nature organizes itself.

This falls into and perpetuates the old either/or trap. Any "science of complementary pairs and the complementary nature" must be able to show *both* how the parts of a system operate in context *and* how they coordinate to produce collective emergent effects. It must see these ancient contrarieties as complementary.

Understanding Multifunctional Dynamics Scientifically

If it can be grounded in science, the philosophy of complementary pairs offers a very attractive way around two great stumbling blocks of contemporary science: either/or thinking and one of its most bittersweet by-products, the artificial separation of reductionist and emergentist approaches to understanding. But if we are really to take this task of scientific grounding seriously, where should we begin?

Well, for starters, we must certainly provide evidence of "real" complementary pairs in some relevant, testable manner. This is not trivial. Remember that Immanuel Kant tacitly held that complementary pairs weren't real, but rather mental constructs that human minds somehow generate to make sense of the world. On the other hand, both Hegel and Marx thought that complementary pairs were actual entities, the endless process of thesis~antithesis acting as the fabric of reality itself. The science of complementary pairs will do well if it can shed some scientific light on this old debate.

Recall also that the philosophy of complementary pairs conceives of complementary aspects as idealized polar tendencies of an as yet to be identified "multifunctional dynamics." If multifunctional dynamics is crucial to the whole notion of complementary pairs, then it also lies at the heart of the complementary nature. Again, the aim is to provide ways~means to study multifunctional dynamics scientifically.

Let's concentrate on these two areas as we begin the search for a scientific grounding of the philosophy of complementary pairs: demonstrating real complementary pairs in a scientific context, and elucidating multifunctional dynamics. A big question is: Is there a previously established, empirically grounded scientific theory that can actually handle both aspects, one that demonstrates "real" complementary pairs and provides insights into how this world of idealized polar tendencies and multifunctional dynamics might be instantiated in nature?

Throughout Movement 2 we'll show that coordination dynamics provides plenty of tangible evidence of complementary pairs. Furthermore, coordination dynamics offers a theoretical~empirical paradigm with which to understand the essential multifunctionality of living things. It may also offer insight into why people dichotomize in the first place. So why do human beings split their world into pairs? How does the human brain~mind really work? If convincing answers to these two basic questions were found, it might significantly alter the way we think about our very existence. And it might well alter the way we behave in situations when it seems that there is no option but to polarize and choose between a rock and a hard place.

A BRIEF DIVERSION INTO QUANTUM MECHANICS

Erwin Schrödinger
(1887–1961)

I am—naturally enough—in love with "my" great success in life (viz. wave mechanics), reaped at a time I still had all my wits at my command (1926 at the age of 39) and therefore I insist upon the view that "all is waves." Old-age dotage closes my eyes towards the marvelous discovery of "complementarity."

letter to John Synge, from Walter Moore, *Schrödinger, Life and Thought* (1989)

Werner Heisenberg
(1901–1976)

The more I think of the physical part of the Schrödinger theory, the more abominable [*abschaulich*] I find it. What Schrödinger writes about *Anschaulichkeit* [visualizability] makes scarcely any sense, in other words I think it is *Mist* [bullshit].

letter to Wolfgang Pauli (1926)

The Wave~Particle Debate

To set the stage for all of this, though, it will be helpful to consider briefly a historically related example of how science handled a very fundamental paradox in the field of physics: How could physical reality, the fabric of nature, be *both* fundamentally continuous (wavelike) *and* discontinuous (particle-like)? It took a very special person to provide an account of that. His name, of course, was Niels Bohr. In our opinion, Bohr achieved this stunning enlightenment via a childlike enthusiasm coupled with a masterful stroke of deep and contemplative insight. Bohr put aside some basic assumptions, an activity we favor: "How wonderful that we have met with a paradox," said Bohr. "Now we have some hope of making progress."

In the early days of quantum mechanics, battle lines were quickly drawn in the struggle between the advocates of wave theory and those of particle theory. On one side was Erwin Schrödinger. A self-confessed philanderer, Schrödinger formulated the mathematical theory he called "wave mechanics" during a tryst in the Austrian Tyrol over one snowy Christmas holiday with a woman of undisclosed identity. On the other side of the fence was the much younger Werner Heisenberg, an Eagle Scout type. He formulated another, different kind of mathematical theory he called matrix mechanics that was particle-based. It explained how radiation interacts with matter using quantum jumps. Waves we associate with a continuum; jumping particles with discontinuity (continuous~discontinuous).

It was Bohr who finally provided a comprehensive interpretation of quantum mechanics, in which wave theory and particle theory are reconciled, as an "either/or~both/and." For Bohr, although the two descriptions of wave and particle behavior are mutually exclusive to an observer, both descriptions are necessary for the full understanding of atomic properties. The reason is that whether light behaves as a wave or as a particle depends on the choice of apparatus for looking at it. Thus, if light photons are observed with a particle detector, the photons behave as a stream of particles, like machine-gun fire. However, if light photons are observed with a wave detector, the photon behaves as a wave, like an ocean wave. What is going on here? Which is the *real* photon, the wave photon *or* the particle photon? But doesn't this sound a lot like Movement 1? There we even mentioned the complementary pair, wave~particle. And now we are telling you that at least in the case of quantum mechanics, wave~particle is not just a philosophical complementary pair. Wave~particle duality is a phenomenon from hard science—verifiable and so far yet to be "disconfirmed." Bohr concluded that the two sets of available evidence regarding light collected from the particle detectors and wave detectors, each having a consistent comprehensive theoretical explanation, were not contradictory, but rather *complementary*. What possessed Bohr to claim something so seemingly counterintuitive? Surely, light is *either* a wave *or* a particle! If this wasn't bad enough, Bohr went even further. At the atomic level, he asserted, the "state" of a system is undefined, having only the potential to take on certain values and probabilities. It is not a thing itself that is measured in quantum mechanics experiments, but the probability of a thing's occurrence. This idea was a radical one, and remains a difficult one to grasp intuitively even today. When we have an opportunity to look at a meter, such as an electricity meter, we normally assume we are measuring stuff flowing through it. But in quantum mechanics experiments, the "meter" actually measures *probabilities!* If after reading this last bit, you are baffled and mystified, don't panic. The great and charismatic theoretical physicist Richard Feynman was fond of claiming that no one, or at best only a couple of people in the world, could actually comprehend quantum mechanics.

That shouldn't stop us from trying to get the gist of this puzzling subject, though. According to quantum theory, all the possible states that an electron or photon can take on are spread out over all of space~time. Only when a measurement is made is the electron's state defined. It is thus the *interaction* between the measuring device and the entity being measured that collapses the probability wave function. A single electron, as the historian of science Arthur Miller pithily comments, changes from being "potentially everywhere to definitely somewhere." (Note the complementary aspects.)

Understanding the world of the incredibly tiny has some truly exotic aspects. For instance, "No elementary particle is a phenomenon until it is measured" is a mantra that the cosmologist John Archibald Wheeler is fond of reciting. This dependence of what is observed on the choice of the measuring apparatus made Einstein very unhappy. Nature shouldn't be like that. But for Bohr it was a welcome and inescapable new feature of nature that had far-reaching consequences.

We will return to the quantum measurement problem later on because the analogy of the "potential to actual" transition of quantum mechanics will be crucial for how coordination dynamics envisages the creation of new, functionally meaningful information. For now, we simply remind you that this set of ideas with its central tenet of complementarity came to be called the Copenhagen interpretation of quantum mechanics. "In the drama of existence," said Bohr, "we play the dual role of actor and observer." How about that?

You might be wondering what these diversionary musings on atomic physics and human knowledge have to do with our story. The everyday person, says the cynic, cares very little about this "dual nature" of the quantum mechanical microcosm. After all, no such ambiguity appears in classical science, in which objects that matter are at scales far removed from the tiny dimensions of Planck's quantum, right? When it comes to understanding biology, the "everyday" physics laid down by Isaac Newton surely prevails.

But this is just the point. If the twentieth-century achievements of relativity and quantum mechanics showed us anything, it was that Newtonian physics was limited. Remember, even the great Isaac Newton was waylaid with the intoxicating grips of either/or thinking: He was firmly convinced that observation was more fundamental than hypothesis. For Newton—and almost every scientist since seeking to understand the nature of anything from the level of genes to cognition and beyond—if two descriptions of the same phenomenon are mutually exclusive, then at least one of them must be wrong. Considering the amazing success of quantum mechanics, mightn't this be a debilitating and prevalent oversight in the life sciences and society at large?

What might happen if the scientific community at large were to opt for a paradigm change from such either/or thinking to either/or~both/and thinking? If you think about what happened when such thinking was accepted into the physics of the twentieth century, try and imagine what might be discovered if it was accepted and employed into the (maybe not so) mundane physics of biology.

Beyond the Quantum Scale

Quantum physics holds out the hope that a savior in the form of some new scientific genius will come along one day and find a way to describe the microworld of particles that will extend easily and without paradox to the macroworld of human beings. But what if we look at it in a different way? What if the "paradox" of contraries like wave~particle in quantum mechanics were also to exist in our descriptions of the macroworld of human brains and human beings? Just imagine, for example, if experiments, analysis, and theory showed that the human brain—a macroscopic object—was capable of producing two apparently contradictory, mutually exclusive behaviors. And what if it turned out that this wasn't exotic, but an aspect of the human brain's normal mode of operation, essential to its function? What if this same complementary nature was also seen to be a crucial aspect of human behavior? Might that not cause us to pause? What if such behavior, as is the case in quantum mechanics, was to be understood scientifically and described mathematically, and led to deeper insights about the nature of living things and how they come to know themselves and the world they live in? What if human brains, both individually and collectively, were shown to be embodiments of the complementary nature? Consider that to some extent they must be, if human brains are responsible for creating, perceiving, and expressing dichotomies in the first place.

What if there was a scientific theory that attested directly, without the slightest trace of vitalism, to the complementary nature inherent in human brains and human behavior? Might it tell us why our perception of the world appears to partition things into pairs? Might it also tell us that there is a more enlightened way to approach nature—a deeper reality that goes beyond superficial appearances, ourselves included? And finally, what are we up to with all these rhetorical questions? Well, imagine how exciting it would be if the answers to those questions were yes, yes, yes, yes, yes, yes, yes, yes, yes, yes, and yes!

And so we ask one more time with feeling: How might ideas, processes, and things that seem entirely different and separate actually be reconciled as inextricable complementary aspects of complementary pairs—for real? To see how, first appreciate that very little in life, even at the most elementary molecular level, happens unless two or more individual things come together.

Just as the author Graham Swift says in his marvelous Booker Prize-winning novel *Last Orders*: "Things come together in this world to make things happen, that's all you can say. They come together." But things don't just come together, do they? They must also be able to move apart again. This is the complementary nature at perhaps its most basic. How can things both come together and split apart? How, in other words, are things, processes, and ideas *coordinated*? And how can they be studied scientifically?

THE DEEP PROBLEM OF COORDINATION

Coordination represents one of the most striking, most taken for granted, and yet least understood features of nature. Imagine a living system whose component parts and processes, on whatever level of description one chooses to examine, didn't interact with each other and with their surrounds. Such an entity—whether a cell, organ, organism, system, or society—would possess neither structure nor function. Howard Pattee, one of the deepest thinkers of our time, a true conceptual biologist, goes so far as to say that he "does not see any way to avoid the problem of coordination and still understand the physical basis of life." We agree. If the first half of the twentieth century belonged to physics and the second half to unraveling the secrets of the gene, the hottest topic of the twenty-first century is going to be the problem of coordination—from molecules to macromolecules to organs, from individual human brains all the way to economies, societies, nations, and eventually (especially if Richard Branson has his way) *between worlds*.

How are complex living things coordinated? Coordination is not just matter in motion. It is a kind of functional order~disorder that evolves in both space and time that can be seen practically everywhere and anywhere we look. Coordination is the regulatory interactions among genes that affect how an organism develops and how some diseases like cancer occur. It is the coherent adaptive responses of organisms to constantly varying environments, like the coordination among nerve cells of the brain that underlie our ability to attend, perceive, think, decide, act, learn, and remember. Coordination is the miracle of a child's first word, the fingers of the pianist playing a concerto, or people working together to achieve a common goal. Look around you: Everything we are and everything we do is an expression of coordination!

The word "coordination" is frequently used and sometimes abused. It is found in all kinds of contexts. Listen, for example, to the philosopher Patricia Churchland, writing in the journal *Science* a few years ago on "Self-Representation in Nervous Systems" (we have used boldface for emphasis):

> The most fundamental of the self-representational capacities probably arose as evolution stumbled on solutions for **coordinating** inner body signals to generate survival-appropriate inner regulation. The basic **coordination** problems for all animals derive from the problem of what to do next. Pain signals should be **coordinated** with withdrawal, not with approach. Thirst signals should be **coordinated** with water seeking, not with fleeing unless a present threat takes higher priority. Homeostatic functions and the ability to switch between the different internal configuration [sic] for fight and flight from that needed for rest and digestion require **coordinated** control of heart, lungs, viscera, liver, and adrenal medulla. Body-state signals have to be integrated, options evaluated, and choices made, since the organism needs to act as a coherent whole, not a group of independent systems with competing interests.

Or listen to Stuart Kauffman, an internationally and self-proclaimed guru of "complexity theory," talk eruditely about life:

> For what can the teeming molecules that hustled themselves into self-reproducing metabolisms, the cells **coordinating** their behaviors to form multicelled organisms, the ecosystems, and even economic and political systems have in common?

And later, talking about "genomic systems":

> Were such systems too deeply into the frozen ordered regime, they would be too rigid to **coordinate** the complex sequence of genetic activities necessary for development.

It's pretty clear from these quotations that a lot of science is tied up in the word "coordination." So who, one asks, is doing all the coordinating? In their attempt to answer this question, philosophers and scientists alike have had a tendency to conjure up "homunculi"—sentient entities thought to be located inside a cell or other living system that are in charge of controlling how the cell or system coordinates its activities. The question that always remains is, if it is a little intelligent being who controls the coordinating of a system, who or what coordinates the little intelligent being, and where does the little intelligent being come from?

Seldom is it inquired what the necessary and sufficient conditions might be for coordinated structure~functions to arise in the first place. Words like "coordination" and its closely related kin—binding, grouping, orchestration, communication, integration, and so forth—are used almost as throwaway words, standing in for what one doesn't know or doesn't wish to explain further. But how are we to understand words like "coordination" and its kin? Are there laws and mechanisms that govern how patterns of coordinated activity evolve in space~time? If so, what form do they take, and how might these laws and mechanisms ultimately pertain to a scientific grounding of the philosophy of complementary pairs and thus to a deeper understanding of the complementary nature?

Now this may seem like a strange bunch of questions. Surely, the coordination of living things is far too complicated to ever be comprehensively described by scientific theory~experiment. But whether the last statement is accurate and informative depends a great deal on how one approaches the describing. If focus is placed only on "the stuff" of nature, then living things are made of ordinary matter, and are at some level of description just a large collection of atoms. In this description, the known laws of physics and chemistry apply, and the job of understanding, possible in principle but impossible in practice, is theoretically complete.

However, this approach seems to miss the point entirely. Just as a computer is much more than a large number of transistors and wiring circuitry, so a living thing is much more than a large number of constituent parts. Knowing what the parts of the computer are made of does not by itself tell us anything about what computers do. Likewise, the key to understanding a living thing lies in understanding how things interrelate and behave and not only in knowing what its parts are made of.

Basically, the problem of coordination is how individual coordinating elements and ongoing processes of any system at any level of description come together to form coherent patterns of behavior, along with the form that such coherence takes. All around and inside us are coordinated patterns of behavior that are dynamical, that change over time. If our insides were inert and motionless, we'd be dead. However, sometimes the changes taking place are so slow that they don't look to us as if they are changing. We call these coordinated patterns biological *forms* or *structures*—like nerves, muscles, bones, and skulls. This domain has traditionally belonged to the anatomist, who studies life from a structural aspect. But organisms are not just pieces of matter. They are quite literally matter in motion—animate forms. If, as the saying goes, morphology is the "royal way to function," then it must also be viewed from a dynamical frame of reference.

On the other hand, sometimes things change so quickly, so zippingly and flashingly relative to our normal frame of reference, that we speak of their coordination patterns as functions, like a neuron firing, a muscle contracting, a bone resisting, a heart beating, or a head nodding. This realm has traditionally been that of the physiologist, who studies life from a functional aspect.

But structure~function is a complementary pair: There are neither functions without structures nor structures without functions. Further, a structure at one level may act as a function at another. But by now it should come as no surprise that historically speaking it is the structural aspects like genes, proteins, cells, muscles, bones, nerves, organs that are thought of as foundational, to be the "parts," the components that compose living organisms. Yet structures are composed of coordinated spatial~temporal patterns of activity in their own right. Each is highly dependent on the others for its integrity and viability. This dual character of coordination, how things can be individual parts at the same time as coordinating together as a collective whole, seems like a paradox no less bewildering than the complementary nature of wave~particle.

HISTORICAL ROOTS OF COORDINATION DYNAMICS

Aharon Katchalsky
(1914–1972)

The possibility of waves, oscillation, macrostates emerging out of cooperative processes, sudden transitions, prepatterning, etc. seems made to order to assist in the understanding of integrative processes ... particularly in advancing questions of higher order functions that remain unexplained in terms of contemporary neurophysiology.
"Dynamic Patterns of Brain Cell Assemblies" (1974)

Over the last 25 years or so, due to the efforts of people working in many fields, a multilevel science of coordination has emerged—coordination dynamics. The main aim of coordination dynamics is to understand the coordination of living systems. This has been and is still being accomplished through extensive theoretical developments, experimental evidence, and field observations carried out in research institutes and laboratories around the world. Early forerunners whose work inspired the theory and research program of coordination dynamics include the biophysicist Aaron Katchalsky; the physiologist and biomechanist Nicolai Bernstein; the behavioral physiologist Erich von Holst; the neurobiologists Charles Sherrington, Roger Sperry, and Hans Lukas Teuber; and the developmental biologists Conrad Waddington, Viktor Hamburger, and Paul Weiss.

People from many disciplines have contributed to the seeding~flowering of coordination dynamics. A far from inclusive list includes the theoretical physicists Hermann Haken, Gregor Schöner, Armin Fuchs, Gonzalo De Guzman, Mingzhou Ding, Tom Ditzinger, Andreas Daffertshofer, Til Frank, Peter Tass, Rudolf Friedrich, Gottfried Mayer, Yoshiki Kuramoto, Jürgen Kurths, and (especially) Viktor Jirsa; the applied mathematicians Ian Stewart, Paul Rapp, Steve Strogatz, Richard Rand, Phil Holmes, Bard Ermentrout, Martin Golubitsky, Tim Kiemel, and Nancy Kopell; the physical chemists Ilya Prigogine, Agnes Babloyantz, and Gregoire Nicolis; the theoretical biologists Robert Rosen and Howard Pattee; the neuroscientists Gerry Edelman, Emilio Bizzi, Sten Grillner, Karl Friston, Giulio Tononi, Olaf Sporns, Tom Holroyd, Steven Bressler (a former student of Walter Freeman), György Buzsáki, Erol Başar, Wolf Singer, Charlie Gray, KJ Jantzen, Doug Cheyne, Hal Weinberg, and the late Francisco Varela and his associates; the experimental psychologists Michael Turvey, Betty Tuller, Carol Fowler, Claire Michaels, Richard Schmidt, Claudia Carello, Robert Shaw, Guy van Orden, and William Warren; and developmental psychologists and kinesiologists including the late Esther Thelen, Linda Smith, Beverly Ulrich, Jane Clark, Jill Whitall, Mary Ann Roberton, Geert Savelsbergh, David Sugden, and many others.

There have also been extremely significant contributions from conceptualizers and synthesizers who have cut across fields, and hence have appeared sometimes on the fringes of coordination dynamics. But in fact, their contributions have been central. These include the late Arthur Iberall, Eugene Yates, Peter Kugler, and Arnold Mandell. Having its origins in studies of the coordination of movement, generations of movement scientists continue to expand the horizons of coordination dynamics. A small sample includes Anatol Feldman, Peter Beek, Reinoud Bootsma, Eric Bateson, Benoît Bardy, Lieke Peper, Eliot Saltzman, David Goodman, Ken Holt, Bruce Kay, Tim Lee, Richard Carson, Winston Byblow, Nia Amazeen, John Jeka, Pier Zanone, Julien Lagarde, John Buchanan, Romeo Chua, Stephen Wallace, Olivier Oullier, David Ostry, Stephan Swinnen, John Scholz, Chuck Walter, Kevin Munhall, Mark Latash, Jean Jacques Temprado, Karl Newell, Dagmar Sternad, Jeff Summers, Ramesh Balasubramaniam, Paul Treffner, Philip Fink, and Patrick Foo. The philosophers Andy Clark, Peter van Gelder, Pim Haselager, Eunice Gonzalez, Tony Chemero, Fred Keijzer, Teed Rockwell, Evan Thompson, Alicia Juarrero, Onno Meijer, and (especially) Maxine Sheets-Johnstone have provided a careful reading of the basic elements of coordination dynamics and addressed its implications for a deeper understanding of mind, body, brain and behavior in their own work. These are only a few of the many researchers with ties to the transdisciplinary, multilevel science of coordination dynamics. Scientists in the social, behavioral, and economic sciences, the brain and cognitive sciences, developmental biology and even molecular biology, have embraced some of the central ideas of coordination dynamics and are now elaborating them both conceptually and empirically.

From its earliest beginnings, coordination dynamics has stressed the coevolution of real organisms coupled to and acting in real environments, a view captured in the term "embodied cognition." Applied fields such as robotics in the hands of people like Randy Beer are now using an approach long advocated by coordination dynamics. Its central tenet of informationally coupled dynamical systems is being used to tackle problems such as path planning and obstacle avoidance. In medical contexts, the concepts, methods, and tools of coordination dynamics are being applied to various brain and developmental disorders and to the enhancement, rehabilitation, and recovery of function.

As emphasized above, coordination dynamics is the culmination of a great deal of work on the part of many people from many fields, spanning many years of research and development. The work is ongoing; it is work-in-progress. Like a good beer that takes time and variety to brew to perfection, coordination dynamics is a genuine blend of experiments and theoretical insights that have been successfully formulated mathematically and implemented computationally.

A broad range of coordinative phenomena have been studied and explained with the concepts, methods, and tools of coordination dynamics. At the same time, coordination dynamics is constantly being elaborated on and extended in numerous ways. In order to provide a less technical description of it, only the most essential concepts and principles of coordination dynamics will be explained here. We don't want to get too bogged down and lose sight of the target, which is to see how coordination dynamics demonstrates real complementary pairs, and how it then has the capacity to help reveal the complementary nature in a novel and useful way. The goal, remember, is to ground the philosophy of complementary pairs in coordination dynamics, a science that belongs to the complex everyday world of human brains and human beings.

WHAT IS COORDINATION DYNAMICS?

Coordination dynamics, the science of coordination, is a set of context-dependent laws or rules that describe, explain, and predict how patterns of coordination form, adapt, persist, and change in natural systems. Through an intimate, complementary relationship between theory and experiment (theory~experiment), coordination dynamics seeks to identify the laws, principles, and mechanisms underlying coordinated behavior among different types of components in different kinds of systems at different levels of description. It explicitly addresses coordination within and between levels in such systems. For instance, it offers a means to relate phenomena occurring "inside" a system (like what is going on in your brain while you are reading this page) with phenomena that are going on outside that system (like the movements of your eyes as you scan this page).

The whole~part paradox is also directly addressed by coordination dynamics. Regardless of whether the individual parts of an organism are simple and homogeneous or complex and heterogeneous, there are a lot of them. Moreover, these parts are made up of parts, which are made up of parts, and so on. Of course this stretches all the way down to the strange world of subatomic waves~particles, *and beyond*. As we said in the Prelude, a whole is a part and a part is a whole. All these enumerable parts must be interrelated, and this pattern of interrelatedness evolves in time. In science, we use words like "coupling" and "interaction" to refer to such patterns of interrelatedness.

Coordination dynamics aims to characterize the nature of the functional coupling in all of the following: (1) within a part of a system, as in the firing of cells in the heart or neurons in a part of the brain; (2) between different parts of the same system, such as between different organs of the body like the kidney and the liver, or between different parts of the same organ, like between the cortex and the cerebellum in the brain, or between audience members clapping at a performance; and (3) between different kinds of things, as in organism~environment, predator~prey, perception~action, etc. Note the complementary pairs already appearing in all of the examples given here.

In coordination dynamics, coupling is invariably "informational" in nature. By this, we mean that information is actively used to coordinate things. This is what distinguishes coordination dynamics from other theories of self-organization that may include measures of information but do not actually use *meaningful* information as a basis for emergent self-organization.

Somehow, under "open, nonequilibrium conditions" (a mouthful we'll unpack a little later) interaction gives rise to both simple and complex spatial~temporal patterns. Such patterns themselves have a dynamic, which may be simple and repetitive, complex and irregular, and even a mixture of both (simple~complex). Starting from specific contexts and circumstances, coordination dynamics tries to discover the key variables and parameters that describe these emergent coordination patterns, and also the rules or laws that give rise to them—the pattern dynamics.

Coordination dynamics describes and explains numerous observations of the way things come together in space~time and, in a complementary way, how they split apart. But it goes a couple of steps further. First, it shows concretely how tendencies for the parts to remain separate coexist with tendencies for the parts to work together as a coherent whole. This coexistent mix of dynamic opposing tendencies appears to be the basis for how human brains operate, and is characteristic of the way living things behave in general. The ubiquitous tendency of the human mind to parse reality into binary pairs is explained in terms of the metastable coordination dynamics of the brain in which coexisting tendencies (integration~segregation, part~whole) occur at the same time. Second, coordination dynamics suggests that a unique blend of self-organizing integration~segregation tendencies gives rise to, in fact *creates*, new functional information. Once created, this functional information can modify, guide, and direct the dynamics. Both the phenomenal coexistence of inextricable yet opposing tendencies and the unity of information and dynamics in living things point to a scientific grounding for complementary pairs and strongly indicate the complementary nature of human existence. How this is achieved and why it is important will occupy much of the rest of this book.

THE MAIN IDEAS OF COORDINATION DYNAMICS

The essential elements of coordination dynamics can be communicated via a small set of main ideas, each supported by a number of related concepts and facts. We consider this an open and evolving list, something to be extended over time rather than to be regarded as set in stone. As coordination dynamics continues to develop as a theoretical~empirical research program, and as more fields begin to embrace problems of coordination, further insights are certain to follow. There is still a great deal to learn. What follows is a brief synopsis of the core ideas that aims to capture what is known so far about the essential nature of coordinated behavior at several different levels of organization. Later on we will delve a bit more into the background concepts, methods, and tools needed to understand them. But keep your eye on the bottom line here. The proposed scientific basis of the complementary nature rests on coordination dynamics and the insights it provides into the workings of human brains and human behavior.

1 Self-Organizing Coordination

Much evidence now exists showing that basic patterns of coordination can, under certain conditions, arise and change spontaneously in a *self-organized* fashion. Why is the concept of self-organization so central to the science of coordination? One reason is that when people see patterns in nature or in themselves, they tend to assume that there must be some kind of centralized causal agent (the homunculus again) responsible for creating and orchestrating these patterns. No matter at what level one looks, if observed behavior of a system is at all orderly, one naturally assumes that something or someone must be orchestrating this behavior. How could it be otherwise?

Ironically, this common assumption is surprisingly akin to the anthropomorphic explanations of ancient cultural mythos. Thunder, for example, must have a cause. Many ancients believed it was the roar of a god. In ancient Scandinavia, the word for thunder literally meant the Norse god Thor's roar. This human tendency to attribute events in the world to some deus ex machina, some ghost in the machine, is a difficult one to overcome, even in the savvy, computer-enhanced technological world of today. Witness the intelligent design movement in the United States, which seeks to undermine evolution.

Self-organizing systems, however, have need for neither homunculus-like agents located inside a complex system nor any kind of cosmic instruction from the outside ordering the parts around, telling them what to do and when to do it. Rather, under certain conditions, many individual coordinating elements interact with each other as well as with their surroundings, and literally organize themselves into dynamic patterns. Organization occurs without an organizer, coordination without a coordinator, switching without switches, etc. Coordination as a self-organizing process means not only that there is no ghost in the machine, but that there is also no ghost and no machine. Self-organizing systems are like having an orchestra that plays without a conductor.

Impossible, you say? Well, actually, it's not as far-fetched as it seems. On February 13, 1922, the First Symphony Ensemble of the Moscow Soviet, the Persymphense, amazed its audience of music lovers by performing without a conductor. The pieces played were not simple, either. They included the Third (*Eroica*) Symphony and a violin concerto by Beethoven. It is said that the orchestra's performance was so exquisite and the sounds so harmonious that professional musicians and critics alike left the concert totally mystified. For an entire decade, the enormously popular concerts confounded audience members, both public and professional, all seeking to unravel the mystery of the conductorless orchestra. How was this accomplished?

Usually, the orchestra is arranged on the stage so that everyone can see and follow the directives of the conductor. The Persymphense, however, was arranged so that everyone could see and hear everyone else. For example, the strings sat around in a circle, some with their backs to the audience. Granted, from the audience's frame of reference, it wasn't the most satisfactory arrangement for a group of performers, but what a strange and fantastic feat! The orchestra could dispense with the conductor because of their ability to interact locally and to cooperate with each other, much like a rock band or jazz combo but on a much grander scale.

The fact that complex coordinated patterns of behavior can emerge due to local interactions and without centralized control is, in the eyes of coordination dynamics, a self-organizing process. Just as new qualities of matter arise when a group of atoms behave as a single particle—the so-called Bose-Einstein effect—new *functional* states of coordination arise on a given level of organization when different components such as molecules, genes, neurons, and muscles aggregate together to form a single coherent entity. Such fleeting pockets of coherent activity arise in many contexts due to both short- and long-range informational coupling among participating components—in the case of the Persymphense, the players watching and listening to each other.

2 Dynamic Patterns *and* Pattern Dynamics

Like any scientific theory~experiment worth its salt, coordination dynamics tries to find key variables that capture the behavior of some system at some level, and to identify factors that influence those variables. The central observable phenomena of coordination dynamics are patterns of behavior, different identifiable configurations of relationships that appear again and again on multiple levels and in different contexts. Such patterns of behavior are characterized by "coordination variables," functional analogues of the collective variables or order parameters used in physics.

Context-dependent patterns of behavior emerge, persist, and often change over time or as a result of changing circumstances, and so possess a so-called "pattern dynamics." On a given level of description, this pattern dynamics may be formalized in terms of equations of motion of the coordination variables that capture the observed behavior of the natural system under study and even predict new patterns.

In a nutshell, in complex systems of organisms and their environments, coordinated patterns of behavior are formed by the process of self-organization. Such context-dependent patterns and their evolution in space~time are described by nonlinear dynamical laws. The collection of these laws or rules, the means by which they are obtained empirically, as well as their implications and interpretations are all part and parcel of coordination dynamics.

Self-organizing dynamic patterns and their emergent pattern dynamics compose one of the main complementary pairs of coordination dynamics (dynamic patterns~pattern dynamics). For coordination dynamics in particular and complex systems in general, one of the biggest scientific challenges is to discover relevant coordination variables and their pattern dynamics in a particular system, level, and context. In Movement 3, we shall provide at least two strategic ways to go about this particular task. But before we move on, let's discuss dynamic patterns and pattern dynamics as individual aspects in a little more detail.

Dynamic Patterns: Multifunctionality and Functional Equivalence

Consider some of the attributes of living things that tend to resist traditional explanations yet are inherent to coordination dynamics. Living things, from genes to brains, are multifunctional. Once again, this means that the same function can be generated by different sets of components, and different functions may be expressed by the same set of components. An example of multifunctionality is one's signature, which is normally written using a pen in one's hand. With a little motivation, most of us can also produce a recognizable signature with a pen attached to our big toe!

Multifunctionality means that very different components can be self-organized to achieve the same functional goal (in this case, making a signature). What about the contrary but complementary aspect, that is, the same components can be self-assembled to do different things? The very different activities of speaking and chewing, for instance, use many of the same pieces of anatomy (jaw, lips, tongue, glottis, etc.) yet are organized in task- or function-specific ways. Such functional equivalence is a crucial aspect of coordination dynamics.

As for different functions being expressed by the same structures, it is now well known that in nervous systems, the same neural circuitry can reconfigure itself or self-assemble according to current conditions to fit the needs of the organism. These circuits, identified and studied by neurobiologists, are called "pattern generators." Note the language of this last term and think about the logic it carries: "If there is a pattern, there must be a pattern *generator*." However, in coordination dynamics this is not an a priori assumption. In coordination dynamics, *laws* of pattern generation complement *mechanisms* such as pattern generators and vice versa. Coordination dynamics inquires into the necessary and sufficient conditions for generating and maintaining a given pattern of activity. According to Allen Selverston, one of the world's experts in the field, even though neural patterns observed experimentally can be highly similar, none of the so-called pattern generators uses exactly the same mechanism. A similar sentiment has been expressed very recently by the neuroscientist Eric Jorgensen. In his words: "the religious study of the wiring diagram will never lead to enlightenment." Context matters. Simply altering the secretion of the neurotransmitter dopamine can reverse the responsiveness of a cell.

A related example from genetics is the phenomenon of pleiotropy: The smallest change in a single base pair in DNA can result in the expression of very different molecular forms that in turn may result in a variety of developmental diseases and syndromes. Thus, very different phenotypes for different traits can emerge as a consequence of tiny changes in the same gene. Just as a single word can have multiple meanings depending on context, so pleiotropy endows a single gene with the ability to express multiple functions. Along with its partner polymorphism (the fact that there are literally millions of variations in the DNA sequence between any two individuals), pleiotropy is the sine qua non of *normal* genetic function. Multifunctionality is an essentially nonlinear property of all living things, from genes to cells to neural circuits to behavior. From the present perspective, multifunctionality is a natural phenomenon, and is understood in terms of multiple coexisting tendencies in the underlying coordination dynamics. Such coexistence of states and tendencies in a complex system is to be seen as a common, if not universal event.

Recruitment~Annihilation

No matter at what level one looks, one can't help but marvel at the ease with which living things are able to switch flexibly from one pattern of coordination to another. Parts and processes can be *both* selectively engaged or "recruited" *and* disengaged as the needs of the organism and the demands of the environment change. For instance, we observe this flexibility in coordination dynamics at the neural level, where some neurons can join a group to accomplish a function, at the same time as other neurons are departing to join other groups to do other things. Likewise, when a person is unfortunate enough to suffer a stroke, parts of the brain not typically used, say, for gripping a golf club or remembering a word, may be recruited to compensate for the deficit.

Coordination dynamics has demonstrated quite conclusively in theory and practice that *both* recruitment *and* annihilation processes go on all the time and at the same time. This is due at least in part to the redundancy inherent in complex living things. Redundancy means that there are more than enough degrees of freedom available for the system to fulfill its functions in more than one way. Although a source of complication in the design of robots and the bane of engineers, redundancy is a wonderful and essential source of flexibility for living things.

Pattern Dynamics: Multistability, Instability, and Bifurcation

Recruitment~annihilation, switching, and other such complex behaviors arise as a result of the inherent and essentially nonlinear nature of coordination dynamics. Stability and instability are complementary aspects that play key roles in all of them. Stability is important because an entity that is stable and robust can resist perturbation. Stability allows a coordination pattern to adapt to changing circumstances without breaking apart. Stability guarantees that regardless of where the system starts from and regardless of the multitude of paths or trajectories it takes, it will nevertheless arrive at the same goal. From a conceptual point of view, stability is intimately connected to "functional equivalence," the ability to achieve the same goal through different paths and with a variety of effector elements. There are many different routes to Rome, and many ways to get there. Conceptually, the multifunctional nature of living things may be understood scientifically in terms of the multistable nature of self-organizing coordination dynamics.

On the other hand, spontaneous switching among coordination patterns is not necessarily and certainly not only attributed to "switches" located somewhere inside that are turned on and off like a light switch. In coordination dynamics, switching and pattern selection arise as a consequence of *dynamic instability*. The most straightforward, intuitive explanation of instability is that a coordination pattern that had been stable up to some point in time can no longer be sustained under the current conditions and context. The pattern is said to have become unstable. To satisfy the circumstances it finds itself in, the system has to switch and do something else: a new coordination pattern must form. Numerous studies have demonstrated that fluctuations from both internal and external sources—often considered unwanted noise or variability in conventional theories—actually aid in the process of recruitment~annihilation and in switching off and on coordination patterns. Coordination dynamics says that if better ways are out there to fit the circumstances and context of a given coordination pattern, fluctuations will help the system (and us) discover and explore them.

Bifurcations constitute a rudimentary decision-making process in coordination dynamics: One dynamic pattern of coordination is adopted or "selected" over another. This is not a switch, per se, but a qualitative change that arises due to the intrinsic nonlinearity of the pattern dynamics. As a basic switching mechanism for all complex systems, bifurcations are found in all systems and levels, from gene regulatory networks to individual neurons to brains and human societies. Like complementary pairs, bifurcations are ubiquitous.

3 Functional Information Flows

The third main idea is that coordination dynamics deals in the currency of functional information. That is, coordination dynamics is primarily concerned with *informationally* based dynamic patterns and *informationally* based pattern dynamics. One might even say that the mathematical equations of coordination dynamics describe how functional information flows in time. Typically, the variables that capture coordination are context- and task-dependent informational variables. They capture the way elements of a system keep each other informed. What could be more important to a system's function than information about how its parts are coordinated?

Now, one has to be careful using the word "information" because it has multiple meanings. Ordinary spoken language, words on a page, messages transmitted over a telephone, images on a TV screen, and so forth are all thought to be forms of information. Many biologists define information in another, less familiar way. They say that "the molecule is the message" (i.e., the molecule is the information). For instance, the sequence of base pairs in the helical structure of DNA is considered to be akin to a series of words that are said to specify, through some pretty sophisticated molecular machinery, the amino acids that form each protein.

Hardly any description of a biological system comes without using the terminology of the information sciences. Words like *input, output, receiver, receptor, code, messenger, encoding, readout, transcription, translation,* and a host of others are indispensable to modern scientific description, and all are related to the word "information" and the "information sciences."

As is the case with the word "coordination," the word "information" is sometimes used in rather vague ways and to stand in for processes that are not so well understood. And again, it is too often assumed that some kind of homunculus, some internal "intelligent operator" must be controlling and coordinating the system as a puppeteer controls a puppet. Perhaps, as the role of self-organizing processes and their intimate relation to functional information become more appreciated and better understood, this centralist way of thinking will diminish. Self-organization and functional information are the complementary cornerstones of coordination dynamics. Like the pillars linked by Leonardo da Vinci's arch, together they make a strength.

What *Is* Functional Information?

Functional information is information that is *meaningful* and *specific* to any kind of coordinated activity. Operationally, information is functional to the extent that it produces an observable effect on a coordination pattern. If it does not produce an observable effect upon a coordination pattern, it is not considered to be functional information. Though it was not part of Darwin's original theory, the concept of functional information is intimately tied to the idea of natural selection. That's because, ultimately, information is functional if it helps an organism survive in its world. Just as the heart evolved to pump blood, organs like the brain evolved to support many kinds of coordinated functions that rely on information creation and information processing.

But the concept of functional information goes way beyond that. Information is functional if it allows people to communicate. We humans call our spoken functional information "language." Nonverbal body language, like gesturing, can also be used to communicate, and as it can impart an observable effect on a coordination pattern, it can be thought of as functional information. Information is functional if it allows people to drive cars, eat, speak, and make love. Information is functional if it helps people learn and remember. Functional information can take many forms, and many forms can realize functional information. Form and function are a complementary pair (form~function). As Samuel Beckett once said of James Joyce's writings: "Form *is* content. Content *is* form."

Sometimes the effects of functional information are direct and easy to see, as when coordination is destabilized and qualitative changes in behavior occur. We will show you how that works shortly. In other situations, delays may ensue between cause and effect that render the presence of functional information more difficult to detect (at least from the scientist's point of view, though not necessarily from an organism's). Development and aging are like this: What was functional information for an organism at an early stage of development may not be functional information later on, and vice versa. Likewise, what wasn't functional information early on in life may become highly relevant at later stages of development, sometimes to the detriment and even demise of the organism.

Returning to molecular biology for an example, certain genes, so-called functional polymorphisms, are known to code for enzymes that metabolize the neurotransmitter dopamine in the prefrontal cortex. People with an unusually active form of the enzyme COMT (which results in lower synaptic levels of dopamine) are at risk for schizophrenia. This is a manifestation that may not appear until late adolescence or early adulthood, when dopamine innervation of the prefrontal cortex is known to change. It thus appears as if the "susceptibility" gene found in schizophrenia alters the microenvironment in a manner that becomes functionally relevant later in life, when the symptoms of the disorder blossom.

Here, as in other examples, functional information lies in the coupling between intrinsic (e.g., genetic) factors and extrinsic factors, such as the surrounding environment (e.g., dopamine levels). Internal and external conditions often have to be just right for functional information to express itself. Context is crucial. For instance, the same gene products can lead to entirely different cell types depending on interactions with surrounding proteins.

According to coordination dynamics, functional information transcends the medium through which parts and processes communicate. This is not fancy talking nor idle speculation. Empirical evidence shows that functional information may be conveyed by mechanical forces, light, sound, smell, touch, intention, and memory. Thus, coupling or "binding" among coordinating elements and processes is mediated by functional information as well as by conventional forces. Functional information is "forceless," though paradoxically it can play a "forcing" function (see number 6 below).

Functional information may be structural or topological rather than material in nature, and can cause qualitative changes in the dynamics of the coordinating parts. To reiterate: "Bound" or coupled coordination patterns in coordination dynamics are informational. Information that changes these coordination patterns is what we call functionally meaningful information. The newborn baby turns toward its mother's voice. The gannet closes its wings before diving into the sea to catch its dinner. The fly prepares to extend its legs for a smooth landing on a piece of cake.

All these coordination patterns are based upon functional information, a patterned interaction between coordinating elements and processes that typically cuts across organisms and their environments. This is the big difference between things that we think of as being alive, like people, plants, and animals, and things like sand dunes and clouds that can be understood as self-organizing and dynamic, but are not alive in the usual sense of the word. This is also why coordination variables are very important to identify, because they constitute the functional information that a system uses to coordinate itself. In coordination dynamics, coordination variables are never arbitrary and context-free; they are always meaningful to a system's structure~function and therefore always context-sensitive.

This does not mean that every context requires a different coordination variable, nor does it mean that the same functional information cannot be used in different contexts. On the contrary, the same coordination variable has been shown to capture coordination among different kinds of things in many different systems and levels. Also, the same coordination dynamics has been shown to capture how the coordination of these different systems and levels evolves in time.

The laws of coordination, like ordinary physical laws, are largely matter-independent. We'll clarify this shortly. Put another way, they apply to all kinds of matter and media. The difference is that the laws of coordination are function-specific and context-dependent. They may be said to govern, or at least describe and occasionally predict, *the flow of functional information*. The key hypothesis is that context-dependent coordination dynamics, which stresses and attests to the enormous diversity of nature, is *complementary* to the "universal" context-independent "first principles" of physics that aim to unify nature. Nature cannot be understood without both.

To say that the laws of coordination are abstract and informational does not at all mean that these rules are not physically embodied, that they have no material instantiation. Laws of function are always materially embodied: Dynamic coordination patterns can only be recognized if they are embodied in material structure. This is the essence of complementary structure~function. It would have been impossible to identify coordination laws in the first place if they were not realizable in our physical world. By definition, the discovery of context-dependent laws of coordination rests on distilling them from the numerous and complex ways these laws are realized in the physical world as events (law~event).

Just as human beings can achieve the same goal in different ways, different mechanisms can realize the same law. It's worth remembering that it was precisely the experimentally established *matter-independence* of the acceleration of falling bodies that provided the most powerful stimulus for the general theory of relativity—its extension to coordinate systems that, relative to each other, are in nonuniform motion. Coordination dynamics, though realized by living matter, is matter-independent in the same sense. That is, the same concepts and principles apply regardless of which particular elements are coordinating.

4 Information Creation

The fourth main idea of coordination dynamics is that it offers an explanation for the *origin* of functional information. In many scientific fields, ranging from molecular biology to cognitive science, the concept of information processing is central. But where does information come from? How is it created, and how is the origin of information related to how information is processed? Surprisingly, fewer have raised the question of the scientific origins of information than you might think. One exception is the eminent physicist John Archibald Wheeler. Wheeler thinks reality itself is based on information. His catchy phrase "it from bit" captures the notion that understanding physical reality always requires a binary yes or no decision. Here's how he puts it (pardon the pun):

It from bit symbolizes the idea that every item in the physical world has at bottom—at a very deep bottom in most instances—an immaterial source and explanation. What we call reality arises in the last analysis from the posing of yes-no questions and the registering of equipment-evoked responses. In short, all things physical are information-theoretic in origin.

As otherworldly as such a position might seem, the cosmologist Wheeler is not alone. The very same sentiment is expressed by Carl Friedrich von Weizsäcker, a nuclear physicist who was a protégé of Heisenberg:

Within the frame of quantum theory I cannot think of any more fundamental physical object than a simple alternative. And this alternative is in the sense of being fundamental far beyond space. Space has to be built up from such alternatives not the other way around.

Think about it this way: If I have a coin in my hand and ask you whether it is a head or a tail, the answer could be said to rest on a choice between alternatives. In this manner of speaking, choosing between any two alternatives, no matter what they are about, always requires one bit of information. Especially in the fields of mathematics, a bit is often coded as "1" and "0," and is instrumentalized as the "on" and "off" of a switch, and finally as an "opening" and "closing" of a gate. Note the complementary pairs here: 1~0, on~off, opening~closing. According to quantum mechanics, out of a universe in which quantum indeterminacy rules, nature selects an alternative—a photon is not a photon until it is measured. Such a measurement constitutes an unsplittable bit of information. One can split an atom, but one cannot split a bit. Quantum mechanics implies that the process of observation and measurement *creates information.*

The way this is done in practice is that scientists and engineers build special detectors (usually at considerable expense) to observe quantal events. To track the minuscule phenomena of the subatomic world, these detectors must be exquisitely sensitive. The slightest change in the detector is intended to result in a gain of information about the microscopic quantal state. Practically speaking, the quantal detector is said to be prepared in a "metastable state." It is the transition from this metastable state to a more stable state that conveys the essential information regarding the otherwise undetectable submicroscopic quantal events. The late quantum measurement theorist, H. S. Greene (who worked with Paul Dirac and other founders of quantum mechanics) describes it this way:

> It is the observable transition between this metastable state and a more stable state that conveys the essential information concerning a sub-microscopic event that would otherwise go undetected.... The functional material of the detector must be macroscopic and in a metastable state which allows the quantal interaction to become manifest at the macroscopic level.

This is how contemporary physics views the creation of information in nature's microcosm. Note that in the information theoretic view of information just outlined, the context or the meaning of the information is irrelevant. A mathematical bit of information is exactly a mathematical bit of information, regardless of context.

What then of coordination dynamics? In coordination dynamics information is *functional* and refers to bound or coupled coordinative states that arise as a result of self-organization. Thus, coordination dynamics views the creation of information as a bifurcation, a kind of flexible, dynamical instability. A coordinating system has the capability of operating in a special regime where the slightest nudge will kick it from one coordination pattern into another, alternative pattern, thereby changing the system's behavior qualitatively, and thereby creating information: ... *the solution turned from black to blue. Now we know more about the solution, namely that it can exist both as black and blue forms. I the observer am now in possession of new information about the solution*. Note that we have to be careful how we use the words "creating information" and what we imply by them. Information is not created in the sense of having a creator. Nor do we assume there to be an independent observer. Rather, information emerges as a result of the inherently nonlinear, functionally self-organizing dynamics. Notice that we are talking about the system generating and conveying information itself, rather than an observer reading a meter in order to gain information.

Analogous to quantum mechanics, the necessary and sufficient condition for the emergence of information in coordination dynamics is metastability. However, in coordination dynamics, we don't refer to metastability as a "state." The reason

THE MAIN IDEAS OF COORDINATION DYNAMICS

is straightforward. The metastable regime of coordination dynamics *does not contain states*, stable or unstable! Rather, in the metastable regime it is competing *tendencies* that are coexistent and complementary. One tendency is for coordinating elements and processes to maintain their independence, while the other tendency is for coordinating elements to behave as an integrated unit.

In coordination dynamics, metastability is therefore not a state, but rather a disposition to behave. Only when the system switches into and out of a state or tendency is functional information created~destroyed. In coordination dynamics, coordination states and dispositions are not arbitrary and bitlike, though they may be describable in such terms. Rather, they are functional and context-dependent. By residing in its metastable regime, coordination dynamics provides the system with a mechanism for the creation and annihilation of informationally meaningful coordination patterns. The stability of information over time is guaranteed by the coupling between individual coordinating elements and constitutes a kind of dynamic memory, about which we'll have more to say later on.

To summarize, when it comes to information creation, both quantum mechanics and coordination dynamics rely on the mechanism of metastability. Quantum mechanics achieves this using intelligent agents, i.e., the physicists who build and operate devices and who use those devices to observe quantal events, thereby "gaining information." Wheeler himself recognizes this; indeed, it is central to his story. In contrast, in coordination dynamics, metastability arises as a coordination pattern of an evolutionarily constrained, self-organizing coupled dynamical system. In coordination dynamics, it is not necessary to propose sentient observers scrutinizing a system in order to gain information. For the human mind (which, among other miracles, invented~discovered quantum mechanics), functional information creation lies in the brain's metastable behavior. And for groups of humans, for the "collective mind," functional information lies in the metastable way human beings behave in a social and cultural context.

Finally, although both quantum mechanics and coordination dynamics may be interpreted in light of complementary pairs, we should note one further, obvious difference with respect to the issue of complementarity. It is helpful to remind ourselves of what Niels Bohr had to say about this subject:

The use of certain concepts in the description of nature automatically excludes the use of other concepts, which however, in another connection, are equally necessary for the description of the phenomenon.

In quantum mechanics the concepts of light as wave and light as particle are mutually exclusive but not contradictory. Both are needed, both are necessary. However, in the microworld of quantum mechanics, nature denies the possibility of observing both at the same time. In the case of everyday, human-scale coordination dynamics, two apparently contradictory tendencies—integration and segregation—may very well exist at the same time, though their effects, of course, are observed sequentially (see, for example, figure 2.16). In fact, according to coordination dynamics, opposing tendencies *must* coexist to make possible the creation of functional information.

So there you have it. The concept of information does not exist in classical, Newtonian mechanics. Hence it is very strange indeed to say, as some do, that "Newton rules biology," a discipline that is as information-rich as one can imagine anything to be. Information is, however, central to quantum mechanics, but requires an observer (though in actual fact, the measurement itself is an objective process). Coordination dynamics introduces and champions the concept of functional information, and shows that it arises as a consequence of a coupled, self-organized dynamical system living in the metastable regime where only tendencies and susceptibilities coexist. These coexisting dynamical tendencies, like the multiple narratives of a James Joyce novel, are what constitute one of the deepest aspects of the complementary nature.

5 Origins of Agency

Nowadays it is quite commonplace to hear talk about "emergence" in everything from condensed matter physics to the behavior of ecosystems to the economy and the mind itself. Novel behavior is thought to emerge from interactions among a large number of autonomous agents. This view is sometimes called the Santa Fe approach to complexity—after the Santa Fe Institute, a think tank in New Mexico that focuses on a mathematical~philosophical approach referred to as agent-based modeling.

In the Santa Fe approach, there is heavy reliance on computer simulations of so-called complex adaptive systems. The *agents* of agent-based modeling are hypothetical, and represent any element~process that can be found in large numbers and that is capable of interacting with others of its own ilk and with its environment. These hypothetical agents can be neurons of a nervous system, genes of a genome, traders in the stock market, birds of a flock, fish in a school, etc.

Although agent-based modeling sounds as if the agents themselves are central to theory, they are not, which is quite ironic. In fact, in agent-based modeling, the agents are considered to be—as one authority said—mindless. Putting it bluntly, the agents are as dumb as doornails; they do not possess agency at all. Think once again of water molecules in cloud formations. Storms can seem to "rage" with behavior even though the water molecules remain, well, just water molecules. As a metaphor of the agent-based modeling approach, the water molecules would be the agents we are describing, the element~process found in large numbers. In computer simulations of complex adaptive systems, all an agent does is alter its output based on its input, which may come from other agents and the environment. Ecological systems, economies, and human minds are imagined to emerge from a collection of mindless elements.

Paradoxically, real coordinated behavior of living things seems to be full of *real* agency (i.e., sentience) and goal-directedness (intention). From the very beginning, moments after birth, a baby's first glance at and scanning of its mother's face, "the turning toward," seems primed with meaning. And it can be confidently asserted that none but a handful of human beings have any doubt whatsoever that it is *they* that direct the motions of their own bodies. As Maxine Sheets-Johnstone says, agency is a matter of doing, and doing is a matter of coordinating elements. But even if "free will" is illusory, as some have argued, where does this sense of agency and directedness come from? Where does the "I" of awareness come from, and how does it work? These questions lead us to the fifth main idea of coordination dynamics: *Spontaneous self-organizing coordination tendencies give rise to agency*. Meaningful information is the joint product of a coordinated system of parts and processes that spans organism and environment. *Coordination establishes meaning*. An important corollary of the fifth main idea is that a most fundamental kind of meaningful information, the conscious awareness of self, springs from the ground of spontaneous self-organized activity.

As Sheets-Johnstone says, it is tempting to think of agency and intentionality in a "logocentric" way—that sentient agents follow "rules of action" that can be stated in words or as a series of if/then statements. In this line of thinking, agency is akin to a chess player playing a game of chess: "If my opponent does that, then I must do this, and so on . . ." But this is putting the cart before the horse. Among the many problems with this metaphor is a complete disregard for the nonlinguistic origins of intentional action that reside in primitive movements that go back before we were able to speak, and even before we were born. The basic spontaneous movements we are born with consist of a large repertoire of self-organized, coordinated actions, like making a fist, kicking, sucking. These activities appear to be built-in, but they are by no means rigid and machine-like. They constitute an "intrinsic dynamics" that enables coordinated activities to happen before we even know how to make them happen, or realize we are making them happen. Likewise, recent work by Rafael Yuste and colleagues at Columbia University shows that the neocortex is full of intrinsic dynamics: spontaneous activity that is far from random, but instead consists of transient preferred patterns. These self-organized dynamic patterns in the brain likely constitute the foundation upon which cortical and cognitive function is built. Broadly speaking, they are a realization of the complementary pair ordered~random and the metastable coordination dynamics of the brain.

As a fetus moves spontaneously within the womb, it eventually discovers arms and legs that flex~extend, a mouth that opens~closes, a body that bends~twists. In other words, the neonate begins to make sense of itself as a living thing. It comes to realize, through its own movements and the sensations they give rise to, that these movements are in fact its own. To see this, if a string is attached to a baby's foot that also moves a mobile, the baby, watching the mobile and feeling its tug, very quickly comes to appreciate that it is its own kicking movements that are causing the mobile to move—in ways that it likes. It is as if the prelinguistic baby is saying to itself: "These movements belong to me! *I am doing this!*" From spontaneous self-organized behavior emerges the self—"I am," "I do"—and from there a huge range of potentialities ("I can do"). "I-ness" thus arises from spontaneity, and it is this "I" that may be said to direct human action.

In sum, the fifth main tenet of coordination dynamics says that evolving processes of self-organization in real organisms coupled to real environments lie at the origins of conscious agency. This is not just a claim based on phenomenological experience, crucial though that is. Coordination dynamics also provides a scientifically testable mechanism: *metastability*. In the metastable regime of coordination dynamics, functionally meaningful information is created. This information can take the form of conscious agency ("I-ness") and hence is capable of steering a system's behavior.

To the extent that it is aimed at understanding the spontaneous self-organized emergence of patterns that form *without* the necessity of agency, coordination dynamics is like other nonequilibrium, nonlinear theories such as Ilya Prigogine's theory of dissipative structures, Arthur Iberall's theory of homeokinetics, and Hermann Haken's theory of synergetics. Coordination dynamics rests on the shoulders of these intellectual giants who recognized and established the central role of self-organization in nature long before institutes for studies of complexity arrived on the scene.

On the other hand, coordination dynamics is a theory of directed self-organization in which *both* spontaneous pattern formation *and* agency coexist and complement each other. In coordination dynamics, agentlike entities are not mindless but meaningful by virtue of the very self-organizing processes that created them.

6 Functional Information Modifies Self-Organized Coordination Tendencies

What does functional information do, and how does it do it? In coordination dynamics, functional information plays a dual role. On the one hand, it may assemble and *stabilize* coordination patterns under conditions in which they might otherwise be unstable and susceptible to global changes—such as undergoing a qualitative transition from one pattern to another. On the other hand, functional information can *destabilize* coordination patterns in order to fit the needs of the organism to the current demands of the situation (stabilizing~destabilizing). In this case, functional information modulates coordination by switching it to fit the needs of the organism.

For example, a number of years ago we established in a series of experimental and theoretical studies that a person's intention is capable of stabilizing and/or destabilizing behavior, depending on context. Thus, by the force of one's intention (the *strength-o'-one's-will*, if you like), one can keep a particular coordination pattern stable that would otherwise quickly become unstable and switch to a more intrinsically stable coordination pattern. A colloquial example of this is the ability to stay awake hours after one's bedtime by willing oneself to stay awake. Another example is a professional drummer who has learned to stabilize complex rhythms that are impossible for a novice.

At the same time, laboratory studies by us and Tim Lee's group at McMaster University have shown that one's intentions are constrained by one's "intrinsic dynamics," the relative stability of the different patterns in one's behavioral repertoire. To continue our sleepy example above, this second part says that even though you might be able to get yourself up in the morning to go to work—an unquestionably intentional act—you might nevertheless find getting up very difficult because you are naturally a night owl. In laboratory experiments, subjects find it much harder to switch out of a more stable coordination pattern to a less stable one than from a less stable coordination pattern to a more stable one. This makes sense. It's easier to escape from the clutches of a less stable coordination pattern, like a jazz improvisation, than a more stable coordination pattern, like a military march. Mathematically speaking, potential minima are shallower in the first case and deeper in the second. We realize this is a mathematical metaphor, but it will become more than that shortly.

Functional information may arise as a result of a system coupling to its external environment, like a baby's natural tendency to turn toward its mother's face or voice, and it may arise from the internal milieu too, for instance if the baby is hungry. Here again, whether stabilization or destabilization occurs depends on context. In laboratory experiments, if a person produces intentional movements in phase with respect to sound and touch, like repeatedly touching an object in time with an auditory metronome beat, coordination is stable over a wide range of movement speeds. In that situation, touch, sound, and movement stay bound together no matter how fast people move. Functional information stabilizes coordination.

However, if touch and sound do not coincide with each other, as when a test subject touches the object in an antiphase or counterpoint relation to the beat of a metronome, coordination becomes unstable and people switch spontaneously such that intentional movement, sound, and touch cohere together. This kind of simple experiment reveals how different kinds of inputs and outputs are bound together to preserve coordinative function. Functional information is the "force" that binds coordinating elements together in a meaningful fashion. This can be imagined from a physical point of view as a process measured by the energy invested in assembling and maintaining bonds. Later on, we'll give an experimental example showing how functional information "glues" or ties the behavior of human beings together.

Since functional information stabilizes~destabilizes the coordination dynamics depending on context, this makes it all the more important to identify intrinsic coordination tendencies and the variables that describe them. A great benefit of knowing a system's spontaneous, self-organizing tendencies, its *intrinsic dynamics*, is that they inform the system (as well as the scientist) about which coordination variables can be modified and which cannot. This is critical, for example, for learning and development.

Since coordination dynamics is always function- and task-specific, functional information is never arbitrary with respect to the dynamics it directs. Information is functional precisely because it "speaks"—formally and biologically—in the language of self-organized pattern generation. This principle goes all the way down to the genetic level. For example, specific "epigenetic" information is responsible for turning on and off the expression of genes, thus affecting whether a cell becomes a kidney or a heart cell, and ultimately affecting what organs do. Here again, whether information is functional or not depends on its ability to stabilize~destabilize ongoing pattern-generating activity.

7 Relating Levels

The seventh main idea is that coordination dynamics offers a way to understand how different levels of structural~functional organization are related. What do we mean by "understanding"? One important meaning of "understanding" is the ability to comprehend how different levels of coordination are themselves coordinated—the smaller with the larger, the faster with the slower. For example, to understand the cell, one must accurately describe the cell's behavior and how it is affected by external influences. But further, one must characterize all the intracellular components~processes that interact in time to produce cellular behavior in various contexts, and so on down, at smaller and smaller scales of magnitude.

THE MAIN IDEAS OF COORDINATION DYNAMICS

Put this way, you might think that it is hopeless to truly understand anything! But reduction of a system down to its "fundamental" elements~processes is only one (albeit important) way to gain understanding of a system.

By now it should come as no surprise that coordination dynamics avoids a purely reductionist or purely holistic approach to relating different levels. Coordination dynamics literally sidesteps the issue of such top-down versus bottom-up, either/or approaches. Rather, a constructive~reductive approach is adopted, in which no level is any more or less "fundamental" than any other. Likewise, there is no absolute "macro" or "micro" level in coordination dynamics. After all, what is macro at one level may be micro at another. For coordination dynamics, the crucial strategic point is that for any system one must first choose a particular level of interest. This should be an informed decision, and there is no recipe for it. This is coordination dynamics at its most pragmatic.

The first step at such a chosen level of interest is to identify relevant coordination variables and their dynamics, equations of motion whose parameters alter the stability and change of observed coordination patterns over time. The second step is to go to the next level down and identify the individual coordinating elements and determine *their* dynamics. This is not trivial, because the individual coordinating elements~processes are usually quite complex themselves, and seldom exist as isolated entities outside the context of the functioning whole.

The final step is to derive the coordinative level dynamics from knowledge of the coupling among individual elements~processes. Such couplings are *essentially* nonlinear. They are the reason that, in the words of the Nobel laureate Phil Anderson, "the whole is more than and different from the sum of the parts." Nonlinear couplings sit behind what some people call "emergence." In coordination dynamics, they allow for a connection to be built across different levels of description in a systematic fashion.

In short, the complete coordination dynamics on *any* chosen level of description always requires at least three adjacent levels: (1) the specific and nonspecific internal~external boundary conditions and control parameters that establish the context for particular behaviors to arise; (2) the individual coordinating elements and their dynamics; and (3) the coordinative pattern level and its dynamics sandwiched in-between. In this way, coordinated behavior at one level may be said to be constructed from the interactive couplings among individual component~processes at another level. We call this complementary scientific strategy "constructing up~reducing down" or just construction~reduction. By starting with a coherent description of behavior on one level and by focusing on adjacent levels "above" (boundary conditions, parameters that set the overall context) and "below" (individual coordinating elements~processes), the behavior of the whole may be seen as "emerging" from the nonlinear interactions among the subsystems. But there is no mystery behind this use of the word "emergence"—just a lot of hard work!

The science of coordination, in essence, is about how interactions both within and between levels evolve in space~time, that is, their coordination dynamics. The relation between levels is not only vertical (as is typical of hierarchies) but also horizontal. One must look up and down as well as side to side. The strategy is context-dependent, and seeks commonalities among nature's forms and functions, yet level- and system-independent. It can be applied at any level and in any system. It is not magic, and it is not easy. But the benefits of a deeper understanding of coordination and hence the complementary nature seems well worth the effort.

8 Coordination Dynamics as a Language of Living Systems

It might sound a bit presumptuous to say that coordination dynamics provides a vocabulary and a language for understanding coordination in living things. Isn't the classical mechanics of matter and motion enough? Can't we just apply that to biology, as in the fields of biomechanics and biophysics? Certainly there has been a massive amount of work in these areas. The question remains, however, is this enough?

Think of how quantum mechanics transformed the language of the physical world at the beginning of the last century by replacing the concept of mass particles that had definite position and momentum with probabilities and wave functions. In the same way, the vocabulary of nonlinear coordination dynamics offers a way to describe and explain how everyday coordination forms, adapts, persists, and changes in living systems. Newton, ironically enough, saw the "self-motion" that God gave animals "beyond our understanding without doubt," even though he understood a great deal about the motion of inanimate bodies. For living things that rely on functional information and self-organization, it seems necessary to *complement* the mechanics of motion with the dynamics of coordination.

As a mathematical toolbox, the mathematical theory of dynamic systems becomes a science of coordination—coordination dynamics—when it is filled with content, namely with the key variables and parameters involved in specific coordinative processes on specific levels of description. As a science of coordination in living things that deals in the currency of informationally meaningful variables, coordination dynamics provides an explanation and interpretation of complementary pairs that is neither metaphorical nor (solely) quantum mechanical.

Although the devil is always in the details, we cannot help but remark again how ubiquitous informationally based coordinated activity is: within~between genes and proteins; within~between different regions of the brain; within~between parts of the body; within~between organisms and their environments; within~between people, social structures, and institutions. Although every system and subsystem has many, many distinguishable aspects, an intriguing possibility exists that what we learn about the coordination dynamics of one field, system, and level may aid in understanding all the others. This is an ambitious but realizable goal. Many of the most deeply puzzling phenomena confronting modern philosophy, science, and technology in the past, such as emergence, nonlinearity, multifunctionality, interaction, and context, are open to theoretical~empirical understanding in coordination dynamics. Thus, more than just offering a novel language, coordination dynamics provides shared principles and activities that may help connect people working in different fields.

FOUNDATIONS OF COORDINATION DYNAMICS I: SELF-ORGANIZING PATTERNS

In trying to convey the key ideas of coordination dynamics, we have hardly said a thing, at least not directly, about complementary pairs. This, despite pondering at the beginning of Movement 2 whether there might be a way to ground the philosophy of complementary pairs in science. The time has now come to address the intimate connection between complementary pairs and coordination dynamics.

To begin, let's examine some of the main foundational concepts and tools of self-organizing dynamical systems in a bit more depth. As we go, we will identify the complementary pairs entailed therein, and by so doing highlight some that appear to be central to our candidate science of the complementary nature. This exercise is vital in another way: We will need to delve into the basic concepts, methods, and tools of coordination dynamics in order to understand how complementary pairs form and change. A bit of work lies ahead, but this, after all, is the heart of Movement 2. It is where we "end our speeches" about how important coordination dynamics is to the complementary nature and show you how it works.

We intend to show you how complementary pairs arise, how they manifest their multilfunctional dynamics and, most importantly, how they may coexist as metastable (integrative~segregative, individual~collective, cooperative~competitive) tendencies of the human brain~mind. As a new principle of how the human brain~mind works, metastable coordination dynamics holds a key to comprehending the complementary nature, so bear with us as we share with you the rather tortuous background behind it.

Individual~Collective

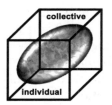

Let's start with a word that's been bandied about quite a lot in this book, namely "self-organization." Self-organization refers to the spontaneous formation of coordination patterns in open systems. An open system is one that is capable of exchanging energy, matter, and information with its surroundings—just the opposite of a thermos, which is designed to be a closed system. Self-organization is nature's way of handling complexity in open systems that contain a large number of multiple, often heterogeneous elements interacting nonlinearly with each other and their surrounding environment. Think of birds flocking, fish schooling, bees swarming. The individual coordinating elements of all of these dynamical collectives are capable of adapting to the very patterns of behavior they play a part in creating.

On the one hand, the dynamical coordination patterns observed in space~time are created by the collective behavior of the individual coordinating elements (the birds, the fish, the bees, the neurons of our brains). On the other, the behavior of each individual element is constrained by the collective behavior of the whole (the flock, the school, the swarm, the brain). Anyone who has watched a school of fish twist and turn, dodge and dart around obstacles when being chased by a predator can appreciate the subtle, complementary relationship between individual behavior and the patterned behavior of the collective (individual~collective).

Inevitably, when individual parts and processes come together to form a coupled dynamical system, novel patterns of behavior arise. For example, consider the puzzling, almost ghostly coordination of two or more pendulum clocks hung on a clock shop wall, no two having exactly the same design, yet somehow ticking perfectly in time together. The weakest coupling, whether through vibration in the wall or displacement of the air, enables the clocks to become mutually entrained, that is, *coordinated without any coordinator at all.*

Such emergent timing reflects the collective behavior of all of the relevant individual coordinating elements working together as a collective whole. As a result of self-organization, the potentially large number of microscopic degrees of freedom are reduced to a much smaller set of relevant coordination variables referred to in physics as "collective variables" or "order parameters." They are given these names by physicists because they describe the collective, ordered state(s) of the system. In general, the number of these collective variables is much smaller than the number of coordinating elements in the system, hence bringing about a tremendous compression of the information needed to describe the system's behavior. In TCN, we call these collective variables "coordination variables" to help us keep the focus on the emergence of coordinated patterns of behavior that are invariably functional, context- and task-dependent.

Note that the names are interchangeable: "coordination variable" = "collective variable" = "order parameter." In line with our newfound appreciation of complementary aspects and complementary pairs, coordination patterns are seen as collective individuals and individual collectives. Here the value of the complementary pair individual~collective to coordination dynamics is immediately evident, as is the value of coordination dynamics to the understanding of the complementary pair individual~collective.

Macro~Micro

Because the order parameter forces the individual electrons to vibrate exactly in phase, thus imprinting their actions on them, we speak of their "enslavement" by the order parameter. Conversely, these very electrons generate the light wave, i.e. the order parameter, by their uniform vibration.

The Science of Structure: Synergetics (1984)

Hermann Haken
(1926–)

Why does pattern formation actually occur in nature? One reason is that near instabilities events on the macroscale occur on a vastly different timescale than events on the microscale. In the theoretical physicist Hermann Haken's famous "slaving principle of synergetics," faster individual microscopic elements in a system become "enslaved" (entrained) to much slower varying macroscopic "collective variables" (= "order parameters" = "coordination variables"). As a result, all the individual coordinating elements of the system no longer behave independently, but appear drawn into an orderly spatial~temporal pattern. The consequence is that billions of molecules cooperate to create just a few macroscopic patterns and structures that evolve in space~time.

The slaving principle, or "center manifold theorem" as it's called by mathematicians, has been used to mathematically describe the cooperative effects that arise spontaneously when ordinary matter takes on novel properties, as in lasers and superconductors, or when new forms of organization among water molecules arise, as in weather patterns. In such situations, the macroscopic and microscopic levels are well defined and the timescales of events on each level are well-separated. That is, microscopic events occur orders of magnitude faster than the slower macroscopic ones, and the microscopic coordinating elements are identical.

On the other hand, the multilevel scheme of coordination dynamics described by main idea 7 seems more suited to biological systems. One reason is that the many individual coordinating elements like cells, neurons, and people are themselves complex and heterogeneous. This is important, because such complexity and heterogeneity can significantly shape the way things are coordinated, as we will soon show you. Another reason is that in complex systems like the brain for instance, the definition of a part, a region, or an area is not always clear nor the boundaries between putative parts well defined. Furthermore, in living systems "macro" and "micro" levels are relative terms, and event timescales are not necessarily well separated.

Recall the basic idea that to understand how any self-organized coordinated pattern arises, one needs information about at least three levels. Remember again, in coordination dynamics levels are relative. On the "upper level" are the constraints and parameters that act as boundary conditions on the potential coordination patterns the system may produce. On the "lower level" are the individual component elements~processes, each possessing their own intrinsic dynamics. *In between* lie the coordination patterns themselves, which typically evolve a bit slower than the individual coordinating elements do. Under the given constraints, this "middle layer"—a coordinative effect—is formed from the interactions among the elemental subsystems. Yet at the same time, as figure 2.1 shows, the evolving coordination pattern itself is a source of the ordering among the very same subsystems.

Imagine a traffic jam. Now, the boundary conditions on a traffic jam are quite varied compared to your typical physical system. They might include one of the lanes being under repair, or a president's visit has led to the erection of security barriers and restrictions, or an accident has occurred, or perhaps (unhappily for the commuters) all three conditions are present at the same time. The traffic jam itself is an emergent, slowly—sometimes *very* slowly—changing coordination pattern created by the behavior of the cars and their drivers. In this scenario, they act as the individual coordinating elements under these usually unpleasant boundary conditions. As a coordination pattern, the traffic jam "enslaves" the individual coordinating elements, i.e., the car~drivers, clearly constraining their behavior. The traffic jam persists even though cars entering and leaving it are always changing.

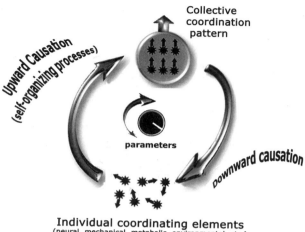

Figure 2.1 The reciprocal causality of self-organizing coordination dynamics. Collective coordination patterns arise from the interaction among variable subsystems and processes yet reciprocally constrain the behavior of these coordinating elements.

This is an unfortunately lucid metaphor, since almost everyone has been in a traffic jam, and *everyone* in a traffic jam feels like they are slaves to it.

Another close-to-home metaphor concerns the fleeting thought patterns of human beings. If it is necessary for you to convince yourself that your own thoughts are fleeting, go try the following: Get an egg timer. Sit down in front of a wall, legs crossed. Take some deep, cleansing breaths, set the timer on five minutes, and then simply *silence your thoughts*. If you start thinking about something, just let it go and silence your thoughts again. Go all five minutes—*no cheating*! After you finish this experiment, come back to this paragraph and continue.

In a given context a human being is never separated from the world. Thinking arises as spontaneous, self-organized patterns of brain activity created by interactions among myriad interacting neurons and neural assemblies. According to the principle of reciprocal causality, thinking, in turn, modifies the activity of the very neurons and neural assemblies that create it. We can all agree that at certain times in our lives, our thoughts certainly *feel* like a storm, hurricane, or traffic jam. Hurricanes self-organize without a merciless hand, and so can mental storms.

It is surely an intriguing possibility for science that though the details obviously differ, the same concept of self-organization may aid understanding of such different things as lasers, the weather, traffic jams, and how we think. In coordination dynamics, thinking involves the creation of meaningful information, fundamentally a coupling between the intrinsic dynamics of a system and its environment. Later on we will show exactly what form this coupling takes in the case of the human brain and how it gives rise to meaning.

Notice in the multilevel scheme of coordination dynamics, the distinction between macro and micro disappears. This is because the terms "micro" and "macro" are relative. What is micro to a physiologist may be considered macro to a molecular biologist, for example. Likewise, the distinction between top-down and bottom-up perspectives also disappears. In coordination dynamics, there is no ultimate preferred route or direction to understanding.

Instead, the idea is to view coordination as patterns of interactions that arise as a result of influences from *both* above *and* below (above~below). The aim is to understand the relation within~between adjacent levels, not to reduce or induce to some ultimate "fundamental" lower or upper level, as is done in strong reductionism and strong emergentism. This may sound as if we are talking about the philosophy of complementary pairs again. But here we aren't talking about complementary pairs directly; we are talking about coordination dynamics.

Control Parameters~Coordination Variables

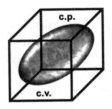

So what causes coherent coordination patterns to form in systems composed of large numbers of individual coordinating elements? And what causes those patterns to change? In open, nonequilibrium systems, naturally occurring environmental conditions and intrinsic, endogenous factors may take the form of *control parameters*. In self-organizing systems, control parameters provide a way for a system to discover the patterns that are available to it. Technically speaking, a parameter is a control parameter if, when continuously varied, a threshold value is reached that causes the coordinated behavior of the system to change in a qualitative way. When a control parameter crosses a critical value, instability occurs, thus leading to the formation of entirely new or different coordination patterns.

Think of the dramatic, qualitative changes or "phase transitions" that occur when liquid water turns to solid ice or to steam. As a thought experiment, think of putting a pot of water on the stove, and warming the stove by turning one of its dials to "1." At this value of the adjustable control dial, the ensuing temperature level makes the water warm, but that's it. Predictably, each time you increase the number on the dial, from "1" to "2" to "3," the water becomes warmer than at the last setting.

Of course, if you put a thermometer in the water, you can be a little more scientific and note the increase in the water temperature with each new setting of the stove dial. But for the sake of argument, that's it. For this stove, you can wait all day, and you will simply have warm water. Finally, however, when you set the adjustable stove dial on "4" and wait, all of a sudden, the water begins to boil—a qualitatively different behavior of the stove-pan-water system. In it, the water doesn't just get hotter, which is a quantitative change. Instead, it goes through a qualitative physical transition from liquid to gas, from water to steam. If you now adjust the dial back to "1" again, the water eventually stops boiling. The behavior has reversed. The dial on your stove is a handy adjustable element that leads the system through qualitative behavioral transitions from one state or phase to another.

In this example, and in the contrary case of water freezing to ice, the control parameter is *temperature*—the stove dial being our way to conveniently adjust it. Although the ice~water~steam example is nice because it is easy to visualize and part of common experience, the qualitative changes or "phase transitions" that occur in living systems are much more complicated. Compared to the topological changes that have been shown to occur in gene regulatory networks, for instance, the transitions from ice to water to steam are very simple indeed. And although the concept of a control parameter is clear enough to us human beings, much work needs to be done to identify the control parameters that lead a human being, organ, cellular network, or cell into spontaneous, qualitative behavioral transitions, such as those that would prevent or induce disease.

A central tenet of self-organizing systems in physics is that control parameters are *nonspecific*: They do not prescribe or contain a code for the emerging structure. Such nonspecific control parameters do not dictate to the system how it is supposed to change or what specific pattern should be generated. For example, the threshold temperatures necessary for water to boil or for ingredients to become a cake don't in any way inform or dictate to those ingredients *how* they should become a cake. This concept seems completely comprehensible in the case of water boiling or cake baking. Note, however, how provocative it would be if one were to propose that the role of genes was to lead a viable system to its phenotypical expression without informing or dictating to the cellular milieu *how* it should become an organism!

In coordination dynamics, control parameters refer to naturally occurring environmental variations and~or specific types of functional information that move a system through its patterned modes of coordination and cause them to adapt and change. Conversely, control parameters may play a stabilizing role, sustaining a pattern that would otherwise undergo a global change such as a phase transition. So in coordination dynamics, control parameters can be both specific and nonspecific and can both cause and prevent change depending on the situation!

Coordination variables that capture dynamic coordination patterns in complex systems and control parameters that lead the system through qualitative transitions are key complementary aspects of self-organizing coordination dynamics. As aspects of a complementary pair, you can't have one without the other. Scientifically speaking, one does not really know whether one has a control parameter unless and until its variation leads to qualitative changes in patterns of behavior. Qualitative or topological change in a system is a sign that some control parameter has crossed a critical threshold. On the other hand, qualitative change is the key to identifying the relevant coordination variables. In a complex living system, many variables may be changing in one way or another, but the ones that change qualitatively are the ones that carry the key information.

We cannot overemphasize the key methodological aspect of coordination dynamics of choosing to focus upon a specific level of description. Once chosen, one must "look up" a level for the boundary conditions, and "look down" a level for the individual coordinating elements. Understanding the entire paradigm depends on this key notion, captured in a saying coined by Arthur Iberall, one of the pioneers of complexity science, that "the overseer always seems to be faced only by king and by peasant." In the paradigm outlined here, although every level may seem awesomely complicated in its nearly uncountable details, we are never faced with more than a small number of interacting levels. This makes it possible in principle and occasionally in practice to seamlessly stitch levels together.

Let us wrap up this section with a brief speculative example from the field of neuropharmacology. We know that within the human nervous system, the moment to moment concentration of certain chemical neurotransmitters may alter a person's mood and cause mood swings. If we were to describe the behavioral patterns corresponding to human mood in terms of coordination variables, and mood swings in terms of coordination dynamics, then concentrations of neurotransmitters like acetylcholine, serotonin, and dopamine, as well as drugs such as alcohol and caffeine, might well qualify as control parameters. Most of us know all too well that a steady and prolonged increase in the amount of alcohol ingested can eventually lead to spontaneous, qualitative changes in one's behavior. Now think of this phenomenon in terms of the current discussion on control parameters: The alcohol level in the blood doesn't prescribe, order, dictate, orchestrate, or program the drinker's behavior. Viewed as a nonspecific control parameter, alcohol can precipitate specific, qualitative behavioral changes. The framework and language of coordination dynamics thus offers a novel and possibly important way to think about how neurotransmitters, neuromodulators, and drugs might work.

A central tenet of pharmacology is that neurotransmitters, neuromodulators, and exogenous neuroactive drugs work in a highly *specific* manner, via *specific* binding to *specific* receptors—leading to *specific* cellular consequences. A pretty *specific* tenet. And this undoubtedly happens. But if we view them as control parameters, we can now also envision neurotransmitter, neuromodulator, and neuroactive drug actions as *nonspecific* control parameters that alter the dynamical landscape of the nervous system, stabilizing~destabilizing ongoing coordinative interactions among neurons and receptors, and so forth. This is a different way to think about pharmacodynamics, one in which the dynamic fluctuations of a *nonspecific* chemical control parameter can lead to quite *specific* qualitative (and quantitative) changes in the nervous system. Considering its possible ramifications for the field of medicine, we believe this concept warrants serious consideration.

Similarly, coordination dynamics might also offer new ways to understand general anesthesia, which remains one of the great mysteries of medicine. For many years, research on general anesthetics has centered on the perplexing problem of how small, rather innocuous, *nonspecific* molecules can lead to such dramatic yet predictable qualitative changes in sensation and awareness. New evidence and mathematical modeling by the Australian medical physicist David Liley and colleagues indicates that the onset of anesthesia may be understood as the bifurcation of a dynamical system.

A related puzzle exists in the case of neuroactive substances like lithium and nitrous oxide that are too small in a molecular sense for their actions to be satisfactorily explained in terms of the usual specific ligand~receptor binding paradigm of pharmacology. Notice that when placed within the context of coordination dynamics this puzzle doesn't seem quite so mysterious—though it suggests there is still a *lot* of work to be done!

Cooperation~Competition

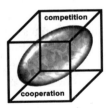

Cooperation and competition are undoubtedly universal processes that underlie pattern formation and change in natural systems. Using the field of fluid mechanics for its illustrative value, we can use convective motion as an example of a collective effect, a widespread cooperation among the many individual fluid molecules composing a fluid's pattern. Always side by side with this cooperativity, however, is competition. In fluid mechanics, the competition is between how forcefully the fluid system is driven (i.e., by heat, pressure, or whatever other gradients apply to it) and how well the system holds itself together, its viscosity. This ratio of driving to dissipative forces in fluid mechanics is called the "Rayleigh number," after the great physicist John William Strutt (Lord Rayleigh).

The general issue, once again intuited by Arthur Iberall, is whether energy sweeping into some field of "atomisms" can be internally absorbed or dissipated by those atomisms. If not, the field becomes unstable and some change in status (phase, macrostate, dynamic pattern) must occur. Notice the message is always the same: Once one of the two forces (driving~dissipation) overcomes the other, instabilities arise, leading to rich and eventually irregular dynamic patterns. All cases of self-organized pattern formation, from the rising smoke plume of a smoking match to the birth of stars, arise as a result of the interplay of such cooperative~competitive processes.

Of course, a crucial and often difficult task remains to identify the physical mechanisms involved, which predictably are far harder to discern for brains and people than for fluids. In the case of the coordinated behavior of brains and other kinds of social organizations, self-organization can arise not only due to competition among conventional forces (mechanical, electrical, chemical), but also among different kinds of functional information. Later on we shall present specific examples of how functional coordination, ranging from neural groupings in the brain to people interacting with each other, depends on the coexistence of cooperation and competition.

Stability~Instability

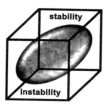

Offsetting the tough, nitty-gritty labor of identifying relevant coordination variables~control parameters is an amazing fact that applies to all self-organizing systems: Regardless of the details, *dynamic instability* is the generic mechanism underlying spontaneous self-organized pattern formation and change in *all* systems coupled to their internal~external environments.

The reason for this is that near critical points of instability, in order to adjust to current conditions—a new value of a control parameter—individual coordinating elements must reorganize themselves in a new and different way. Though it's tempting to talk of the formation of qualitatively different coordination patterns as being due to purely cooperative effects, always lurking behind apparent cooperative phenomena are complementary competitive ones—*a tension of opposites*. In coordination dynamics, competition exists between the new, qualitatively different coordination patterns that are forming, the coordination patterns being left behind, and the tendency for the individual coordinating elements to simply do their own thing.

Inevitably, the interplay of cooperative~competitive processes results in fluctuations in the coordination dynamics. Fluctuations are always present in complex systems, constantly challenging the system's stability under the current conditions, and nudging it into producing new, more adaptive behavioral patterns. Think of the fluctuations of a tightrope walker's pole while considering how stability~instability works as a complementary pair.

Summarizing Self-Organizing Patterns of Coordination Dynamics

All self-organizing processes in living things are of potential relevance and interest to coordination dynamics. As in any discipline, it takes some effort to learn its language and appreciate its methodology. But keeping the concepts presented so far in mind, you may begin thinking and practicing coordination dynamics, at least on an experiential level. Try the following exercise yourself: When you notice some strange or unusual pattern(s) of behavior or events in your life or in your laboratory, ask yourself the following kinds of questions:

How do I know something is coordinated when I see it? What do the coordination patterns actually look like? What might the key coordination variables and control parameters be? What exactly changes in the coordinated behavior in question, and what factors seem to be causing these changes? Are the observable behavioral changes smooth and predictable, sudden and unpredictable, or both depending on the situation? Are these factors very specific or nonspecific to the patterns of behavior that emerge, or are they both? What might the individual coordinating elements be? What is the nature of their interaction? When and where are interactions occurring, and how do they evolve in space~time?

Answers to these questions, regardless of the context and level at which coordinated patterns appear, can be quite useful and insightful. Naturally, answering these questions is a lot harder than asking them. Yet it certainly helps to ask the right questions. An important idea to remember in coordination dynamics is that even though the physical mechanisms of pattern formation are undoubtedly different across fields, systems, and levels, the basic principles of coordination are the same.

Intimately tied to the principles we have discussed thus far is a fundamental role for complementary pairs such as individual~collective, macro~micro, coordination variable~control parameter, qualitative~quantitative, cooperation~competition, and stability~instability. That is quite a few complementary pairs already, even in this preliminary explanation of coordination dynamics.

Although many if not all of the phenomena and principles of coordination dynamics seem to entail complementary pairs, the search here is for the scientific *underpinnings* of complementary pairs, not their mere enumeration. To get to the heart of that, we need to provide some further basic concepts and tools that will enable us to directly observe the behavior of real complementary pairs. So get comfortable, and take the next sections slow and sure.

FOUNDATIONS OF COORDINATION DYNAMICS II: PATTERN DYNAMICS

A central tenet of coordination dynamics is that the dynamic patterns that emerge and change due to self-organizing processes have a rich and essentially nonlinear pattern dynamics. What exactly does this mouthful mean? Let us take it from the beginning:

"Pattern dynamics" means that one has a set of coordination variables **cv** ($\mathbf{cv}_1, \mathbf{cv}_2, \ldots, \mathbf{cv}_n$) that evolve in time according to some dynamical law. To refresh your memory, a variable is just something about a system that changes for whatever reason. It can be anything from the moment-to-moment velocity of a pendulum, to the monthly size of the fish population in the local lake, to the hormone levels in your bloodstream.

Dynamical laws are usually expressed as sets of mathematical equations stipulating the temporal evolution of all the values that a coordination variable can possibly take on. Therein lies the rub: To do the science, one must determine what the *relevant* coordination variables are in the first place, and attempt to identify their dynamics on a given level of description. This search for relevant coordination variables cannot be overstated. In any system, many things change over time. But not everything that changes over time is relevant for understanding the system's behavior.

Generally speaking, the term "dynamical systems" refers to physical systems that change in time: the weather, the organism, the climate, etc. The term is also given to sets of mathematical equations used to study, understand, and communicate how a system behaves. Such sets of mathematical equations are referred to as "models."

Hence, a model of Los Angeles's rush-hour traffic usually refers to a set of mathematical equations run on computers whose behavior is intended to simulate real Los Angeles traffic. Both the real Los Angeles traffic itself and its models are examples of dynamical systems, the former physical, the latter a virtual rendering (physical~virtual). These days, some type of dynamical system model exists in almost every field of research and at practically every scale of observation, from the minute to the grand.

Of course, the availability, speed, and power of contemporary computers, from the supercomputer to the now ubiquitous personal computer and all the way to the palmtop, have greatly expanded this activity. Hence, not only do we describe naturally occurring physical systems as dynamical systems, we also use model dynamical systems to help us understand them, and even to predict their behavior in novel situations.

It is not our aim in the present work to try to do justice to the fascinating and very active field of mathematical dynamical systems research and its numerous technical applications. Fortunately, there are books upon books on this topic, not to mention all the information about them that's available on the Internet. Here, we will provide only the most elementary concepts and tools necessary to comprehend coordination dynamics in its simplest form. It is our goal in this section to communicate the essential mathematical ideas descriptively, without getting caught up in equations.

Nevertheless, there is no getting around the fact that every field of inquiry has its own terminology, its own jargon. Such terminology is one of the main reasons that communication between scientists and nonscientists, and even between those practicing different scientific disciplines, can be so difficult. We are keenly aware of this problem, and have made an effort here to describe the concepts as comprehensibly as possible. We will pick up on this problem again in Movement 3. For now let's begin to describe dynamical systems and some of its basic terminology.

Dynamical systems come in many varieties. They can be simple, complex, linear, nonlinear, deterministic, stochastic, etc. Most of the systems we are interested in are living dynamical systems and are quite likely to entail all of these different varieties in one way or another, and at one time or another. The mathematical tools of nonlinear dynamics can be applied to anything that evolves in time. In the elementary case treated here, we consider how a coordination variable (abbreviated as **cv**) varies as a control parameter (**cp**) changes in a hypothetical dynamical system.

Different Ways of Looking at the Same Dynamics

In the following section we introduce some basic ways to visualize the behavior of dynamical systems. Though they emphasize different aspects, all of these visualizations are equally valid ways of expressing certain fundamental properties of dynamical systems, such as stability~instability. Used together, they provide a pretty complete picture of what it means for a dynamical system to "behave," regardless of what kind of system it is and the level at which it's observed.

Convergence~Divergence of Dynamical Trajectories

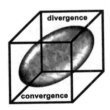

First, consider how a hypothetical coordination variable, **cv**, might evolve in time. Figure 2.2(a) shows a number of different expressions of this behavior. In the first graph at the top left, you see lines starting at different initial values of **cv**, some above the horizontal axis and some below it. Notice that over time (proceeding from left to right), they all converge toward the horizontal axis or **cv** = 0. In the parlance, the different initial values of **cv** are called "initial conditions." If one is doing a real experiment, perhaps these initial conditions are the values of the real experimental **cv** at the beginning of different experimental runs.

FOUNDATIONS OF COORDINATION DYNAMICS II: PATTERN DYNAMICS

Different Ways to Look at the Same Coordination Dynamics

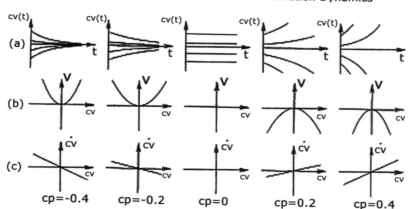

Figure 2.2 Ways to observe how things behave: (a) Trajectories, (b) Potential landscapes, (c) Phase portraits, (d) Bifurcation diagram (courtesy A. Fuchs).

The lines in figure 2.2(a) are called "trajectories," and are akin to the streaks left in the sky from passing jets. In fact, the many different graphs in figure 2.2 represent five different experimental situations as one reads from left to right. For instance, reading the five top graphs of figure 2.2(a) from left to right, you can see that the groups of trajectories converge quite quickly in the first graph furthest to the left. They converge a little more slowly in the second graph. In the third, middle graph, they don't converge at all, but remain constant. Moving on to the right, in the fourth graph, the trajectories slowly diverge from one another. In the fifth graph of figure 2.2(a) farthest to the right, they diverge from one another quite rapidly.

Obviously, if we want to understand how a pattern of behavior as expressed by the coordination variable **cv** forms and changes, we need to know what is influencing it. What is making the trajectories of the *same* coordination variable behave so *differently* in the five top graphs? Well, to begin with, you might notice that the initial conditions of cv are the same in each of the five cases, so *that* isn't it. In fact, the different ways these five collections of trajectories behave depend only on the value of a control parameter, **cp**.

At the risk of redundancy, we remind you once again that a parameter simply means something in the system under study that is adjustable, and that when adjusted causes the coordination variable to change. We call it a *control* parameter because adjustments of this parameter can lead the coordination variable **cv** through qualitative changes in the way it behaves. In the leftmost graph in figure 2.2(a), the trajectories converge to a single value. In the rightmost graph, the trajectories are diverging from one another. It doesn't get much more qualitatively different than that!

In figure 2.2(a), when the control parameter **cp** is less than zero, all the trajectories converge toward the same place, the horizontal axis. It's as if you were living in the time of Julius Caesar: No matter where one might start, whether from ancient Hibernia or the far reaches of Asia Minor, all routes lead to Rome. In the language of dynamics, such a place is called a "fixed point" because when it is reached, the pattern captured by the coordination variable comes to rest there—it no longer changes with time. When all the trajectories from all initial conditions converge to this point and stay there, it is called a "stable fixed point." In the mathematical field of dynamical systems, this point is also called an "attractor."

A little note: It is a bit unfortunate, definitely complicating, and certainly trying that sometimes the very same behavioral phenomena have different names depending on whether one is a mathematician, physicist, biologist, philosopher, or poet. There is arguably very little about human behavior that is consistent and easy to comprehend, and this isn't helped much by the "Tower of Babel" problem. Our hope is that we can make you aware of the concepts, and mention the other terms when necessary, so that you can try to keep clear of the traps of terminology.

Fixed Points, Potential Landscapes, and the Skateboarder Analogy

Now, right about here in many other primers on dynamics comes an example of a swinging pendulum that over time eventually comes to rest. That's a good physical example, but as we are ultimately interested in more complex systems including living organisms, let's use a slightly more vivacious analogy to make the same point. A skateboarder jumps from any point on the side of a deep empty circular swimming pool that is funnel-like (figure 2.3). Putting no extra effort in, the skateboarder just rides the board as it climbs and drops, climbs and drops, up and down the sides of the pool. Finally, the skateboarder comes to rest at the bottom of the pool, and just stands there.

Subsequently, if pushed slightly up any wall of the pool and released, the boarder drops and climbs, drops and climbs, but eventually returns to the same point at the bottom. In the language of dynamical systems, the bottom of the pool would be called a "fixed-point attractor." It is as if there is a giant skateboard magnet (called gravity) at the bottom of the pool that pulls the board and boarder toward it.

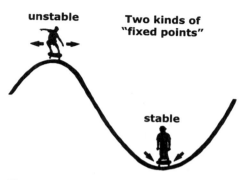

Figure 2.3 Stable and unstable fixed points and the skateboarder analogy.

Of course, what we call this point doesn't tell us *why* the boarder has come to rest there. Yet, this is what *seems to be happening* when we observe the skateboarder. And we are so accustomed to such behavior that we usually don't even question it. But a curious child might very appropriately ask why the skateboarder seems to end up in the same spot no matter where the skateboarder begins. "Well, it's the lowest point, and that's where gravity takes us," we explain with confidence. And there it is: When we drop something, it falls to the floor, as if the earth were a magnet, as if the object we dropped was caught in the pull of the earth's "attraction." This is an intuitive description of why we might speak of something as an attractor, even though that description in itself tells us little or nothing about *why* the attraction is the way it is, nor how it is manifested.

The skateboarder's motion, an actual flesh-and-blood system (figure 2.3), can be modeled in terms of a coordination variable, **cv**, that varies in time, along the lines of the graphs shown in figure 2.2(b). In this analogy, the control parameter **cp** determines the shape of the "pool." Imagine that the pool is made out of some sort of flexible material, like rubber, and we are able to repeat our skateboarder experiment a number of times with shallower and steeper shaped pools, as dictated by the value of **cp**. When the control parameter is zero in figure 2.2(a) (middle graph), the pool would be flat. Of course in that scenario it isn't really a pool anymore—it looks like a circular rug (not so interesting for the skateboarder).

Furthermore, when the control parameter becomes greater than zero (figure 2.2(b), fourth and fifth graphs to the right), we have flipped the flat rubber pool inside out. As a result, the pool is now a little hill, which grows steeper as we increase the value of the control parameter further. From **cp** = 0 and higher, we will certainly observe a dramatic change in the skateboarder's behavior. No matter where you gently push the boarder up the hill and let go, the skateboard moves away from the midpoint of the pool that used to be its bottom. Why? Because it is now the top of the hill! Now, all the possible trajectories flow away from the horizontal axis, as if they are trying to get away from the fixed point, as if they were being repelled. This means that the fixed point is now unstable. Thus it is called a "repeller." But since the skateboarder always ends up moving away from the top of the hill, why do we still refer to it as a fixed point?

To answer this, let's make it more dangerous and thrilling for our skateboarder. This time, the skateboarder is towed to the top of a very long, steep hill by helicopter, and set gingerly down at the very tiptop. Let's say that the summit is very pointy, and barely has room for the skateboard. Our intrepid skateboarder has been given instructions to keep both feet on the board and just stand there. And indeed, there our boarder stands, and nothing happens—no motion, at rest. And it is possible in principle to be at rest in that position, at the fixed point at the top of the hill, and to exhibit no movement over time. Our skateboarder is momentarily fixed, and thus must be standing at a "fixed point"!

However, a slight nudge, a gust of wind in any direction, a momentary tremor from too much coffee, and skateboard and boarder begin to roll down the hill, accelerating away from the unstable fixed point, never to return, except by helicopter again. This is why it is called an "unstable fixed point" or "repeller" (attractor~repeller). It is a fixed point because any dynamic object placed there can remain there at rest. But it cannot say there forever because, without conscious effort on the part of the skateboarder, any tiny perturbation will set the object in motion.

This is true of tightrope walkers and of any sort of situation where we are trying to balance something, like balancing a broomstick vertically in the palm of your hand. Yes, even broomsticks are useful to the inspired scientist! That you are able to balance the broomstick tells you that you have found (and been able to stabilize) a fixed point. That the slightest nudge will break the balance and make the broomstick fall to the floor tells you that it's an unstable fixed point. One can also think of *learning* in these terms, like learning to walk, to balance, to do all kinds of balancing skills as in circus acts—as the ability to stabilize inherently unstable systems.

Attractors and Repellers in Dynamical Systems

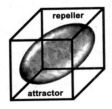

Fixed points are the skeleton upon which dynamical systems are built. They enable one to make predictions about the system's behavior. The route that a coordination variable takes toward a fixed point, called a "transient," can also be very informative, because it tells one what is going on with the system before it settles into an attractor. In our continuing example, we can tell something about the skateboarder's transient in each case by dipping the wheels in paint before each run, and analyzing the paint trail left by the wheels as the skateboarder slowly wheels to a stop at the bottom. We are confident that any skateboarder you ask to help you do this will gladly comply.

Attractors come in different types. They can be fixed points, in which all initial conditions converge to some stable state, as in the case of our skateboarder, or a pendulum. But attractors can also be periodic, exhibiting preferred stable rhythms or orbits on which the system settles regardless of where it starts, like beating hearts, breathing lungs, and walking feet. For more physics-class-worthy mental images, think of an ice skater spinning, the earth's spin, and also planetary orbits. Think of all your biological rhythms, from brain waves and heartbeats, to breathing, to menstruation. In all of those familiar systems, the "attractor" is some stable rhythm, a stable cycle.

There are even attractors called "strange attractors"—"strange" because their quasi-cyclical behavior is chaotic and their moment-to-moment behavior is difficult if not impossible to predict. Take the weather, for example. Although we know that in some parts of the world a hurricane season happens every year (repetitive, cyclical, predictable aspect), it is impossible to say exactly when and where a hurricane will emerge (chaotic, unpredictable aspect). Strange attractors exhibit "deterministic chaos," a type of irregular behavior resembling random noise, yet often containing pockets of quite orderly behavior (chaos~order).

The foregoing discussion on skateboarding wasn't to load you up with another metaphor, but to help you visualize what the next series of graphs represent. For these diagrams, a good way to model the dynamical behavior is as a ball moving through a hilly landscape. To move from our skateboarder metaphor to our mathematical model, we simply get rid of the skateboarder (figuratively speaking) and imagine a little rolling ball instead, as shown in figure 2.2(b). In fact, now that we have explained what is going on, we hope that you can focus on the behavior in a quite intuitive way. The model allows one to study the dynamical system in detail by manipulating parameters on a computer rather than on a skateboarder in the field.

In the landscape pictures, minima and maxima correspond to fixed points. Figure 2.2(b) shows five examples. On the left, regardless of where the ball starts (its initial conditions), it will always arrive at the bottom of the hill. And the speed with which it runs down the hill depends on how steep the hill is. This is exactly equivalent to how fast the trajectories of our dynamical system converge on the fixed point, as in the trajectories of figure 2.2(a).

The trajectory graphs in figure 2.2(a) and potential landscapes in figure 2.2(b) are describing exactly the same behavior only in different ways. The differences have to do with exactly what it is about the dynamical behavior we are discussing. The minima in the landscapes are just like the bottom of the skateboarder pool—they are stable fixed-point attractors. Similarly, the maxima of the landscapes, just like the tiptop of skateboarder hill, are unstable fixed points. The slightest little bit of noise or the tiniest fluctuation will cause the little ball to run away, as if being repelled from the tiptop.

In between these two circumstances, *between* the stable and unstable, is the case where the land is perfectly flat. No matter where the ball starts, it will just stay there. This condition is called "neutrally stable." No changes here. From this, one can appreciate intuitively that there always needs to be a little wrinkle or curvature in the potential landscape to make the ball move toward or away from this neutrally stable spot.

Figure 2.2(b) shows how the potential landscapes behave as a function of the control parameter **cp**. The **cp** is indeed influential because it can lead the system from behaving like a valley containing a stable fixed point to behaving like a hilltop containing an unstable fixed point. Notice that as the **cp** changes, the valley on the left becomes less steep until it flattens at **cp** = 0, and then turns into a hill when the **cp** increases further. In the language of dynamical systems, the behavior that evolves in time can be stable or unstable depending on the value of the **cp**.

Phase-Plane Trajectories and the Stability~Instability of Dynamical Systems

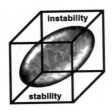

Yet another valuable way to visualize dynamical behavior is to plot the behavior of the coordination variable **cv** against its own rate of change. This is often written mathematically as **cv** with a dot on top of it, or spoken of as **cv-dot**. Think of **cv-dot** as a quantity that tells you how fast the behavior changes in time. When the skateboarder is standing at the fixed point, **cv-dot** = 0. No change in the behavior over time. If **cv-dot** is negative, the skateboarder is slowing down. If **cv-dot** is a positive value, the skateboarder is speeding up.

The graphs displayed in figure 2.2(c), also called phase-plane trajectories, are important because they not only highlight fixed points as do the trajectory paths and potential landscapes just discussed, but can also be used to determine how stable the fixed points are. A set of phase-plane trajectories like those in figure 2.2(c) is often called a "phase portrait." Fixed points, as we have discussed, are where the behavior of the system does not change, that is, where **cv-dot** equals zero.

In our example, one can see that all the fixed points in figure 2.2(c) lie where **cv-dot** crosses the horizontal axis. It just so happens in this simple case that the value of **cv**, since it's at the origin, is also zero. In other situations, the **cv** could easily be shifted to other values along the horizontal axis. In terms of our metaphor, we can use the phase portrait not only to model skateboard in motion–to skateboard at rest, but also scenarios where the behavior switches from one pattern of motion to another.

What determines the stability of a given coordination pattern? When the slope intersecting the horizontal axis is negative (i.e., for all values of **cp** < 0), the fixed points are stable. That's because if the ball is perturbed away from the fixed point, it returns quickly. How quickly depends on the value of **cv-dot**. When the slope is positive (i.e., for values of **cp** > 0), the fixed points are unstable. That's because if the ball is perturbed from the fixed point, it speeds away. One of the useful features of phase-plane trajectories is that they provide an easy way to evaluate the system's stability. The steeper the slope of the line, the more or less stable or unstable the fixed point.

It is important to keep emphasizing that all of the different ways to visualize the dynamics are equivalent. In figure 2.2(a), how fast the trajectories converge or diverge tells us how stable or unstable the fixed points are. In figure 2.2(b), the minima and maxima of the potential landscapes locate the fixed points, and the curvature of the landscapes tell how stable or unstable the fixed points are. In figure 2.2(c), the values of **cv** where **cv-dot** equals zero locate the fixed points, and the direction and steepness of the slope of the curve tell you whether they are stable or unstable fixed points, and how stable~unstable those fixed points are.

Taken together, these are quite useful ways to visualize how coordination patterns evolve in time. We owe these methods for visualizing dynamical behavior to the great French mathematician, Henri Poincaré, the very man who said that science is not about things, but the relations among things. Now we have an inkling into why he said it.

Bifurcation Diagrams and Quantitative~Qualitative Changes

The last row of graphs in figure 2.2(d) illustrates the last main concept needed in order to finish this part of our little tutorial. It shows all the fixed points, stable and unstable, as the control parameter **cp** changes. Notice, however, it does not show *how* stable the fixed points are, unlike the graphs in figures 2.2(b) and 2.2(c). Instead, it just provides information about the system's qualitative behavior. For most journeywork, that is a lot of what one needs. The picture in figure 2.2(d) is called a bifurcation diagram. It shows that when the control parameter reaches a critical value (**cp** = 0 in our example), the system's behavior changes qualitatively. That is, it goes from being stable and attractive to being unstable and repulsive. Below the critical point, all the fixed points are stable (solid circles). Only unstable fixed points remain as **cp** increases beyond **cp** = 0 (open circles).

Embedded even in this simplest of examples is a powerful message. A control parameter can change continuously (as **cp** does here) without changing the qualitative behavior of the system at all. In this range (**cp** < 0) it may lead to *quantitative* changes in the system's behavior as evaluated, for example, by measures of stability. But when the **cp** reaches a critical threshold value, an abrupt change in the coordination variable occurs due to loss or exchange of stability. Such qualitative change turns out to be crucial, because it is at these "critical points" or "thresholds" or "bifurcation points" or "instabilities" that switching occurs and the process of self-organization reveals itself.

Fluctuation~Determination

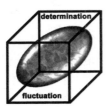

Fluctuations, it's worth repeating, are always present in natural systems. They stem from all sorts of sources. To treat fluctuations mathematically is a formidable task, but one may still get an intuitive feel for their importance from our simple pictures. Fluctuations are intimately connected to the notion of stability, so let's briefly discuss how stability is measured in dynamical systems.

In figure 2.2(b) the curvature of the potential landscape dictates how fast the system will run down the hill to its attractive fixed point. The steeper the slope, the more stable the attractor, the less time it takes the system to settle. Stability in this case is aptly measured as a "relaxation time." As the curvature decreases, it takes longer and longer for the system to approach the fixed point. This is called "critical slowing down" and is a major sign of dynamic instability. The system is nearing a critical point: One coordination pattern is becoming less stable and is beginning to give way to another, qualitatively different one. Critical slowing down, measured as an increase in relaxation time, is thus a predicted signature of qualitative change.

Looking back at figure 2.2(b), it is apparent that the more stable the landscape, the less effect random fluctuations will have upon the system. Likewise, it is equally apparent that as the system becomes less stable, the effects of fluctuations will be enhanced or amplified. Sure enough, this phenomenon is called "enhancement of fluctuations" and, like "critical slowing down," is a telltale sign of self-organizing processes at play. Fluctuations can be thought of as incessantly testing or probing the stability or robustness of a system, on the one hand challenging the stability of current coordination states, and simultaneously provoking the emergence of new ones.

Notice we have not said a word about whether the outcomes of self-organizing processes are good or bad, positive or negative. This depends. In the context of an organism's development and maintenance, cell division and proliferation is necessary for viability. But cell proliferation in a disease like cancer occurs when a cell develops emergent behavior that from the cell's point of view may be very stable, but of course renders the organism fragile. Thus, dynamical concepts must always be evaluated in the context of the system's own properties, context, and the problem at hand.

Bistability~Bifurcation: The Heart of Multifunctionality

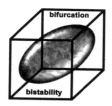

Since the beginning of this little primer on foundations of coordination dynamics, we have already discussed no fewer than eleven complementary pairs: individual~collective, macro~micro, control parameter~coordination variable, cooperation~competition, instability~stability, convergence~divergence, attractor~repeller, difference~similarity, things~relations, qualitative~quantitative, and fluctuations~determinism. Though they are not unique to coordination dynamics, all seem to fall out of it naturally. Are there even more?

In the following slightly more complicated example, you will see that there are indeed many more. There is no need to keep track of them at the moment, because in Movement 3 we will provide you a list of the complementary pairs of coordination dynamics, along with short descriptions of each one. But perhaps the most important point about the upcoming section is that it provides the first hint, only the beginnings perhaps, of the mathematical underpinnings of the complementary nature. For example, polarized extremes (complementary aspects) may be interpreted in light of the ubiquitous phenomenon of bistability in nonlinear systems, and the dynamics of switching from one extreme to the other may be seen as a classic bifurcation.

Although this rich mathematical phenomenology provides intuition, how it is grounded in empirical science, in particular in terms of how brain and behavior are coordinated, will have to wait a little longer. So turn now to the curves in figure 2.4, which, though a bit more complicated, are exactly parallel to the ones we have already explained. For this reason, we'll cut to the chase and just describe how things behave in the plain language of nonlinear dynamics.

Figure 2.4(a) shows trajectories from different initial conditions converging on fixed points. Reading from left to right, one can see that initially those trajectories that begin with initial conditions above and below the **cv**-axis converge to the same fixed point at different rates determined by the first two values of the control parameter **cp**. The system's behavior is said to be "monostable," meaning it has just one stable way of behaving, regardless of where it starts.

Mathematically speaking, in this hypothetical dynamical system a stable fixed point exists for all values of **cp** < 0. As the control parameter rises above zero, however, we see the strangest of things. At the third value of the **cp**, the trajectories begin to split apart! Those that started above the line converge on a new fixed point above the **cv**-axis, and those that began below the line converge on a different fixed point below the **cv**-axis. This means that there are now two possible stable fixed-point attractors for the same value of the control parameter!

How can the same circumstances produce two different outcomes or solutions? The reason is that under these conditions, this particular dynamical system's behavior is "bistable." More than that, the system has switched from being monostable (one stable fixed point) to being bistable (two stable fixed points). This

Different Ways to Look at the Same Coordination Dynamics

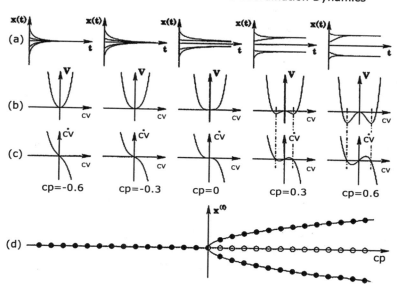

Figure 2.4 Ways to observe how things behave: (a) Trajectories, (b) Potential landscapes, (c) Phase portraits, (d) Bifurcation diagram (courtesy A. Fuchs).

sounds familiar—something that can act both as a single thing and a dual thing: Complementary pairs are like that. But what is the significance of such bistability in coordination dynamics?

Well, for one thing, neurons themselves—as the basic structural~functional units of the brain—happen to be bistable. In fact, neuronal behavior can even be multistable, that is, display several different patterns for the same control parameter values. In its so-called rest state, the neuron is quiescent. However, when subjected to an increase in synaptic levels of an appropriate neurotransmitter, stimulating current or a brief dose of an exogenous drug, the neuron's membrane permeability changes, and the cell "depolarizes." If the depolarization reaches a critical threshold value, the neuron will start to fire, and can fire repetitively.

This change from resting to periodic neuronal spiking can be and has been understood as a dynamical bifurcation. In the experimental scenario, one may interpret the amount of the neurotransmitter, applied current, or drug as a control parameter that increases the resting potential of the neuron until, at a critical point, the quiescent state disappears and the neuron's activity becomes oscillatory—*the coordination pattern changes*. Without such action potentials, neurons would not be able to communicate with each other, and without such communication the brain~mind could not function. Without bistability~bifurcations, the brain~mind could not switch, nor could choices and decisions be made. Bistability (and in general multistability) is a vital, *essentially* nonlinear property of coordination dynamics that underlies the multifunctional nature of living things.

The Importance of Symmetry~Symmetry Breaking

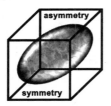

Moving on, it can be seen in figure 2.4(b) that the potential landscape starts to deform as the value of **cp** approaches zero. The single minimum becomes flatter and flatter until at a critical point two minima appear. This is called a "symmetry-breaking instability" because, before the bifurcation occurs, the landscape is perfectly symmetric around the vertical *V*-axis. Before the bifurcation, only the pattern (represented by the value of **cv**) at the origin is possible. When **cp** > 0, the previously stable coordination pattern around the origin becomes unstable and the system can exist in either of two alternative coordination states, patterns, or modes.

It is easy to see that which pattern is selected will depend on the slightest little fluctuation nudging the system into one coordination mode or the other, thereby breaking the symmetry of the potential landscape around the vertical axis. In this particular case, chance fluctuation determines which of the system's two possible behaviors is adopted. This phenomenon is called a "symmetry-breaking fluctuation." Later on, we'll show that although such symmetry breaking is important for the dynamics of both brain and behavior, it doesn't necessarily work the same way as shown in figure 2.4(b). Nevertheless, the concepts of symmetry creating and symmetry breaking are vital to understanding the process of pattern formation and change.

As in figure 2.2(c), the next set of graphs in figure 2.4(c) display phase-plane trajectories (plots of **cv-dot** against **cv**), which convey crucial information about the location and stability of fixed points. On the left graphs of figure 2.4(c), the negative slope at the intersection of the horizontal axis means that the fixed point is stable, like a pendulum (or our infamous skateboarder) coming to rest. As before, the single fixed point loses stability when **cp** $= 0$. In fact, as the **cp** increases in

FOUNDATIONS OF COORDINATION DYNAMICS II: PATTERN DYNAMICS

value, the previously stable fixed point becomes unstable (read out as a positive slope through the intersection), and two new stable fixed points are born.

The layout of the fixed points, remember, is the skeleton that supports the coordination dynamics. So what are the fixed points in the diagram on the far right of figure 2.4(c)? There are three: A positive slope intersecting the **cv**-axis where **cv** and **cv-dot** are both equal to zero indicates an unstable fixed point (like the skateboarder on the hilltop). That's one. Flanking this unstable "repeller" is a pair of stable fixed points where the slopes intersecting the **cv**-axis are negative. That makes three fixed points altogether.

Last but not least, we come to the bifurcation diagram in figure 2.4(d). It shows how the system behaves as the control parameter **cp** varies continuously. Solid circles indicate that a fixed point is stable for all values of the control parameter less than zero. Then, at a critical point, a bifurcation occurs. Two new stable states are spawned, like a fertilized cell splitting in two, and the formerly stable states lying along the horizontal axis now become unstable (open circles). This is called a pitchfork bifurcation after the easily identifiable shape of the bifurcation diagram.

Foundations of Coordination Dynamics: Some Take-Home Messages

For a practitioner of coordination dynamics, where a dynamical system "lives" in its parameter space determines which behavioral patterns are observed. We saw that nicely in the pictures: For a whole range of values of the control parameter, nothing much happened—see figure 2.4(d). Then the system's behavior changed qualitatively, so the parameter space displays all the values of the observed variable **cv** as a function of all the values that the control parameter (**cp**) can take on, including those that lead the system through qualitative changes and those that do not. Thus, in order to be able to determine a system's full behavioral repertoire, one must find the relevant control parameters and use them to explore the full parameter space. From the point of view of a dynamical system itself, where it happens to be located in its parameter space determines how it behaves. Appreciate that in living things, some values of control parameters might not be so good for the system's survival. Some values can even lead to a system's demise. That's important information for the scientist~practitioner to know, for example, in order to prevent disease.

To the left of the vertical axis in figure 2.4(d), the dots lying along the line indicate the system's behavior remains qualitatively the same regardless of changing values of the control parameter. Then, when the value of the control parameter reaches a critical value, new things suddenly happen. Two stable coordination states or patterns appear where formerly there was only one. In this region of parameter space, the system is bistable: two modes of behavior are possible. Which one actually occurs, of course, depends on the initial conditions—the basin of attraction—as shown in figure 2.4(a).

Bistability and bifurcations are complementary, *essentially nonlinear* properties of coordination dynamics. They confer on the system both stability (i.e., the same patterns are stable over a wide range of **cp** values) and flexibility (i.e., the ability to switch among stable patterns or modes of operation). Stability~flexibility is a complementary pair of the first order, necessary for life, matter, and mind. At least in coordination dynamics, one does not exist without the other. Notice once again what an important role complementary pairs seem to play in the function and interpretation of coordination dynamics. Notice also how the essentially nonlinear features of bistable (multistable) states and symmetry-breaking bifurcations offer preliminary mathematical and visual intuition into why complementary pairs might be ubiquitous and dynamical in the first place.

To ground complementary pairs in science demands more than mathematical phenomenology. Nevertheless, as the great physicist Eugene Wigner remarked, mathematics has an "unreasonable effectiveness" in capturing the behavior of nature and, we might add, the nature of behavior. As our story unfolds, coordination dynamics will also seem "unreasonably effective" in capturing the essence of the complementary nature.

In closing this brief tutorial, a caveat or two is in order. One is that the mathematical theory of dynamical systems contains many more subtleties than have been mentioned here. For example, qualitative changes or bifurcations come in a number of forms, some with exotic names and properties. Bifurcations from rest states to periodic states, from periodic states to quasiperiodic states and from quasiperiodic to chaotic behavior are all possible, as well as mixtures thereof. All of these are excellent examples of the kind of behavioral complexity that can emerge from even a quite simple, low-dimensional nonlinear dynamical system consisting of only one or a few relevant coordination variables and control parameters. Complementarily, such low-dimensional dynamics represent one way for very complex systems to self-organize in order to produce both simple and complicated behavior.

Many different kinds of stability exist, too. A system can be locally stable, globally stable, marginally stable, and structurally stable. The trajectories, landscapes, phase portraits, parameter spaces, and bifurcations of real systems are typically far more complicated than those illustrated here. Nevertheless, the same basic concepts, methods, and tools apply, transcending such complications. Such notions apply no matter what it is that is evolving in time, and no matter what level is currently under investigation—keeping in mind, of course, the relation between levels.

Thus, when confronted with a system they don't know much about, practitioners of coordination dynamics wonder what the relevant control parameters might be and which patterns of connectivity and coordination they alter. They will try to determine under what conditions coordination patterns are stable or unstable, and use the results to identify relevant coordination variables. By inquiring whether the dynamics change smoothly or abruptly in different contexts and for different values of the control parameter(s), they will generate dynamical landscapes of the system under study, asking what causes the topology of these dynamical landscapes to change and what forms the new layouts take, and so on. These are just a few leading questions that experiments can begin to be designed around.

Many challenges face the practitioner of coordination dynamics. One is that there are many different kinds of coordination possible in nature to wonder about, and on many different levels too, with many possible coordination variables and control parameters. Another clear and present difficulty in studying coordination scientifically is its invisibility-by-familiarity. In the words of the prolific British writer G. K. Chesterton, "the more a thing is huge and obvious and stares one in the face, the harder it is to define." As we mentioned in Movement 1, complementary pairs are like that. Coordination is definitely like that. It is quite easy to take coordinative processes for granted exactly because they are so essential to our own moment-to-moment viability.

FROM DYNAMICAL SYSTEMS TO COORDINATION LAW

Erich von Holst
(1908–1962)

... a kind of neural cooperation that renders visible the operative forces of the central nervous system.

The Behavioral Physiology of Man and Animals (1939)

Prerequisites for a Coordination Law

Having explained the elementary concepts and tools of self-organizing dynamical systems, we now wish to bring them to life. The goal now is to describe a real coordination law, and show how it helps us ground the philosophy of complementary pairs in the scientific method. It will take a little work to build up the pool of concepts necessary to understand the coordination law, and to see how it provides grounding for the philosophy of complementary pairs.

In the same way that we have sought to present the basic concepts of dynamical systems, we must now discuss the concepts of coordination science in both theoretical and empirical ways, and highlight some of the most important phenomena and constraints that coordination laws must account for. Along the way, we'll talk about studies of the human brain and behavior that provide valuable evidence supporting the theoretical framework of coordination dynamics. As before, we suggest you take these next sections at a comfortable pace, letting the big picture sneak up on you. The endgame is to enfold the concepts of coordination science with those of dynamical systems in the form of a bona fide coordination law.

Let's start by thinking about coordination at its most basic. It is clear that very little happens in mind, life, and matter—from the human bond to the chemical bond—unless two or more separate things come together. So how do things come together and when they do, what's the nature of their coordination? As students of the complementary nature, we must also inquire into how things break apart, how such bonds are broken, and what the nature of *that* process is. Thus, for a coordination law to be useful, it must adequately explain how things bond and break, how they converge and diverge, how they integrate and segregate. A plausible scientific explanation of such integration~segregation seems a necessary prerequisite for any worthwhile coordination law.

Along with the prerequisite of explaining integration~segregation, laws of coordination dynamics should also predict and provide a parsimonious account of different kinds of coordinated behavior. Additionally, such laws should embrace rather than deny the notoriously difficult-to-understand properties of living things, like interaction, nonlinearity, emergence, and context-dependence. In other words, they need to capture something of the essence of coordinated behavior amid the great diversity of biological forms. Moreover, as a candidate science of complementary pairs, coordination dynamics has an extra burden: It has to demonstrate that complementary pairs are not mere mathematical phenomenology, rich though that may be, but are actually grounded in empirically testable reality. Last but not least, since it is the human mind~brain that polarizes and synthesizes, dichotomizes and entertains contradiction, coordination dynamics had better have something to say about how our brains are coordinated. What kind of a brain can split the world into pairs yet is able to capture everything in between?

To answer this question, a central one in the study of TCN, we need an entry point. Fortunately one is to be found in the work of the great German behavioral physiologist Erich von Holst, who spent his entire life studying coordination in a wide variety of creatures, ranging from worms all the way to human beings.

Absolute Coordination~Relative Coordination

In a masterful survey called *On the Nature of Order in the Central Nervous System* (1937), von Holst noted the prevalence of two basic kinds of coordination. (Surprise, surprise!) The first kind he called "absolute coordination," in which the individual coordinating elements of a system under study are locked together in time. We all know what absolute coordination is. It is what Olympic synchronized swimming teams try to achieve, as do cheerleaders and most military marching formations. In nature, absolute coordination can be seen in a host of different contexts. One dramatic example is the beautifully synchronized flashing patterns of fireflies. Another is when everyone starts clapping rhythmically at a rock concert or a big rally. Phase-locking in the brain, neurons that fire with the exact same rhythm, is a further example of von Holst's absolute coordination. Such synchronous oscillations in the brain have been proposed by many scientists, including the Nobel laureates Francis Crick and Gerald Edelman, to be a crucial biological mechanism for binding. "Binding" is a term for the highest level of brain integration that is our sentient human awareness.

The second kind of coordination patterns that von Holst identified and studied he called partial or relative coordination. Here the coordinating elements lock in only transiently, and then break away from one another again, sometimes spontaneously and sometimes as circumstances change. Relative coordination is like a child walking hand in hand with a parent on the beach. The parent must slow down and~or the child must add a step so that they can continue to walk together. This "coming together and breaking away" waxes and wanes as the pair progresses up the beach.

Relative coordination is obviously less regular and more random-looking than absolute coordination. One reason not much has been said about relative coordination is that we human beings like things to be crisp and orderly when we are trying to make sense of the world. We may even tend to *impose* order on the observations we make, a bias that can create a blind spot. It turns out, however, that relative coordination, despite a lack of rigid observable order, provides a way to think about the dynamics of fleeting neural assemblies in the brain that last long enough to give rise to and propagate "mind." Understanding relative coordination thus becomes central to understanding complementary pairs and their dynamics.

One might wonder whether these two simply stated forms of coordination are *all* that our coordination law is trying to explain. It might not seem like that much on first blush. Yet, just as the secret and diversity of life are tied to the helical arrangement of only two thin strands of DNA and its combinations of complementary base pairs, so the secret and diversity of coordination may lie in the numerous blends and transitions among absolute and relative coordination.

So how is the complementary nature instantiated in the human brain, and how does coordination dynamics account for it? To anticipate, a quick yet accurate answer is that complementarity rests on a subtle and dynamic interplay of *tendencies* in the human brain. To explain this interplay, coordination dynamics as a science needs to extend beyond the currently widely held view of "states." This is not a trivial step, given their ubiquitous and vernacular usage, as in mental states, bodily states, perceptual states, brain states, emotional states, etc. Most physicists would never dream of a world without the concept of state and the definition of state variables. Our very descriptions of matter rest on this terminology. The less well-understood world of biology and psychology, however, is on far shakier ground when it uses these words, because the relevant state variables are seldom defined. So that's the first task of coordination dynamics: come up with relevant state variables! Then, after that, if it turns out that the brain~mind cannot be adequately captured purely with a state description (and its attendant notions of stability, etc.) but also requires the concept of *tendencies*, this fact will need to be reflected in our coordination law. With these considerations in mind, let's take a moment to ponder how the brain~mind might actually work.

HOW THE BRAIN~MIND WORKS

Virginia Woolf
(1882–1941)

Examine for a moment an ordinary mind on an ordinary day. The mind receives a myriad impressions—trivial, fantastic, evanescent, or engraved with the sharpness of steel. From all sides they come, an incessant shower of innumerable atoms; and as they fall, as they shape themselves into the life of Monday or Tuesday, the accent falls differently from of old.... [Life] is a luminous halo, a semitransparent envelope surrounding us from the beginning of consciousness to the end.

Mrs. Dalloway (1925)

So how does the human brain~mind work? Given the enormous complexity of the human brain, both in terms of the detailed workings of its trillions of cells and synapses as well as its rich repertoire of behaviors, this question might seem a bit daunting, if not totally outrageous. But it is an important question to ask nevertheless. Although the complexity of the brain is mind-boggling, there is also a danger that the more we know about the details (which literally fill hundreds of scientific journals these days) the less we are able to understand Virginia Woolf's "ordinary mind," an understatement if there ever was one.

Informational flooding is found not only in the neurosciences. Starting in the last part of the twentieth century, especially with the advent of the digital computer, many fields of research~development have simply had too much data to process. Were an individual or group to pursue the strategy of forgoing synthetic analysis until all possible data are collected, they might *never* get the chance to understand their subject~object! Faced with the reality of information flooding, we must at least try to discover~invent laws and principles based on abstraction from a restricted selection of data, so that we may arrive at understanding in a more expedient fashion. Newton did not give us the laws of motion of a falling leaf. But he did give us the laws of motion for how things, as abstract mass points, move and fall. As we all know, falling leaves are very complicated events! They tumble and rotate unpredictably depending on many factors—their form, the season, the weather, and the wind. When it comes to brain research, though, there is an even older problem that complicates modern efforts to expedite the process of understanding the brain~mind. Would you be surprised if we tell you that it has to do with the interpretation of contraries?

Local~Global, Integration~Segregation

The last couple of centuries of neurology, psychiatry, and brain research have been steeped in controversy about how the human brain works. The reason? You guessed it. There are *two contrary theories* of brain function. One theory is based on the notions of anatomical *localization* and *segregation*. It stresses that the brain consists of a vast collection of distinct regions each localizable in the cerebral cortex and each capable of performing a unique function or set of functions. The other school of thought looks upon the brain not as a collection of specialized centers, but as a highly integrated organ that works by a kind of "mass action" and is capable of forming gestalt-like patterns. In such a holistic view, no single function can be the sole domain of any unique part of the cerebral cortex. This line of thinking contends that the brain works in a *global* and *integrated* fashion.

Like most either/or debates, these opposing views of how the brain works—one local and segregated, the other global and integrated—have shed more heat than light. Caricatures though both may be, they are alive and well. In modern parlance, researchers debate whether the brain is segregated into its parts or integrated as a whole. They debate whether information is represented in a modular, category-specific way or in a distributed fashion in which many distinct areas of the brain are engaged, each one representing different kinds of information. What do you know—it's contraries again!

A reasonable question is: Why couldn't the brain be both? Instead of viewing global integration and local segregation as conflicting processes and theories, why not view them as complementary? Might such a subtle blend of integration and

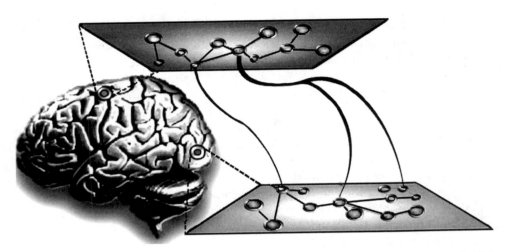

Figure 2.5 Coordination dynamics of the brain. Distributed neural assemblies—each blob within a neural region is itself a coordinated functional unit—interact in time via long-range cortical connections (adapted from Varela et al. 2001).

segregation hold the key to understanding the brain~mind, as well as complementary pairs *and* the complementary nature? That sounds promising, but what might it look like? And even if we did have both segregation and integration in the brain, how could we explain it?

An image of how neural segregation and neural integration are conceived of in the brain is shown in figure 2.5. Here one sees two hypothetical neural assemblies that are dynamically linked by virtue of long-range, reciprocal connections. In the language of coordination dynamics, the coupling between assemblies is two-way. Each talks to and affects the other. How they communicate and what they communicate about is shaped by context and experience.

Experience alters the brain by modifying synapses, junctions where neurons connect and communicate with each other. Experience alters a lot of other things too. Synaptic modification just happens to be a primary means of neural communication and the focus of much neuroscientific research. Since a single cerebral neuron can have as many as 1,000 synaptic connections, the possibility for experiential modification is enormous—practically infinite, in fact. We'll return to this topic again in Movement 3 when we discuss how people learn.

Notice in figure 2.5 that each neural group within a segregated cortical area (represented by the blobs inside the trapezoids, perhaps amounting to about 1,000 neurons or so) is also a coordinated entity in its own right and thus perfectly able to function independently as a segregated entity (echoes of "a whole is a part~a part is a whole"). Indeed, the evidence for functional segregation of local areas on many levels of brain organization is considerable. However, when the *mind* is working (pondering whether to buy these shoes or those, attending to what the shop assistant is saying, remembering what the professor said about coordination, and planning the next summer vacation), individual neural collectives work together as an integrated sentient whole. How is this accomplished?

Oscillation~Rhythm

It is no coincidence that coordination dynamics deals largely with nonlinearly coupled oscillatory processes. Oscillations and rhythms—"brain waves"—are ubiquitous in the brain and have been known, since the time of Hans Berger in the 1930s, to occur in different frequency bands. Different parts of the brain oscillate at different rates. Until recently, however, the functional significance of these rhythms has remained obscure. For example, the so-called "theta rhythm," which varies roughly between 3 and 5 Hz, dominates in the hippocampal formation, a structure~function strongly implicated in the formation of new long-term memories. Likewise the thalamus, a structure~function sometimes called the "organ of attention," possesses an intrinsic rhythm that hovers around 10 Hz. In fact, when one spans the spontaneous firing rates of neurons in the cerebral cortex, 10 Hz is the dominant firing rate.

The ubiquitous tendency of neural groups to oscillate is not because there are clocks in the brain or anything like that, metering out time like a quartz crystal. Rather, it's because neural groups are composed of populations of synaptically coupled excitatory and inhibitory neurons. The former excite the latter which in turn inhibit the former, causing cycles of activity (excitation~inhibition). How fast neurons oscillate depends on many factors ranging from the low-level membrane biophysics of ion channels, to synaptic delays between neurons, to the intensity of neuromodulatory influences. Just as phase-locked oscillations may arise locally among relatively small segregated populations of neurons, so too such oscillations may arise globally from interactions between participating brain regions—sometimes over long distances (figure 2.5).

The key idea is that multiple oscillations from widely distributed brain regions are coupled or "bound" together into a coherent network when people attend to a stimulus, perceive, think, remember, and act. As units of structure~function, such oscillations are multifunctional: a given oscillation may be involved in multiple functions, and multiple oscillations underlie a single function. Binding is a dynamic, self-assembling process, wherein different specialized areas of the brain engage~disengage each other in time, as in a country square dance. In the simplest case, oscillations in different brain regions can synchronize in-phase, neural populations rising and falling together. They can also synchronize in an antiphase pattern, one oscillatory brain activity reaching its peak as another hits its trough and vice versa (in-phase~antiphase).

Detailed mathematical analysis and simulations of neural clusters by Eugene Ishikevich and Frank Hoppenstadt at the University of Utah show that in-phase and antiphase are the *only* stable solutions in large, synaptically connected clusters of neurons. That's an interesting result because it suggests that out of the many possible phase relations that could exist between different, specialized brain areas, in-phase and antiphase coordination patterns predominate as the stable ones. It will be useful to keep this in mind, as these coordination patterns constitute empirical phenomena that a science of coordination dynamics must explain. Such phenomena may not seem so impressive at first blush, but realize that there isn't any reason why such global synchronization should necessarily occur at all. That brains manage to form *any* globally coherent coordination patterns among billions of neurons with trillions of synaptic connections seems like a miracle! From the viewpoint of self-organizing coordination dynamics, however, it could hardly be otherwise.

Not only does the brain possess different phase relations within and among its many diverse and interconnected areas—how many is unknown—but it can also switch flexibly from one phase relation to another, just as flexibly as a mind can switch thoughts. In the experiments and theory of coordination dynamics, these switchings are understood to be actual nonequilibrium phase transitions in the brain, rapid shifts in brain coordination caused by sensitivity to changes in both external and internal conditions. Recall from the tutorial that in addition to being a sure sign of self-organization, instability is a selection mechanism that leads a dynamical system to the most suitable pattern of behavior for the circumstances at hand. Here again, the science of coordination dynamics aims to show how this comes about.

In Kelso's *Dynamic Patterns* a considerable amount of evidence and theory was presented suggesting that transient, metastable phase coherence within and between specialized areas of the brain represents the rapid creation and dissolution of neural assemblies that last long enough to give rise to the "stream of consciousness." An example of this creation~annihilation process is displayed in figure 2.6, drawn from the work of the late Francisco Varela and colleagues. The figure shows the instantaneous phase difference between two "raw" signals obtained by recording directly from different places in the brain. One can readily see that the two rhythms fall in and out of step, the phase relation (bottom picture) locking in only temporarily and then proceeding to fluctuate all over the place. The beauty of this kind of relative coordination is that it suggests not only a means for the

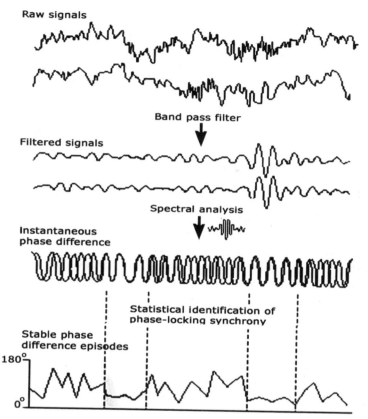

Figure 2.6 Relative coordination in the brain. Two "raw" signals recorded from two different parts of the brain are said to be in synchrony if their rhythms coincide. Notice that the two rhythms go in and out of phase, coming together only briefly and then fluctuating quite randomly. Brief epochs of phase locking are interspersed with brief epochs of phase wandering (adapted from Varela et al. 2001).

rapid formation of cell assemblies ("binding") but also their active dissolution ("breaking apart"). Now why would the brain do that? And for what function? Might this be a way for the brain to create~destroy information?

A typical strategy for handling data like these would be to focus in on only the brief, ordered episodes of phase-locking behavior while ignoring the phase-scattering part (phase locking ~ phase scattering). Reasonable though it seems, such limited treatments miss the rather extraordinary complementary nature of *the ordinary mind*. They fall into the historical either/or habit because they assume the brain to be run predominantly via integration, thereby emphasizing phase locking over phase scattering. Look at the brain waves in figure 2.6 again. Under the reasonable assumption that, to a first approximation, phase scattering reflects neural segregation and phase locking neural integration, the time series on the bottom of figure 2.6 may be characterized by the following sequence: *segregated-integrated-segregated-integrated-segregated*, and so on.

If, instead of artificially partitioning the brain data (or any dynamical data for that matter) into isolated epochs of pure integration and pure segregation, one considers the entire epoch, one discovers that the behavior seems to be *relatively* coordinated. The kind of relative coordination displayed in figure 2.6 is a nice example of what, in *Dynamic Patterns*, Kelso calls "the twinkling metastable mind." Metastability is a crucial feature of coordination dynamics, and holds the key to understanding the complementary nature.

In the metastable brain, brief epochs of phase-locking synchrony like those shown in figure 2.6 do not correspond to stable coordinated *states* of neural integration at all. Likewise, the brief epochs of phase wandering do not correspond to fully segregated behavior. Rather, individual parts~processes exhibit *tendencies* to function independently as well as *tendencies* to coordinate together. From this perspective, the data shown in figure 2.6 represent one unified picture composed of two coexisting tendencies.

Here one cannot help but recall Carl Jung's words that the "conscious and the unconscious [exist] at the same time, that is, conscious under one aspect and unconscious under another." It is unknown whether the segregation~integration of the brain corresponds to the unconscious~conscious of the mind. If true, such a novel correspondence principle would seem a rather more agreeable solution to the mind~body, psycho~physical problem than the usual story of how one-way neural firings are causing mental events.

Metastable coordination dynamics is an entirely new and different conception of how the brain works. Metastability says that individualistic tendencies for the diverse regions of the brain to express themselves as segregated entities *coexist* with collective, coordinative tendencies to bind and integrate as a functional unit. The metastable brain~mind implies that the patterns of the brain are fluid precisely because tendencies to bind coexist with tendencies to break apart. It implies that the patterns of the brain are diverse precisely because tendencies for the parts of the brain to cooperate coexist with tendencies for the parts to compete. In this view of the brain, the view of coordination dynamics and the complementary nature, apartness (segregation) and togetherness (integration) coexist as a complementary pair of tendencies. At the same time, opposing theories and pure states of integration versus segregation appear as polarized and idealized extremes.

This new view of the brain puts either/or brain versus mind, localizationist versus holist, and nativist versus empiricist debates in sharp relief. It suggests how apparently contrasting properties of the brain may coexist and how they may be reconciled. Coordination in the brain is like a Ballanchine ballet. Neural groups briefly couple, some join as others leave, new groups form and dissolve, creating fleeting dynamical coordination patterns of mind that are always meaningful but don't stick around for very long. It is transient coupling~uncoupling tendencies, both locally within individual brain regions and between cortical and subcortical areas, that underlie the workings of the brain~mind and its complementary nature.

HOW BRAIN~MINDS WORK TOGETHER

What then of the coordination of two or more brain~minds? Might they too exhibit this same kind of relative coordination? Might metastable coordination dynamics also capture something essential about human interaction? Human beings work together, play together, dance, sing, and communicate with each other. Moreover, this coordination is accomplished *wireless—through the air*. We know that we can do this, but how? In the previous section, we illustrated how different areas of the brain communicate through long-distance, reciprocal pathways of connection. But what happens if there are no physical wires (mechanical or electrochemical) to do the job?

Figure 2.7 shows two brain~minds "talking to each other." In this case they are not actually talking; all that is going on is that each person is sometimes watching the other and sometimes not. The coupling is very long-range compared to the relatively short distances between coordinating neurons inside a single brain. Here the coupling is mediated by the exchange of visual information. In this simple experiment, pairs of people sit facing each other. Each is asked to move their index finger up and down at their preferred frequency, "as if they were going to have to do this all day."

The pair's movements are recorded in real time by an optical recording device. This produces time series just like those in figure 2.6, except this time the behavioral signals are from finger movements, not brainwaves, though that is possible too. Finger movements and brain activity are obviously related. A few years ago we reported in the journal *Nature* that there is a beautifully tight connection between the two. Visual coupling can be simply manipulated by asking the subjects to open and close their eyes. A key experimental condition is when both subjects start this test with their eyes closed. This means the two people are acting as segregated entities, totally independent of each other. As before, a good measure of this uncoupled, segregated behavior is "phase wrapping."

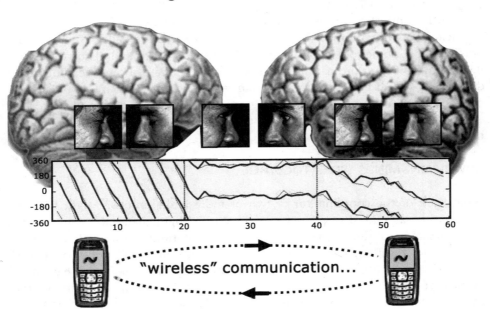

Figure 2.7 Two brain~minds "talking" to each other "without wires." Do they share a common coordination dynamics with that of a single brain whose parts communicate through neuronal connections? (Courtesy O. Oullier.)

Notice in figure 2.7 how the relative phase drifts in a linear fashion. This means that there is no relationship between the movements of one subject and those of the other. Each is moving at a different, self-chosen frequency. However, as soon as the subjects open their eyes, they spontaneously get in step with each other. The relative phase between the two people's movements now stabilizes around zero, indicating that the two people adopt a common frequency and phase and are now literally in-phase. They have formed a coupled, integrated unit. This means that if you know what one is doing, you know what the other is doing too.

Just as in the single brain coordinating the activity of different regions, the interaction that produces functional integration here is reciprocal and two-way: each member of the pair spontaneously adjusts to the other. In this experiment neither person is the boss, although similar studies using bosses and employees, males and females, friends and strangers, and other such variations may be quite informative. Detailed analysis which we won't go into here shows that both people are mutually synchronized. This is a classical example of informationally based self-organization that vanishes, needless to say, when test persons close their eyes again. With eyes closed, the subjects again become uncoupled and the phase relation between them drifts once more.

Evidence from brain imaging studies indicates that neurons in the prefrontal cortex are synchronized when people socially interact with each other. Remarkably, these so-called "mirror neurons" in the premotor cortex of monkeys and in the inferior frontal gyrus of humans are active only when observing another perform the same actions that one is capable of performing oneself (hence the term "mirror"). It seems likely that these mirror neurons will fire in synchrony when people are bonding in the fashion shown in figure 2.7. But let's remember not to get too hung up on specific functions for specific neurons; neurons in the cortex are patently multifunctional.

The significance of this "two brains talking" experiment is that it offers more than just a window into the mechanisms that neurally bind humans together. It also reveals that exactly the same principles of functional integration~segregation apply to two brain~minds unconnected by neuronal circuitry as to one brain that very obviously *is* connected by neuronal circuitry. For coordination in complex systems, coordination dynamics predicts that, provided there is a medium through which functional information can flow, integration~segregation will occur.

Neuronal connections in the brain are an effective medium that nature has provided through the mechanisms of self-organization and natural selection. But the important concept here, as these simple experiments on people reveal, is biologically relevant information. Such functional information can be conveyed by local connections and between distant areas, both by two-way interactions. In this capacity, local~global and integration~segregation are two of the key complementary pairs of coordination dynamics.

So how does coordination dynamics explain all these phenomena? And what are the implications of this different view of the brain for the complementary nature? The empirical data suggest that integration (pure coordination) and segregation (no coordination) in both brain and behavior may be viewed as polar, idealized extremes and that the key to understanding lies in the dynamical interplay of integration~segregation tendencies. As a candidate science of the complementary nature, coordination dynamics must minimally be able to explain both polarized complementary aspects and all that falls in between them. So let's see how it rises to these challenges.

CONDITIONS~ASSUMPTIONS OF COORDINATION DYNAMICS

The derivation of a scientific law, regularity, or principle depends on the validity of the assumptions it employs as well as the conditions under which these assumptions are met. Often the word "assumption" is accompanied by the word "hidden," so let's make our assumptions explicit before we get into the coordination law itself. After that, we will show that it is the interplay between these assumptions and conditions that gives rise to all the natural forms of coordination that have yet been observed. This is the power of a law or principle: With the aid of a few core concepts, one can arrive at a workable comprehension of a system without the necessity of trying to process and integrate every piece of data from the ceaseless informational flood. Detailed knowledge is important, but not if it obscures the principle.

Condition~Assumption 1: Heterogeneity~Homogeneity

The overseer always seems to be faced only by king and by peasant. *That* is our key notion.

"Physical Basis for Complex Systems" (1978)

Arthur Iberall
(1918–2002)

The first condition~assumption concerns the similarity~difference of the constituent elements that compose an individual entity or component. Physical components that make up a complex system like an organism—its "atomisms," to use Iberall's word—are heterogeneous, even though each atomism certainly possesses some of the same constituents. Every cell, for example, contains DNA. Nevertheless, cells of the brain differ within themselves and are different from liver cells and heart cells. Though this subject takes us too far from our present topic, it is indeed a fascinating problem to understand how genes are switched on and off to express the proteins that result in different types of cells. What we wish to state up front here is that, in general, all the individual components of complex living systems represent, in the words of the late theoretical biologist Walter Elsasser, "a kind of heterogeneous order."

Coordination dynamics follows Elsasser's insight and assumes that the individual coordinating elements are heterogeneous in complex living systems, as are the dynamic connections between them. Heterogeneity contrasts with the homogeneous character of the atoms and molecules that make up an inorganic material (like water) and the homogeneous connections between them (like hydrogen bonds). Note that the assumption that individual coordinating elements are heterogeneous doesn't exclude the role of homogeneous elements within that heterogeneity. As a science of complementary pairs, coordination dynamics naturally considers homogeneity and heterogeneity to be complementary aspects.

Condition~Assumption 2: Oscillation~Rhythm

In steady-state systems, the flow of energy through the system from a source to a sink will lead to at least one cycle in the system.
Energy Flow in Biology (1979)

Harold Morowitz
(1927–)

The second necessary condition~assumption concerns time, timing, and rhythm. In living things, processes and events occur on multiple timescales. We assume this is because coming together~apart in time is inherent to animate form~function. Processes go up and down, in and out, back and forth. They have a rhythm to them. Oscillation or rhythm is the cornerstone of life, of all systems that are not at equilibrium. Although rhythmical behaviors can be quite complicated, the principles underlying them possess a beautiful simplicity. Rhythms confer positive functional advantages on the organism, like spatial~temporal organization, prediction of events, energetic efficiency, and precision of control. Rhythms can be both continuous and wavelike, like walking, or discrete events with a beginning and an end, like a golf swing. Although no single golf swing is ever exactly the same as any other, it has an obvious rhythm to it (continuous~discrete).

All this is nice, poetic even, but we would like to understand the property of rhythm in a more rigorous scientific manner. Inspiration comes from Morowitz's theorem. An eminent biophysicist and emeritus professor at Yale University, Harold Morowitz wrote, among other books, *The Wine of Life* and *The Thermodynamics of Pizza*. Morowitz's theorem says that for any process to persist, a cycle of work must be performed. This means that the fundamental processes of living things, including the slowly time-varying ones we call structures, *must* be time-dependent and inherently cyclical. Structures like the cells of the heart and the neurons of the brain, and processes like glycolysis and blood-cell production, are cyclical in nature. They cycle with different characteristic periods, contributing to the wide distribution of timescales that we see in living things. Oscillation, therefore, is a dynamical archetype of all behavior.

One point gets lost all the time in discussions and presentations on oscillations: To be inherently oscillatory does not imply that a process must always be oscillating. Consider the theta and gamma rhythms in the brain that are associated with memory and perception. Memory and perception are not usually thought of as oscillating, though, like attention, they can. The more general point is that a system or a function may have the potential to oscillate without necessarily oscillating itself. An oscillator can cycle as well as be at rest, depending on its parameters (cycle~rest). Think of a mechanical system like a mass attached to a spring. If the mass-spring system is heavily damped, it won't oscillate at all. If it is lightly damped, once set in motion it will oscillate for a long time.

Furthermore, oscillators don't necessarily wiggle back and forth in perfect cadence. Their cycling does not have to be strictly periodic. A limit cycle oscillator, for example, may oscillate with different periodicities, just as long as its limit cycle is stable. An example is the circadian rhythm common to humans and fruitflies. Its cycle is stable even though its period (roughly 24 hours) can vary quite a bit.

In limit cycles, if the initial conditions of a process begin somewhere outside the limit cycle, the cycle will tend to return to some basic periodic orbit depending on how fast the process dissipates energy (the equivalent of Morowitz's "sink"). Likewise, if a process begins inside the limit cycle, the oscillation will grow, depending on how much energy is available (the equivalent of Morowitz's "source"). A steady state is reached when energy input and energy dissipation, source and sink, reach a dynamic balance (source~sink). When this happens, the cyclical process may be said to be "on its limit cycle."

Many different kinds of oscillators and types of oscillation exist. Think of the eddies and vortices that form in the wake of a fluid flowing past an obstacle, like water over stones in a stream. These are periodic structures, alternating back and forth in one direction or another. In the inanimate world, oscillators are the basic components of most man-made machines, from helicopter rotors to pistons to grandfather clocks. Oscillations in chemical reactions, originally discovered by chance, are now widely studied not only for their inherent interest and practical applications, but also because of parallel discoveries of oscillations and cycles in biochemistry and biology.

A good example of a complex oscillation from the field of biochemistry is the glycolytic cycle, which is responsible among other things for the breakdown of glucose for its use in oxidative metabolism ... *cycles of work*. Cyclical processes are part and parcel of all things, permeating processes and ideas, great and small. They are as ubiquitous at the individual level, from cells to organs like the heart, lungs, and brain, as they are in whole organisms and the collective coordination patterns formed by groups and populations—from business cycles in the economy to an audience clapping in a theater. The tendency toward keeping together in time is universal. It is literally what makes us tick. Following Morowitz's theorem, cycles are as basic as anything can be since their roots lie in thermodynamics.

Condition~Assumption 3: Coupling~Uncoupling

Linus Pauling
(1901–1994)

If the structure that serves as a template (the gene or virus molecule) consists of, say, two parts, which are themselves complementary in structure, then each of these parts can serve as the mould for the production of a replica of the other part, and the complex of two complementary parts thus can serve as the mould for the production of duplicates of itself. In some cases the two complementary parts might be very close together in space, and in other cases more distant from one another—they might constitute individual molecules, able to move about within the cell.

"Molecular Architecture and the Processes of Life" (1948)

The final basic condition~assumption of coordination dynamics, following Linus Pauling's work in *The Nature of the Chemical Bond*, is that for coordination to occur and to manifest the wondrous forms it takes, there must be some degree of coupling between individual coordinating elements. In the cell, for example, two different macromolecular "chaperones" bind unfolded proteins together most effectively when they act together as a synergy. Meiosis, a basic form of cell division, involves an intimate coordinated dance between homologous chromosomes. Nothing much happens without coupling.

In general, organisms and their environments evolve on all scales as a single, coupled dynamical system. This is because dynamical systems at all levels both adapt to the environment and change it. The interaction is two-way. In the context of the brain, the Nobel Laureate Gerald Edelman calls these reciprocal interactions and speaks in terms of "re-entrant pathways." In coordination dynamics, such two-way interaction, like cross-talk between components, mutual coupling, and information exchange, is crucial for self-organized pattern formation.

In nature, actual coupling mechanisms are diverse and take on many forms. Things may be coupled by mechanical, electrical, and chemical forces and, as emphasized in coordination dynamics, by functional information. When it comes to people, not only can the five senses couple us together, so can ideas and feelings—a *sixth* sense that sometimes allows us to anticipate events. Regardless of the way things are coupled, for a system to be multifunctional—for multiple states and tendencies to coexist—the coupling *must* be nonlinear. Nonlinear coupling is at the very heart of coordination dynamics and hence of the complementary nature.

A DYNAMICAL LAW OF COORDINATION

We are still mostly ignorant of the laws governing the context dependency and integration of environmental signals into the output patterns.

"Maneuvering in the Complex Path from Genotype to Phenotype" (2002)

Richard Strohman
(1930–)

In the paradigm of coordination dynamics, all the basic forms of coordination and various mixes and assortments of them emerge naturally from three basic conditions~assumptions: (1) the *heterogeneity* of coordinating elements and connections; (2) multiple time scales by virtue of inherent oscillatory processes; and (3) *nonlinear coupling*. Other factors, such as time delays, e.g. due to transmission between brain areas, are known to significantly shape the coordination dynamics but are not essential for present purposes.

Let's now examine a coordination law that has been tested in numerous studies of real-life biological coordination. This law has a relatively long and cherished history in the field, and has been elaborated over the years in a series of theoretical~empirical steps. In figure 2.8, using the same terminology we introduced earlier in our little tutorial, we put it in the most user-friendly and general form possible.

The equation says that the term called **cv-dot**, that is, the rate of change of the coordination variable, **cv**, is a function **f** of three basic factors: the momentary value of the coordination variable **cv** itself; one or more control parameters **cp**; and noisy fluctuations, F. Recall that coordination variables capture the functional nature of behavioral patterns. Such coordination variables can be quite abstract in that they can characterize coordination among the same and different kinds of elements and processes (homogeneous~heterogeneous).

$$\dot{cv} = f(cv, cp, F)$$

Figure 2.8 The archetypal form of coordination dynamics.

cv-dot stands for the moment-to-moment rate of change of the coordination variable's behavior and therefore expresses how coordination patterns evolve in time. That's what we typically want to know and understand: how coordination patterns persist, adapt, and change. Recall from figures 2.2 and 2.4 that a parameter is a control parameter if it leads to abrupt, qualitative changes of the **cv** at a critical threshold value. Below or above critical values of the **cp**, the **cv** may change only gradually or not at all. It's at critical threshold values of the **cp** that instabilities leading to the emergence of new and different patterns of coordination occur. Instabilities, bifurcations, and phase transitions are the stuff of dramatic, qualitative change. Without them life as we know it would not exist.

The last term in the equation, F, is crucial because nature is full of chance happenings—*fluctuations* that provide an essential source of variability. The laws of coordination dynamics are therefore both deterministic and stochastic: How a coordination pattern evolves in time depends upon both the values of one or more control parameters and inherent noise (deterministic~stochastic).

As is true in every scientific field, in coordination dynamics the choice of key variables is typically based on empirical insights. Though it may not be the only bearer of information about how things are mutually coupled, the evidence in favor of *relative phase* as a key coordination variable representing the dynamic linkages among mutually interacting components is pretty overwhelming. Everything from individual neurons to neural groups to different brain regions and cognitive and behavioral function appears to be coordinated phasically. This is important because it tells us that coordination dynamics can exist in different situations and different levels. And generalizability, of course, is an essential criterion for any would-be scientific law.

According to the paradigm of coordination dynamics, understanding at *any* level of coordination requires at the bare minimum a knowledge of three things: (1) the control parameter(s) **cp**; (2) the individual coordinating elements, which may be of quite different kinds; and (3) the patterns of coordination that emerge from component interactions under the influence of control parameters, as expressed by the coordination variable **cv**.

Mathematically speaking, the coordination law contains three different kinds of parameters: One corresponds to control parameters (**cp**) that govern the *strength of coupling* between individual components. This coupling is typically mutual, but it can also be one way, once again pointing to the complementary nature. A second parameter corresponds to *intrinsic differences* between individual coordinating elements. In complex systems, individual coordinating elements are usually neither simple nor purely homogeneous. They may possess inherent time delays and attendant memory. The third mathematical parameter in the coordination law reflects *noisy fluctuations* (**F**) that contribute to the way coordination patterns vary and change.

As we will shortly show, this simple coordination law wraps up or "enfolds" in a single equation many of the basic coordinative phenomena studied so far. It has been shown to capture coordinative relations within~between components of an organism, between organisms, and between organisms and their environments, thereby attesting to what the theoretical physicist Hermann Haken calls its "universal nature." Remember, however, that coordination dynamics, as a set of context-dependent, informationally meaningful laws of coordination for living things, is intended to complement not replace the universality of physical laws.

VISUALIZING A BASIC LAW OF COORDINATION

Clouds are not spheres, mountains are not cones, coastlines are not circles, and bark is not smooth, nor does lightning travel in a straight line.

The Fractal Geometry of Nature (1982)

Benoit Mandelbrot
(1924–)

For reasons of clarity and exposition, we will stick with one of the simplest cases of coordination, namely the mutual interaction between two rhythmically active components. However, this coordination law has been successfully elaborated in a host of different ways and in new directions. For example, people like the brilliant bioengineer James Collins (whom we'll come back to later on in Movement 3) and his mathematical mentor Ian Stewart have extended it to handle not just two, but multiple interacting components. Others such as Viktor Jirsa, Armin Fuchs, and Scott Kelso have extended and derived the law from a biologically realistic model of the brain's structure~function. Empirically verified cognitive influences on behavioral coordination such as intention, learning, memory, and attention have been incorporated. Rhythmical as well as discrete behaviors have been accommodated, and new predictions made. The elementary form of the coordination law has spawned models of speech and visual perceptual dynamics. Indeed, the coordination dynamics paradigm has inspired dynamical approaches to cognition, development, and emotion. We'll return to some of these developments in Movement 3.

VISUALIZING A BASIC LAW OF COORDINATION

To see how the coordination law captures how coordination forms and changes, observe how the key coordination variable *relative phase* behaves as a function of control parameters. In the lingo, the relative phase is written as "phi" and abbreviated by the Greek letter Φ. We follow Φ-dot (the change in relative phase over time) plotted against Φ—just as we did in figures 2.2(c) and 2.4(c)—as the coupling between the coordinating elements is changed. The format is exactly the same as in the phase-plane trajectories described in the dynamics tutorial, but the graphical terms now reflect an actual, bona fide coordination variable, and a control parameter that reflects the heterogeneity of the coordinating elements. What, *more stuff on the graphs*? Yep. But don't worry. Here is a brief debriefing to keep you on top of the game.

Reading the Pictures, Following the Flow

In the following diagrams, the relationship between the coordination variable Φ is plotted on the *x*-axis and its rate of change Φ-dot is plotted on the *y*-axis for all values of $\delta\omega$ on the *z*-axis. These pictures allow us to study the flow of the coordination dynamics as parameters change. As before, let's take this slow and steady. Take a look at figure 2.9. We aren't going to interpret anything yet. Let's just start by getting acquainted with the terms representing the axes of these three-

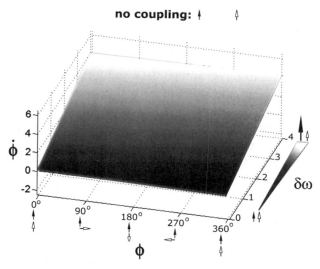

Figure 2.9 No coupling, no coordination. Without coupling, no coordination between components is possible, no matter how much or how little the components differ.

dimensional graphs. The relative phase Φ represents the difference between the "phases" of the coordinating elements. Phases are points on a circle and are expressed as an angular measure with one cycle or period being 360°. (Imagine the rotating hands of a clock, with relative phase being the angular difference between the hands.) Each cycling element (rotating hand) has its own phase, and in principle the two can be quite independent, like the "phase wrapping" that occurs when the two parts of the brain or the two brains are not communicating with each other.

So how does one interpret values of the relative phase or the phase relation? It is simple enough. If the two cycling elements happen to be synchronized, they will move round and round, up and down, or back and forth together in time. In this eventuality, there is no difference between their respective phases, and so the phase of one relative to the other is zero (relative phase, $\Phi = 0°$). This is represented on the graphs right under 0° as the black and white arrows pointing in the same direction. In the language of coordination dynamics, this condition of relative-phase synchrony is called "in-phase." Similarly, if the coordinating elements are exactly in counterpoint, one is all the way up when the other is all the way down. This is represented on the graphs right under 180° as the black and white arrows pointing in opposite directions. In this case, there is a 180° phase difference between the component processes. This "counterpoint" coordination pattern is often referred to as "antiphase."

Summarizing before we move on, the relative phase Φ is the coordination variable in the coordination law and is read out on the graphs as degrees going from 0° to 360°. Underneath each of the degrees are black and white arrows representing the coordinating elements. Notice that at 0° and 360°, the arrows point in the same direction. That's because they are in-phase and go back and forth as a single unit. Rhythmical systems are circular, so 0° and 360° are really the same point, like a snake biting its tail.

At a relative phase of 180°, the arrows representing the individual phases of our coordinating elements point in opposite directions. They are thus said to be antiphase or counterpoint—one is up, the other is down, like your bicycle pedals as you pedal. Of course, the relative phase can take on *any* value between 0° and 360°, whether 37°, 45°, 90°, or 117°. But what we will show you very soon is that out of the very many possible phase relations for a system to adopt, nature seems to prefer *just two*, in-phase and antiphase, to do most of its work.

Φ-dot on the left vertical axis represents the change of the coordination variable relative phase over time, just as in figures 2.2(c) and 2.4(c). It is interpreted in exactly the way already discussed. So what's next? Well, as we have labored to explain, we can't have coordination variables without control parameters. And we can't do coordination dynamics without both. This basic coordination law contains two control parameters: one corresponds to the strength of coupling between the coordinating elements, and the other to intrinsic differences between the components.

In all of the following pictures, the coupling parameter is shown at the top of the graphs followed by a black-and-white arrow with a number of asterisks between them to represent the "coupling strength." Notice that this kills two birds with one stone. We need a control parameter to view how our dynamical law behaves, and we fulfill condition~assumption 3, that to have coordination there must be some coupling between coordinating elements.

Finally, we have the term $\delta\omega$, or delta omega, which represents the heterogeneity parameter. Remember that one of the three built-in conditions~assumptions about biological coordination is that coordinating components are seldom, if ever, identical. In our case, this parameter represents intrinsic differences between the individual coordinating elements, specifically whether or not they differ in their cycling rate or frequency. On the graphs, this is also represented by the sizes of the black and white arrows lying on the right-hand side.

In order to avoid possible confusion, note that because this is a three-dimensional graph, it *looks* as if the right vertical axis is following the phase-plane landscape from 1 to 4, though it actually runs from 0 to 4. When $\delta\omega = 0$, the coordinating elements are the same. That would be like coordinating your two index fingers. Two fingers have very similar shapes, sizes, and intrinsic cycling frequencies when you wiggle them back and forth. Thus, when we study the coordination patterns of two rhythmical finger movements experimentally, $\delta\omega$ will be close to zero. Of course they won't be perfectly identical, but they'll be surprisingly close.

In contrast, the black and white arrows at the top right corner of the graphs are quite different in size when $\delta\omega = 4$. That represents a situation like trying to coordinate two very different kinds of things, like a finger and a leg (or two neural ensembles that tend to oscillate at quite different frequencies; a "genetic clock" with very different transcription rates; a very large person dancing with a very small person). The $\delta\omega$ parameter will become larger as the difference between the coordinating elements grows. Thus, the more the components differ, the more $\delta\omega$ moves from zero. Given weak coupling, this means that rather than coordinating together in some fixed-phase relationship, the two coordinating elements will exhibit a stronger tendency to go their own way—to act as independent entities. In fact, although delta omega has a physical interpretation, one can view it more generally as a parameter that represents the *differences between things and processes*—regardless of what they are made of.

With this preamble over and done with, let's look at the graphs showing how the coordination law behaves. If along the way you get lost or disoriented, just return to the tutorial on dynamics and study the section on the meaning of phase-plane trajectories again. To keep things simple, in all of the following graphs, only one thing is being adjusted—the amount of coupling between components, which reflects the effect of a continuously changing control parameter on the behavior of the coordination variable Φ. Keep in mind, however, that we are always studying how the strength of coupling affects coordination, and how coordination is affected by the degree to which the components differ. If your *complementary* intuition tells you that the more~less the individual coordinating elements differ, the stronger~weaker the coupling needs to be to keep them together, you are spot on.

No Coupling, No Coordination!

Figure 2.9 represents the initial case when there is *no coupling* between the elements. In this case we observe a flat landscape with no fixed points. The gray levels represent the values of the heterogeneity parameter $\delta\omega$, a measure of the difference between coordinating elements that varies from 0 to 4 "units." The message to be learned from figure 2.9 is a tad boring and straightforward: Without coupling between coordinating elements, there can be no coordination, no matter whether the individual coordinating elements are the same or not. Without coupling, individual coordinating elements are free to do their own thing, to act independently. They show no preferred phase relation.

Conception: The Birth of a Complementary Pair

However, when just the tiniest bit of coupling is introduced, one can just detect a delicate wrinkle in the phase-plane surface, representing the emergence of coordination states or patterns (figure 2.10). What formerly was a completely flat phase-plane surface has now become slightly curved. Additionally, a filled and an open dot appear at the phase values of $\Phi = 0°$ and $180°$ at $\delta\omega = 0$, that is, when the coordinating elements are the same. (Note: the black dots at 0°and 360° are equivalent.)

As explained in our little tutorial, these are the so-called *fixed points* of the coordination law. Notice that the birth of coordination involves the creation of both a stable (filled dots at 0°and 360°) and an unstable fixed point (open dot at 180°). The stable attractor is always accompanied by an unstable repeller. One does not exist without the other. *Does that remind you of anything?* The obvious conclusion is that attraction and repulsion are a complementary pair, perhaps one of the most basic complementary pairs to be found in coordination dynamics and all nonlinear dynamical systems. Considering figure 2.10, it is ironic that some scientists

VISUALIZING A BASIC LAW OF COORDINATION

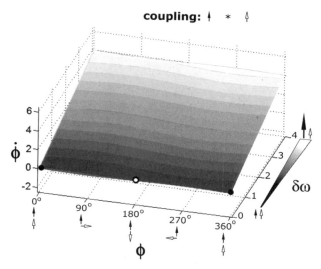

Figure 2.10 Conception. The weakest bit of coupling between individual coordinating elements results in a tiny wrinkle in the dynamical landscape, which corresponds to the birth of coordination, and the emergence of complementary aspects.

tend to overemphasize the study of a dynamical system's stability, while others like to focus solely on its instability. *The either/or mind-set turns up in the darndest places.*

The philosophy of complementary pairs does not go along with this. It says that attractive and repulsive fixed points are complementary aspects of the complementary pair attractor~repeller, and stability and instability are complementary aspects of the complementary pair stability~instability. One might even speculate that the simultaneous creation of an attractor and a repeller in the coordination law hints at the complementary nature of the coordination dynamics, and thus reflects the complementary nature itself!

There is another lesson to be learned from the picture shown in figure 2.10. The fixed-point, absolutely coordinated phase-locked states (filled dots) arise only for the relatively idealized situation in which there are no intrinsic differences among the participating coordinating aspects ($\delta\omega = 0$). Notice that the coupling is too weak to create coordinated states when the coordinating elements differ by just a small amount, i.e., when $\delta\omega > 0$. Even with tiny differences between the elements and weak coupling, no stable forms of coordination exist. We can see from figure 2.10 that there would be only the faintest of tendencies for any coordination to arise in most of the picture, that is, for all values of $\delta\omega$ above zero. In this case,

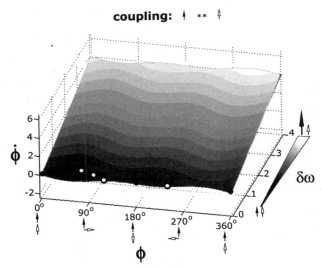

Figure 2.11 Adaptive change and the two Bs—bistability and bifurcation: the creation of multiple coordination states as the coupling is increased.

the individual coordinating elements will not work together; they will just do their own thing.

Bifurcation~Bistability

If the coupling is increased just a little bit more, as expressed by the two asterisks in figure 2.11, a threshold is crossed, and two astonishing events happen. The first (looking at the left side of the graph between 0° and 180°) is that a *whole set* of new fixed points has emerged on the dynamical landscape. These are the three filled dots and three open dots between 0° and 90°. Notice again that stable and unstable fixed points always appear on the scene together.

What this means is that the current level of coupling not only sustains a coordinative interaction between identical coordinating elements (the idealized situation where $\delta\omega = 0$) but also is able to create and maintain a persistent coordination even when the parts are slightly different, or heterogeneous (a more realistic situation). These are the second and third closed and open dots on the left, between 0° and 90°. This is when the heterogeneity parameter $\delta\omega$ is just a bit greater than zero.

Notice how coordination is lawfully adapted to this new situation in which the coordinating elements are no longer absolutely identical. That is, the fixed points spontaneously shift along the Φ-axis to accommodate differences between the elements. The stable value of coordination is now shifted away from perfectly in-phase coordination: The elements~processes are still highly coordinated in time, but phase-shifted by an amount proportional to the differences between the coordinating elements. This also demonstrates that when the individual coordinating elements start to differ, a stronger bond (more coupling) is needed to keep them together.

Once again, coordination dynamics seems to reveal the complementary nature in action. When the individual coordinating elements begin to differ, the collective coordination dynamics adapts to the differences. Conversely, if one were to observe a stable phase relationship between the parts that was not perfectly in-phase or antiphase, this would mean that the individual coordinating elements must be different by an amount proportional to the change in the coordinative phase relation. This has in fact been successfully tested in many experiments, most especially in the beautiful work of Michael Turvey's group at the University of Connecticut and Peter Beek's group at the Free University of Amsterdam in Holland, as well as by others.

When we look at the right side of figure 2.11 (between 180° and 270°), the second major feature to emerge is the spawning of an entirely new coordinative state in which the phase relation between the two individual coordinating elements is antiphase. That is, the individual coordinating elements still keep time with each other, but as one is all the way up, the other is all the way down, and vice versa—they move in equal and opposite directions ($\phi = 180°$). Once again, this sudden appearance of *two* stable states (one at $\phi = 0°$ and one at $\phi = 180°$) where previously there was only *one* (at $\phi = 0°$) is called a bifurcation. There is a splitting into two, two stable states emerging where previously there was only one. The dynamical behavioral potential has become divided, as it were, like a parent cell dividing into two daughter cells.

Notice also that the system can now support two stable coordinative states as well as their complementary unstable partners for exactly the same coupling strength. What does it mean for two stable coordinative states to coexist for exactly the same coupling strength? Recalling the tutorial, it means that now the dynamics is bistable. The "two Bs", bifurcation~bistability, are essentially nonlinear features of coordination dynamics, and also happen to be a complementary pair.

Why do we say that bifurcation~bistability is *essentially* nonlinear? In a linear system if one plots a coordination variable as a function of a parameter on a graph, any change in the parameter will cause a proportional change in the coordination variable. This is what one thinks of in garden-variety cause and effect: A single cause leads to a single effect. Increase one, and the other goes up or down in a strictly proportional, linear fashion. But in the case of nonlinear coordination dynamics, we observe that a single cause can lead to two (bistability) and in general more than two possible effects (multistability).

Thus, in coordination dynamics the usual either/or contrast of "linear versus nonlinear" which crops up quite often in science simply does not apply. As you can see in figure 2.11, both attributes are inherently present in the same system (linear~nonlinear). That is, for a certain range of parameters, coordination states may change and adapt quite smoothly, *as if* the system is behaving linearly. This can be seen for example in the adaptive changes in coordination—shifts in the fixed points that occur in a coupled dynamical system as differences between individual coordinating elements increase. The true signature of nonlinearity, however, is the unexpected and dramatic *qualitative* changes in coordination that occur when a control parameter—here coupling strength—reaches a critical threshold value.

Multistability and qualitative change are *essentially nonlinear* features of coordination dynamics. It means that the entire layout of attractors and repellers changes, as when two or more stable coordination states emerge where previously there was only one or vice versa. The ability to bifurcate provides living things with an essential means of flexibility. Now, there are at least two coordination patterns available for use under the same conditions. Recall from Movement 1 that such multifunctionality is also an essential requirement for complementary pairs. This "more than one way to skin a cat" property of coordination dynamics is characteristic of all complementary pairs and hints strongly of the presence of the complementary nature.

Returning to figure 2.11, notice that for this coupling strength, the emergence of two stable coordinative patterns is possible only when the coordinating elements are nearly identical. When the coordinating elements become more dissimilar ($\delta\omega > 0$), this value of the coupling can only support one stable coordinative state (e.g., left side, second row). Thus, under the same coupling strength a different kind of bifurcation or "phase transition" can be observed from two stable coordinative states (when the components are identical; bottom line) to only one when they are not. This is called a "symmetry-breaking bifurcation," created, as you can see, by just the slightest asymmetry between the coordinating elements.

In the tutorial we showed you how the trajectories of a dynamical system evolve in time and why it's important to see how these trajectories behave. This is what is shown in figure 2.12. All initial conditions near in-phase coordination (i.e., near 0° and 360°) converge and stay there. That's because the state near in-phase is a stable coordination pattern. On the other hand, imagine what happens to the time evolution of our coordination variable near the transition. All initial conditions near the less stable antiphase coordination state (middle part of figure) approach that state and then diverge, switching to near in-phase coordination.

VISUALIZING A BASIC LAW OF COORDINATION

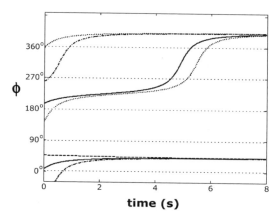

Figure 2.12 A nonequilibrium phase transition. The coordination variable ϕ evolves as a function of time and displays dynamic instability. Initial conditions near in-phase (0° and 360°) converge to a stable unchanging pattern or fixed point (the plateaus). Initial conditions near antiphase (180°) converge temporarily, but then the antiphase pattern loses stability and switches to the more stable pattern near in-phase.

And why is that? You are correct if you thought "dynamic instability." As we now know, such instability is a creative cause of all pattern formation and change in the complementary nature and a key to coordinative flexibility.

Let's summarize this constellation of events briefly before moving on. When the coupling strength among individual coordinating elements is slightly increased but still quite weak, we see the birth of two coordinative states where previously only one was present. The coordination state near in-phase is apparently more stable than the one near antiphase, because the latter disappears when the intrinsic differences among the coordinating elements increase. Of course, when we say "disappear," we mean that a phase transition has taken place, a spontaneous change from a less stable coordinated state to a more stable one. *Pure in-phase and pure antiphase are thus idealized coordination states in a coupled system composed of identical coordinating elements.* This crisp, rather Platonic picture is characteristic of much of the early work on biological coordination, including the HKB model (named after Hermann Haken, Scott Kelso, and Herbert Bunz).

The HKB model assumed that the components being coordinated were identical, and hence is a simpler form of the coordination law described here. Nevertheless, HKB was able to account for early experiments showing that coordination is self-organized in human beings, including the discovery of bistability, transitions from antiphase to in-phase coordination (a pitchfork bifurcation), and hysteresis, a primitive form of memory.

Later experiments found that still more interesting coordinative effects happen when the components being coordinated are not identical. This heterogeneity among coordinating elements is a crucial source of *symmetry-breaking* that changes the entire character of the coordination dynamics. The combination of a little coupling and a little heterogeneity among the individual coordinating elements gives rise to weird and wonderful coordinative effects, including adaptation, abrupt changes, and as we shall soon see, much more.

After over 20 years of detailed study, it is probably time to put the more idealized HKB model of coordination to bed. It has served its purpose well. By explicitly showing that crucial observations about the stability and change of human behavior could be understood in terms of self-organizing dynamical systems, the HKB model stimulated a great deal of empirical research and theoretical development. But biological coordination seldom deals with identical components and pure symmetry. The newer, still quite elementary version of the coordination law presented here not only is able to handle all the previously observed phenomena treated by the HKB model, it also embraces both symmetry and broken symmetry (symmetry creating~symmetry breaking) and leads to altogether novel effects and totally unexpected consequences, both of a scientific and epistemological nature. As we pursue these consequences, it is significant to note that complementary pairs continue to emerge at the center of the action, not only as metaphors and not only as mathematical entities, but as testable, repeatable phenomena in actual biological experiments. This is what we have been after all along: a scientific basis for complementary pairs and the complementary nature that goes beyond metaphor.

Adaptation and Repetition of Themes

Figures 2.13 and 2.14 represent two further increases in the coupling strength, and in both we observe a continuation and repetition of themes: The coordination dynamics adapts by shifting the fixed points to accommodate differences between coordinating elements. Once again, increased coupling strength accounts for sustained coordinated patterns around in-phase and antiphase. Because of the increase in coupling and because the in-phase pattern is inherently more stable than the antiphase one, the number of stable fixed points is greater near in-phase than antiphase. Moreover, it is possible to coordinate individual coordinating elements that are even more different from the ones shown previously in figure 2.11.

In each case, notice that coordination is bistable, but that new fixed points continue to emerge with increased coupling strength (compare figures 2.11 and 2.14). And again, as the intrinsic dynamics of the individual coordinating elements be-

VISUALIZING A BASIC LAW OF COORDINATION

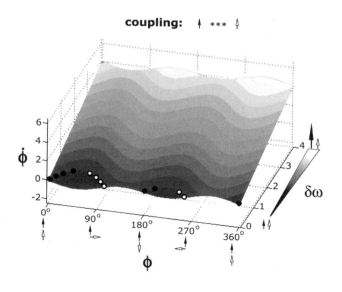

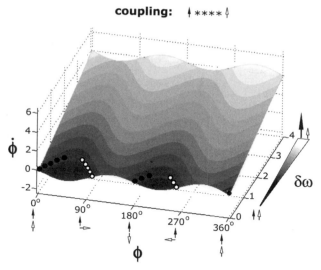

Figures 2.13 and 2.14 As the coupling increases, a bit more of the same. Nature operates with repeated (ancient) themes.

come more heterogeneous, a bifurcation occurs in which the fixed points near antiphase disappear. This means that the system must switch its behavior to the only remaining fixed point corresponding to the near in-phase coordination pattern. A technical note: In phase space, the maximum possible phase difference between individual coordinating elements is 180°. In other words, that's as far apart on the circle that two coordinating elements can be "out of phase" from one another.

As coupling increases even further (figure 2.14), the hills and valleys of the landscape become steeper and more defined, and all themes described before are repeated. One can see that adaptation of the coordination dynamics takes the form of a progressive shift in the fixed points to accommodate intrinsic differences among component elements and processes. Bistability~bifurcation occurs at critical values of parameters: One of two stable coordinated patterns, antiphase, becomes unstable, and the system switches to its only remaining stable configuration near in-phase, from two states to one.

Notice that the coordination dynamics doesn't pit one type of change against another. Changes can be smooth, gradual, and adaptive, as seen in the shift in fixed points. They can also be discrete, discontinuous, and abrupt, as seen in bifurcations from bistability to monostability and back again. The pictures tell the story well, and the story they tell is a real one. All of these graphs of the coordination dynamics represent actual behavior expected or already observed in real biological experiments.

Before continuing, let's take stock of the phenomena up to this point. A little coupling between dynamic elements and processes, and coordination emerges from where there was none. As circumstances change, the coordination dynamics adapts accordingly. We see that the coupling is sufficient to keep the system together despite intrinsic differences between the components. We observe that bistability is a natural and essential property of coordination, which has been confirmed again and again by empirical studies: Living things, even hypothetically "simple" ones like sea slugs, aren't limited to doing one thing. They are multifunctional and possess a repertoire of coordinated behaviors from which their needs may be met.

Bifurcations provide a selection mechanism, the means to decide when one mode of behavior is no longer able to do the job and to switch to one that can. All of these phenomena can be understood theoretically in terms of a relatively simple and easily visualized mathematical law. This is already quite remarkable, but there is more. The broken symmetry version of the coordination law shown in these pictures suggests a scientific basis for complementary pairs and the complementary nature.

VISUALIZING A BASIC LAW OF COORDINATION

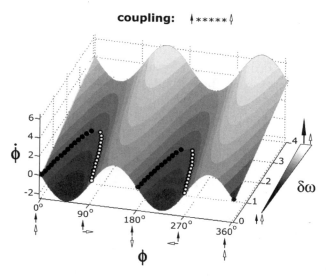

Figure 2.15 Emergence of the "in-between." The territory beyond where the fixed points (filled and open circles) coalesce and disappear shows a new kind of coordination that reveals the essence of the complementary nature.

The Metastable Regime of the Coordination Law

We now draw your attention to the region in the phase-plane diagram where the stable and unstable fixed points coalesce. Notice how they approach each other, "kiss," and then disappear. Now, this in itself isn't that different from the preceding diagrams. As we said, when the fixed points disappear, the system is no longer able to sustain stable coordination states. But in figure 2.15, we want to draw your attention to the beautiful hill and valley regions of the phase portrait where there are no longer any fixed points of the coordination dynamics, stable or unstable. What do these uninhabited hills and valleys represent? In principle and according to our tutorial, one might conclude that since there are no longer any fixed points—no coordination *states*, that is—coordination is no longer possible. But before jumping to that conclusion, look at what is going on.

Physically speaking, as coordinating elements~processes become too disparate (increasing $\delta\omega$), the coupling is no longer able to keep them together. In figure 2.15 (and for that matter from figure 2.11 on), for larger values of $\delta\omega$ the stable attracting and unstable repelling fixed points disappear entirely. Does the disappearance of fixed points cause coordination to drift aimlessly and randomly? That is, above the place where the fixed points merge, collide, and annihilate each other, is coordination possible? The answer is a paradoxical, emphatic yes! And why is that?

Notice in figure 2.15 that after the fixed points have disappeared, the surface of the dynamic landscape still retains its curvature. This means that, despite the fact that attractors and repellers no longer exist, there is still *attraction* and *repulsion* in the vicinity of where they used to be! What is the cause of this weird state of affairs? Before it happened, the individual coordinating elements, despite their differences, were still coupled strongly enough so as to produce fully and absolutely coordinated states. But now, although coupling is still present, it is not sufficient to hold the elements together. When the fixed points coalesce, the individual coordinating elements exhibit only *tendencies* to coordinate together at the same time even as they exhibit *tendencies* to do their own thing.

The key insight is that, despite the disappearance of the fixed points, coordination is far from totally absent (as it was at the beginning in figure 2.9, when the elements were uncoupled and the dynamical surface representing the coordination law was as flat as a pancake). Instead, above the fixed points in figure 2.15 we observe a region of the phase portrait—a huge territory in fact—in which *tendencies* for togetherness and apartness coexist even though *states* are nonexistent (togetherness~apartness; states~tendencies).

Look at what figure 2.15 is telling us. It says that depending on circumstances (parameters, boundary conditions, etc.) we can be in *either* one state *or* the other (either/or). It says that due to the inherent nonlinearity of the law, coordination patterns can even switch from one state to the other. Then it says something quite startling. Out beyond the fixed points, beyond where coordinated states live and where the boundaries between states have disappeared, where there are no longer *any* fixed points of the coordination dynamics, a beautifully articulated dynamical landscape full of potential coordination is still possible!

This region of the dynamical landscape isn't ruled by the either/or. *Both* the tendency of the parts to function independently (segregation) *and* the tendency to couple together (integration) coexist at the same time. Out in this "metastable" regime of the coordination dynamics (*meta* meaning "beyond") lies the grail we are after. As a scientific demonstration of the either/or~both/and, *metastable coordination dynamics grounds the philosophy of complementary pairs in science*.

Now take a look at figure 2.16. It's in the same format as figure 2.12, and it might be useful to compare the two. The latter, for example, shows absolutely coordinated stable states near in-phase, as indicated by the trajectories converging to a flat horizontal line. On the other hand, in figure 2.16 if we follow our coordination variable, relative phase (Φ), as it evolves in time, we can see that it drifts all over the phase space, as if the individual coordinating elements weren't coordi-

VISUALIZING A BASIC LAW OF COORDINATION

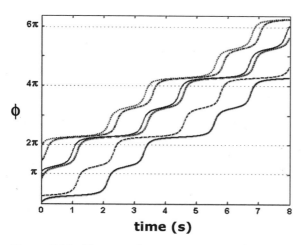

Figure 2.16 The complementary nature of coordination dynamics. This is the territory where all the fixed points (both stable and unstable) of the coordination law have disappeared, as shown in the previous figure. The paths or trajectories of the coordination variable evolving in time stem from six different initial conditions. Notice the plateaus on each curve, indicating a tendency for the components to converge and stay together ("trapping"). This tendency coexists with a tendency to diverge and separate ("wrapping"), allowing the system to visit everywhere in the space of possibilities. The presence of both convergence and divergence attests to the truly nonstationary, transient nature of the coordination dynamics in the metastable regime.

nated. But then it gets locally trapped, in its past as it were, near remnants or ghosts of the coordinated states around in-phase and antiphase.

Notice that coordination hangs around longer near the formerly stable in-phase coordination state (0, 2π, 4π, 6π, etc.) than it does near antiphase (π, 3π, 5π, etc). Notice also how antiphase switches to near in-phase and hangs around there for a while before moving on. If the coordinating elements were independent and completely uncoordinated, figure 2.16 would show a set of six straight lines at a slant sweeping through the phase space with no plateaus or kinks. This would look exactly like the "eyes closed" part of figure 2.7, that is, when there is no coupling between the two people. Likewise, were the coordinating elements absolutely coupled together, all the paths would converge to the same fixed-point attractor (akin to the horizontal line shown in the lower part of figure 2.12) regardless of initial conditions.

The territory beyond where the fixed points disappear is the metastable regime of the coordination law. In it there are no attractors or repellers, no fixed points, no coordination states of the coordination dynamics. Nevertheless there is still attraction to where the fixed points used to be. Thus, in figure 2.16, the relative phase between the coordinating elements drifts in time, but is occasionally trapped near places where the coordinated states used to be, near in-phase and near antiphase. Metastable coordination dynamics thus offers insight into how a global integrative tendency—in which the component parts tend to work together in harmonious binding—is reconciled with the tendency for the parts to function locally as specialized, autonomous individual entities. By living in the metastable regime of its coordination dynamics, the human brain is not stuck in an integrated or segregated state, but is able to exhibit a far more variable, plastic, and fluid form of coordination in which integration and segregation tendencies coexist. This balance between integration and segregation among the individual coordinating elements may be characteristic not only of the human brain, but of all viable coordinating systems—*all living systems.*

As we saw earlier, support for metastable coordination dynamics has been borne out in numerous studies and sophisticated analyses of brain structure~function. For example, novel statistical measures of brain complexity derived by Giulio Tononi, Olaf Sporns, and Gerald Edelman at the Neurosciences Institute in San Diego suggest a balanced interplay between integrating and segregating influences. Such statistical measures, however, are based on the assumption that interactions among neural groups do not change in time (i.e., the assumption~condition of stationarity) and that the anatomical connectivity is fixed.

Here we show how, in the metastable regime, informationally relevant dynamic links among neural populations are transiently formed and, just as important, are dispersed as the stream of perception, action, and memory flows. Metastability appears to be the key principle, and phase gathering~phase scattering the key mechanism underlying the forming~dissolving of neural ensembles. In plain English, it looks as if living organisms require metastability to retain their viability. And here we have a link between the complementary nature and life itself.

MECHANISMS OF INFORMATION PROCESSING AND INFORMATION CREATION

We promised a word or two about information processing and information creation, and will now discuss them within the explicit context of multistable and metastable coordination dynamics. It may come as a surprise to learn that few, if any, scientists talk about information processing and information creation in the same breath. Indeed, few address the issue of information creation at all. Of course, students of the complementary nature would suspect the two to be intimately related.

One picture of their relation that seems quite natural and almost conventional by now can readily be seen in the progression from figure 2.10 to 2.11. In them can be seen a phase transition scenario for the creation of information. In figure 2.10 one stable state exists, and in figure 2.11 two stable states exist, as the result of a changing parameter. Going from one state to two in one direction of parameter change and from two states to one in the other direction correspond precisely to the act of information creation and destruction.

In his important book *Information and Self-Organization*, the physicist Hermann Haken uses the phase transitions discovered in biological coordination by Scott Kelso as a means to quantify information transfer when a biological system switches from being bistable to being monostable. In this treatment, the coordination variable (= order parameter) is seen as a "compressor of information" in the sense that the very many individual parts (which require a lot of information to describe) behave in a coherent, cooperative fashion (which requires only a minimum of information to describe). This approach highlights the important role that self-organization plays in information processing and likewise how important phase transitions are in creating~annihilating information.

Turning now to the metastable regime of coordination dynamics, a similar principle applies. As is repeatedly shown in figures 2.10 through 2.15, depending on context (i.e., coupling and intrinsic parameters), states can turn into tendencies and tendencies can turn into states. Thereby information, in the conventional use of the word ("bit from it" and "it from bit" and all that) can also be created~annihilated in the metastable regime.

Interestingly, this metastable scenario quite precisely parallels the so-called measurement process in quantum mechanics. Like quantum mechanics the creation~destruction of information in coordination dynamics rests on the interaction of a measuring device (e.g., the brain, with its own intrinsic dynamics) and an object being measured (e.g., the world, within which brain~bodies have coevolved and experienced each other through development and learning).

There is yet another intriguing and less conventional scenario for the creation of functionally meaningful information. As we discuss it, keep in mind throughout the significance of self-organization for information processing. In figure 2.16, we saw that the changing lengths of the plateaus suggest that coordination tendencies hang around for variable times, some longer than others, with a probability that quantitatively depends on how far the system is from where the fixed points lie. That is why we refer to these tendencies as "ghosts" or "remnants" of attractors.

In combination with figure 2.15, the picture shown in figure 2.16 suggests a remarkable resonance with the thinking of William James (1842–1910), one of America's most influential thinkers at the turn of the last century. In a chapter of *The Principles of Psychology* James compares consciousness to the flight of a bird, whose life journey alternates between "perchings and flights." "In certain persons," James says, "consciousness may be split into parts which coexist but ignore each other and share the objects of knowledge between them. More remarkable still, they are complementary ... representing relations of mutual exclusion." James's beautiful metaphor of perchings and flights for the stream of consciousness is realized physically by metastable coordination dynamics. His perchings correspond to integrative, phase-gathering tendencies (the plateaus in figure 2.16) and his flights, the drift between "perching" plateaus in figure 2.16, to the segregative, phase-scattering escapes that explore the space of possibilities.

We can now venture a hypothesis about what all of this means: *Thinking*—the creation of information in the mind—is a transient nonstationary dynamic process. It corresponds to a flow of converging "perchings" (integrative phase-locking tendencies between brain areas) and diverging "flights" (segregative decoupling tendencies and individuation of brain areas). Both tendencies are crucial: the former to create thoughts, feelings—information in general; the latter to release individual brain areas to participate in other acts of cognition and emotion. We might add the obvious: To be stuck in a phase-locked state is to be temporarily "trapped in thought," to be depressed in affect, in one stationary state or another, the limited repertoire of the either/or.

Finally, it may be useful to keep in mind that the same principles of information creation~destruction in the metastable regime of coordination dynamics apply not only to the brain~mind level of coordination dynamics, but to any level of coordination dynamics studied. After all, this is what we mean when we talk about a coordination law.

A SCIENCE OF COMPLEMENTARY PAIRS AND THE COMPLEMENTARY NATURE

Let's gather things together, and return to the original objective of Movement 2, namely to put complementary pairs and the philosophy of complementary pairs on a firm scientific footing via the science of coordination dynamics. To begin with, isn't it astonishing that the coordination dynamics of both brain and behavior entails the complementary pair individual~collective? In one complementary aspect, individual coordinating elements *couple* together to form coherent self-organized coordinated states. These are stable states, classic stationary attractors, of a dynamical system and represent its integrated behavior. In the other complementary aspect there is no coordination or interaction among the parts whatsoever. These same coordinating elements are free to do their own thing; they are totally segregated and behave completely independently.

We show that two key factors are necessary and sufficient to produce such integration~segregation of individual~collectives: the strength of the coupling between the coordinating elements and the intrinsic differences between them.

Measures such as how long the system hangs around before it leaves, its "dwell time," open up new ways to calculate the strength of the interaction between the components of complex systems. Reciprocally, how fast the system escapes, or its "flight time" in James's terminology, allows a way to calculate the relative independence of the components, their individual freedom.

Despite the fact that coordination dynamics is formulated in mathematical terms, this isn't just mathematical phenomenology we are talking about. Stable coherent states of coordination dynamics correspond to actual phase- and frequency-locking behavior in nature. Phase synchrony has been proposed to be one of the major mechanisms of neural integration in the brain and more generally reflects von Holst's biologically studied principle of absolute coordination. On the other hand, segregation refers to the independence of individual coordinating elements, and phase scattering or wrapping corresponds to a "breakdown" in binding or coordination. Both "pure" integration and "pure" segregation are mathematically and physically understood through coordination dynamics. One or the other is not enough. An exhaustive account of brain and behavioral coordination requires a subtle blend of both.

In between these idealized extremes lies the huge territory of tendencies and dispositions. This middle ground is full of complementary pairs: individual coordinating elements coexist with the collective (individual~collective). Cooperation coexists with competition, component parts cooperating with each other while trying to retain their autonomy (cooperation~competition). The tendency to converge toward attractive fixed points (phase- and frequency-locked states) and the tendency to diverge from coexisting repelling fixed points dominate the flow of the dynamics (convergence~divergence, attractor~repeller). Qualitative changes (phase transitions and bifurcations) are produced by quantitative variation of parameters, and accompanied by quantitative consequences such as enhancement of fluctuations and critical slowing down (qualitative~quantitative).

And of course, the ringer is that once freed of the fixed points, the traditional building blocks of all dynamical systems, one discovers the metastable regime, where the tendency to integrate coordinating elements coexists with their tendency to segregate. It seems that in coordination dynamics we have found a way at last to pursue a science of either/or~both/and, a scientific grounding for both polarization and reconciliation. The empirically and theoretically founded science of coordination dynamics allows for either states or tendencies and it allows for both states and tendencies, depending on the situation.

As far as we can tell, up to now the scientific basis of complementary pairs and what it means for them to be dynamical has been either metaphorical in character (like Fritjof Capra's "dynamic" interplay of yin~yang mentioned in Movement 1), or rests on an interpretation of how the subatomic world behaves, namely, the

Copenhagen interpretation of quantum mechanics. Coordination dynamics reconciles the classical world of Newtonian mechanics with its forces, masses, and motions and the weird, but highly successful world of quantum mechanics with its probabilistic waves and particles. In coordination dynamics, which deals in the informationally rich currency of coordination variables like phases and amplitudes of brain and behavioral and social activity, we offer an explanation of complementary pairs that is neither purely metaphorical nor purely quantum mechanical in origin. With its built-in, essential nonlinearity, coordination dynamics says that two opposing tendencies like integration and segregation are complementary and coexistent. An exhaustive account of how the brain works rests not on one or the other. Coordination dynamics shows it is a subtle interplay of both.

GATEWAY TO MOVEMENT 3

The great Isaac Newton considered the "self-motion" that he believed God gave animals "beyond our understanding without doubt." In contrast, his immortality lies in his genius for dealing with the complex motions of inanimate masses and forces of the physical universe. Yet it is the coordinated movements of human beings and the brains that give rise to them that have driven and continue to drive the development of a science of coordination, and now of complementary pairs. It is true that when one climbs on a rock and one's feet slip, the force of gravity overcomes the force of holding on. But there is more to climbing than gravity. It seems that for living things that reside at the scale of everyday human existence, we must complement the information-free mechanics of motion with the informationally meaningful dynamics of coordination.

Compelling testimony to the pervasiveness of the complementary nature is that complementary pairs were discovered in coordination dynamics and the human brain when no one was looking for them. In fact, the whole conceptual framework of coordination dynamics begun 25 years ago was achieved without evoking complementary pairs, at least not explicitly so. Furthermore, up to now, brain research has predominantly considered holistic integration and reductionist segregation—its two prize, competing theories—*as contraries.*

Likewise, the study of how the brain perceives the world is still wrapped up in the empiricism versus nativism debate of philosophy, though now couched in the more contemporary engineering terms of feedback (from the senses) or feedforward (due to the brain's own intrinsic activity). In coordination dynamics, not only have we found a way to ground the study of complementary pairs in science, but complementary pairs seem to help advance the very science that explains them—*curiouser and curiouser ...*

One might even wonder if these two insights are themselves complementary? Might the complementary pairs of coordination dynamics provide some insight into lesser known fields, endeavors, and activities, and thereby afford a deeper and wider understanding of the complementary nature? On the other side of the coin, how might coordination dynamics and its potential extensions and elaborations accommodate the many complementary pairs that lie outside its currently practiced form? These are questions we take up in Movement 3.

We are coming to the end of Movement 2. In it, we have tried to provide the gist of coordination dynamics, an ongoing multilevel, transdisciplinary science that aims to understand certain features of informationally coupled, self-organizing dynamical systems. Although we tried to keep the math as much at bay as possible, the tutorial was not sugar-coated. We took you through the gauntlet of nonlinear dynamics, its basic concepts, methods, and measures, for a reason. We believe they provide a feasible, operable means for you to run with them. After this, we hope you won't ever think of stability without instability, cooperation without competition, and so on.

And now, let's run things the other direction for a moment and wax philosophical about coordination and coordination dynamics. It was inevitable that human beings—as evolved, self-organized, informationally rich dynamical systems themselves—would eventually find and explore the principles and mechanisms of coordination dynamics. We suppose that this is the case partly because the theme of complementary pairs is so readily apparent in the history of ideas, as found in the lives and times of the men and women who conceived and studied them.

Based upon our understanding of the coordination dynamics of brain and behavior, it now appears that it is quite literally *in our biological nature* to split the world into pairs. More and more, it looks as if both life itself and the tools we use to study it emerge from the same source: informationally coupled, self-organizing dynamical systems—coordination dynamics, in short. If this is so, it would appear that, like complementary pairs, coordination dynamics is thoroughly implicated in anything and everything that we humans do.

Why then has a lawfully based science of coordination taken so long to emerge? It seems possible that at least part of the reason is because many of the basic concepts and methods necessary for its advancement have been curtailed by either/or thinking. Could there really be a relationship between the phenomena of self-organization in physics and chemistry, for example, and the way people and animals move, or between integration and segregation in the brain, or for that matter, between complementary pairs and coordination dynamics? If so, how could such important relationships be missed for so long?

A contributing factor may be our "self-centered" view of the world. Yes, it is difficult to think of attack and defense as co-occurring at the same time. It violates our sense of causality. Likewise, it is blatantly obvious that we cannot arrive and depart at the same time, either. Once we step outside of this somewhat anthropocentric stance, however, we realize that in a battle or a football game, attack and defense are going on simultaneously in space~time. And people come and go at airports and cocktail parties at the same time all the time. Doors, it seems, can only be either open or closed, but a revolving door can be both open and closed, and for good reasons too.

Eccentric and highly creative people like the architect and inventor R. Buckminister Fuller and the physicist Richard Feynman provide nice examples of counterintuitive thinking (or its lack) in the way we humans go about the business of observing and trying to understand nature. Fuller astutely points out that there is nothing about the earth *itself* or the moon *itself* that will help a person intuit the invisible gravitational bond between them. Perception of both the earth and moon moving together is necessary to even *conceive* of gravity in the first place. Of course, once it is conceived, it seems so incredibly obvious.

Feynman tells the story of how in Michael Faraday's day, electricity was being studied and chemistry was being studied, both in earnest, and important discoveries were being made about each. However, scientists did not realize at that time that they were two aspects of the same thing. In much the same way, there still exists very little in the usual educational settings and paradigms that prepares students to think about how apparently different things may be different aspects of the same thing. Yet this activity is usually a mark of great scientific and intellectual innovation and achievement.

If complementary pairs are as conspicuously present in coordination dynamics as they appear to be, it would seem that coordination dynamics might be a real contender for a "science of complementary pairs." But there is more to this than our feelings. There is also a very compelling reason why this might be so, stemming from the complementary way that human brains, as well as the nervous systems of other organisms, have been shown to work—via a subtle, dynamic blend of opposing yet coexisting tendencies.

A possibly profound understanding of the complementary nature seems to lie in how the human brain works, and the ability of coordination dynamics to successfully explain it. Metastability is a powerful new idea. In it, there are literally no coordination *states*, only fleeting yet coexistent tendencies or modes of coordination pattern persistence~change. The metastable brain creates *functional information*, and by so doing may properly be called brain~mind.

In Movements 1 and 2 we have presented two very different approaches to understanding complementary pairs, one philosophical and one scientific, and introduced the idea that these approaches seem to be intimately related. The philosophy of complementary pairs offers an intriguing interpretation of ubiquitous complementary pairs. Coordination dynamics offers a successful scientific theory~experiment for understanding ubiquitous pattern formation, persistence, adaptation, and change in self-organizing systems that are coupled to and sculpted by information. As individual conceptual frameworks, they both seem quite useful. However, it now seems quite clear that they might be even more useful if they were contemplated as a complementary pair themselves. This is the subject of the final movement, Movement 3.

MOVEMENT 3   COMPLEMENTARY PAIR~COORDINATION DYNAMICS

RECONCILING PHILOSOPHY AND SCIENCE

Louis de Broglie
(1892–1987)

In the nineteenth century there came into being a separation between scientists and philosophers. The scientists looked with a certain suspicion upon philosophical speculations, which appeared to them too frequently to lack specific formulation and to attack vain insoluble problems. The philosophers, in turn, were no longer interested in the special sciences because their results seemed too narrow. This separation, however, has been harmful to both philosophers and scientists.

L'avenir de la science (1941)

The Polarization of Philosophy and Science

Let's put our cards on the table. As is already evident in the words of Louis de Broglie over 60 years ago, *philosophy*—literally, the love of wisdom—and *science*—from the Latin word *scio*, "know"—have become polarized, segregated disciplines. That a physicist like de Broglie should call for the reconciliation of philosophy and science is not as surprising as it seems. After all, his greatest scientific achievement was also a profound reconciliation. In 1923, while still a graduate student at the Sorbonne in Paris, de Broglie introduced the incredible idea that light *particles* (photons) may exhibit *wave* properties (particle~wave). In our book, de Broglie had it right: The split between philosophers and scientists has been harmful, and not only to philosophy and science but to all humanity.

What a strange world it is, where the love of wisdom and the pursuit of knowledge are separate enterprises! Ostensibly, both philosophers and scientists are committed to advancing knowledge and pursuing wisdom. Yet despite this shared goal, their paths have clearly diverged. These days, the two fields are nearly always pursued in different university departments, and nearly always reside in different buildings. The students too are often miles apart, following quite different agendas.

Why this separation? Part of the reason is that many philosophers eschew the lack of philosophical prowess in their scientist colleagues. They feel that scientists, in their zeal to gather and use knowledge as a means of control, ignore the deeper philosophical ramifications of what they are doing. They feel that scientists have by and large become technologists and technocrats, with little time for philosophy, or even worse, have abandoned philosophy all together, as a superfluous and irrelevant undertaking.

The other side of the coin is that many scientists view philosophers as know-it-alls who rarely admit to not understanding something. Or they treat philosophy as if it were a sort of elite academic sport, best played out in more informal social gatherings, over a glass of wine, a gin and tonic, or a cold beer. Along these lines, one of the biggest putdowns of a scientific theory is, "That's no theory, that's just *philosophizing*." For example, the Nobel laureate physicist Sheldon Glashow claims that cosmological string theory is "just philosophy" because it isn't testable. Though this attitude may be shortsighted, it is not uncommon. For most scientists, who tend to verge on the side of positivism, only *testable* hypotheses and repeatable observations are capable of distinguishing what is from what is not.

How is it that we have come to such a strange set of affairs? Can one really have a philosophy without science? Can one really have a science without philosophy? Few would argue against the idea that philosophy and science are two key aspects of human curiosity. Just as the human mind needs a brain, philosophy and science appear to need one another. So what can be done about the rift between them? Naturally, TCN sees philosophy and science as complementary aspects of a complementary pair. As such, they are inextricable, even though common practice treats them as separate.

The Complementary Nature claims that a reconciliation of philosophy and science is not only possible in our age, but necessary. And sooner or later, preferably sooner, someone has to come along and do it. Consider the words of the late Ilya Prigogine, Nobel laureate in chemistry, calling for a reconciliation of physical science and biology: "Almost by instinct, I turned myself later towards problems of increasing complexity, perhaps in the belief that I could find there a junction in physical science on one hand, and in biology and human science on the other." Like Aristotle, Thomas Aquinas, and Immanuel Kant, Prigogine suggests a reconciliation of complementary aspects that have somehow become artificially polarized. As a physical chemist, Prigogine sought a reconciliation of physics and biology, rather than of philosophy and science, but it is exactly the same sentiment as the one expressed so eloquently by Prince Louis de Broglie.

The Effectiveness and Challenges of Reconciliation

If you hold opposites together in your mind you will suspend your normal thinking and allow intelligence beyond rational thought to create a new form.

from "The Art of Genius" (1998)

Niels Bohr
(1885–1962)

Whether consciously aware of it or not, many celebrated thinkers have achieved deep and penetrating insights into the nature of the universe by reconciling complementary aspects. Yet humanity routinely fails to appreciate, never mind adopt, the very mind-set that led to these amazing accomplishments in the first place! Instead, the dominant focus has been placed on practical applications of the great reconciliations—their cash value, so to speak. As a result, the process by which reconciliation was arrived at tends to be missed. And so, even as we benefit from the fruits of reconciled complementary aspects, the means by which these fruits were obtained are routinely overlooked by almost everyone. Richard Feynman had it right again when he remarked that newspapers have a standard line for anyone who makes a discovery, say, in physiology: "The discoverer said that the discovery may have uses in the cure for cancer." But few pay any attention to the details of the discoveries themselves or the means and strategies through which they were discovered.

If one did decide to pursue a reconciliatory mind-set, one would find oneself in eminent company. As we saw in Movement 1, Socrates strove to reconcile truth and falsehood. Plato struggled to reconcile the one and the many. Aristotle strove to find an "enlightened middle ground," a *golden mean*. The Buddha sought reconciliation of happiness and suffering. Jesus tried to reconcile friend and enemy, rich and poor. Thomas Aquinas attempted to reconcile faith and reason. Immanuel Kant sought to reconcile empiricism and rationalism, telos and mechanism, whole and part. Mohandas Gandhi tried to reconcile civility and disobedience, Hindus and Muslims, etc.

Unfortunately, to achieve successful reconciliation and exhaustive description in the face of rigidly entrenched, polarized styles of thinking is easier said than done. Gandhi, for example, was a great inspiration and agent for change, but try as he might, he could not convince the populace of his beloved India to adopt reconciliation as its modus operandi. This frustrated Gandhi no end, as summed up in his famous retort, "An eye for an eye makes the whole world blind!" Ideological polarization and segregation are still present in India 60 years later; only now, India and Pakistan both carry nuclear arms in their arsenals, a danger to us all. We can only hope that the recent efforts of these sister nations to reconcile their differences are successful.

It cannot be overemphasized that in TCN we are not trying to achieve reconciliation of philosophy and science by eliminating the *natural process of polarization*. But we do believe the tendency to polarize should be relegated to a more reasonable position in the overall scheme of things. How might it be possible to spare natural polarization at the same time as ridding ourselves of debilitating polarizing mentalities? Instead of claiming that polarization is "wrong" and reconciliation is "right," we make the case that polarization~reconciliation is a complementary pair whose multifunctional, metastable dynamic is capable of expressing both tendencies. The complementary nature both polarizes and reconciles. The big problem lies in getting frozen into one of these idealized modes, getting trapped in the either/or mind-set.

Likewise, blind overemphasis on reconciliation for reconciliation's sake can lead people astray. As we strive for enlightened reconciliation of complementary aspects as complementary pairs, we are *not* calling for the process of reconciliation itself to be held as the new overarching modus. Our goal is nothing like an "ultimate reconciliation," as it might have been for Spinoza and the neutral monists. Reconciliation itself is only an aspect; it should not be reified as an overarching principle. The clarion call is rather for a general "unfreezing" of humanity's dynamic when it comes to polarizing~reconciling mentalities. Since the overriding modus of the day is polarization, unfreezing the dynamic means that humanity might do well to shift the dynamic in the direction of reconciliation.

So what options are available for those who seek to reconcile opposing, conflicting, or competing aspects—*contrarieties*—that nevertheless seem to be inextricably connected to each other? This is the challenge we take up in Movement 3. We have suggested that complementary pairs and their interpetation have been central to all progress, to all human discoveries and creations throughout the ages. And the science of coordination dynamics has led us to a novel interpretation of how complementary pairs might work and the role they might play. Now let us see how this reawakened interest in the study and appreciation of complementary pairs and their dynamics may lead us to a more harmonious attitude toward ourselves, each other, and the world around us.

INTRODUCTION OF A NOVEL PHILOSOPHY~SCIENCE

Ilya Prigogine
(1917–2003)

I find our period remarkable precisely because some of the questions in social science and in natural science form a kind of confluence. In the past we've seen two other periods in which such convergences occurred: the Greek classical period and the Renaissance. And during both those periods you had people like Plato and Aristotle, or Descartes and Newton, who were philosopher-scientists.

from "Wizard of Time" (1984)

In Movement 1, we described our philosophy of complementary pairs. As many great thinkers have intuited throughout history, human reality~fiction is entrenched in binary oppositions. In philosophical circles this is sometimes called the "metaphysical exigency." Philosophies based upon the tension between complementary aspects are certainly as old as humanity itself. However, it doesn't necessarily follow that just because a subject has been pondered upon for eons all that can be said has been said, or that nothing new can be contributed. In our case, the squiggle character (~), which is used syntactically to express complementary pairs and their dynamics, represents a novel take on the ancient issue of dynamic contrarieties. Remember, the (~) is not a bridge or conceptual glue holding complementary aspects of a complementary pair together, nor does it represent a simple tension of opposites. Rather, the squiggle is a constant reminder that complementary aspects are inextricably related as a complementary pair, and possess an intrinsic dynamics.

In these first years of the twenty-first century, the acquisition of knowledge and its interpretation continue to depend upon the scientific method. As such, a compelling argument in support of our interpretation of complementary pairs would be to demonstrate their existence using currently accepted scientific methods. Thus, complementary pairs shouldn't only be treated as conceptual or metaphorical models, they should also be realizable and observable as bona fide physical phenomena. The chief aim of such work would be to understand how complementary pairs form and dissolve, persist and change, during the normal operation of real world complex~simple systems.

In Movement 2 we introduced coordination dynamics and showed how complementary pairs like individual~collective, integration~segregation, and competition~cooperation seem to fall out naturally from every facet of its theoretical~experimental spectrum. Complementary pairs are not merely present in coordination dynamics. They are so thoroughly entrenched in it that coordination dynamics seems like a good candidate for a science of complementary pairs. Think about it: Is coordination better explained by full phase and frequency locking, as in absolute coordination, or is it *better* explained by partial phase locking and partial independence among the coordinating components, as in relative coordination? And again: Is the brain *better* explained by a large-scale global integration theory or by a localizationalist perspective? Is perception based on input from the senses (empiricism and feedback), or is it based on processes intrinsic to the brain (nativism and feedforward)?

For all such scenarios, of which there are so many, something far more subtle and much more exhaustive seems to be going on. Once one moves away from always attempting to understand things by trying to eliminate one of two idealized descriptions, one discovers a metastable dynamic manifested by behavioral tendencies for both complementary aspects. Tendencies and dispositions are the complements of idealized states. In the metastable regime, there are no immutably stable or unstable states, only simultaneous tendencies for participating elements to act independently and to work together. This is where coordination dynamics appears to offer something new.

Here in the third Movement, we attempt to reconcile the philosophy of complementary pairs from Movement 1 with the science of coordination dynamics from Movement 2. We proceed as we have done all along, by forming a complementary pair, in this case complementary pair~coordination dynamics (for short, CP~CD). CP~CD celebrates the idea that a philosophical interpretation of complementary pairs (CP) is inextricably grounded in tangible, realizable scientific principles and mechanisms (CD), and vice versa. That is, coordination dynamics entails complementary pairs, and complementary pairs entail coordination dynamics. CP~CD is literally a reconciliation of philosophy and science. Employing the syntax of complementary pairs, CP~CD *is* a philosophy~science.

When you think about it, CP~CD seems to be a rather extraordinary development. It means there can be no deep understanding of complementary pairs without coordination dynamics, and vice versa. It means that complementary pairs and coordination dynamics cannot be captured exclusively by the philosophy of complementary pairs or by the science of coordination dynamics alone, by CP/CD as a dualism, or even by CP-CD as a multimodal whole. CP~CD is much more than and different from a simple merger of CP and CD.

Of course, one is still free to focus one's attention (or at least think one is focusing one's attention) exclusively on complementary pairs without coordination dynamics. Similarly, one is also free to concentrate exclusively on coordination dynamics without complementary pairs. Such has been the case pretty much until recently. Before the advent of CP~CD, only a kind of tacit awareness of complementary pairs like stable~unstable, creation~destruction, control parameter~

order parameter, and so forth existed among those working in coordination dynamics.

However, the philosophy of complementary pairs predicts that much better progress will be made by doing what both F. Scott Fitzgerald and Niels Bohr advised: namely, to try to keep *both* the philosophy of complementary pairs (CP) and the science of coordination dynamics (CD) in our heads at once as much as possible. CP~CD helps us do that. Like all complementary pairs, CP~CD is much broader in scope than each of its complementary aspects, CP and CD, alone. That's because CP~CD not only includes all that is contained in the philosophy of complementary pairs and all that is contained in the science of complementary pairs, but also includes the self-organizing, multifunctional dynamic responsible for the appearance of both of those aspects in the first place. And let's not forget that, as a complementary pair, CP~CD must itself have a metastable regime, where simultaneous tendencies for CP and CD are present. Employing the syntax of complementary pairs once again, we can say that CP~CD is a philosophy~science of the complementary nature.

Because both of its complementary aspects deal in complementary pairs, CP~CD is a tricky complementary pair to get one's head around. But once you get the hang of it, CP~CD has the potential to be applied in any field of interest whatsoever. Indeed, boundaries between fields tend to disappear when one adopts a CP~CD view of the complementary nature. A simple reason might be that CP~CD is relevant anywhere one finds complementary pairs. Also, to the extent that it can explain how human nervous systems work, CP~CD might provide not only a unique window into the complementary nature, but also insight into how human beings come to believe such things in the first place.

All this is well and good, but the most difficult job remains, which is to figure out what this reconciliation of CP and CD *means*, how we might study it, and how it might help people make sense of the world and each other. Up front, we should acknowledge that so-called immutable truths will probably continue to change and evolve forever. Here we part with some thinkers, like Nobel laureate Steven Weinberg, who believe physical science is near the end of a journey that began with Kepler, Galileo, and Newton in the sixteenth and seventeenth centuries. Be that as it may, CP~CD, which certainly does not exclude physical science, is still very much at its beginnings. Moreover, we think that it is highly unlikely that we will ever run out of things to discover about the complementary nature. The complementary nature, remember, sees the "universal" laws of physics and context-dependent coordination dynamics as a complementary pair.

This somewhat daunting long-term outlook should be taken as a promising sign. As many before us have noted, confident though humankind may be about its information-gathering activities, there is much more that is unknown, and even more that is unknowable. However, such a humbling prospect seems a small price to pay for what may be gained through a deeper appreciation of CP~CD. If the complementary nature *is* best seen through the window of complementary pairs and their coordination dynamics, then CP~CD is likely to yield new facts and insights in the foreseeable future. Like a bowl whose use depends on both its walls and its emptiness, CP~CD is above all practical, a philosophy~science that we all can use, no matter the sphere of activity.

It is important at this point to rein in our hubris a bit, and remind you that for all of what we think is a promising future, there is a lot of work ahead, as well as some formidable obstacles to overcome. History teaches that fundamental reconciliations have always been very difficult to accomplish. The present undertaking is expected to be no different, possibly even more difficult and controversial. This is because CP~CD is a "super" reconciliation, a sort of reconciliation of reconciliations that includes those discovered in the past and those yet to be discovered in the future. We have to admit that such a comprehensive reconciliation, though worth the effort, may be out of reach. The lifework reconciliations of both Thomas Aquinas and Immanuel Kant serve as cogent examples. No one said it was going to be easy!

Keep in mind, however, that although CP~CD certainly qualifies as a work-in-progress, it wasn't exactly started yesterday. The science aspect alone represents many years of research and development, with conceptual and empirical roots extending into the brain, behavioral, cognitive, and social sciences, all clothed in the physical concepts of self-organization and the mathematical tools of nonlinear dynamical systems. And as we have noted more than once, the philosophical contemplation of complementary pairs has been with us in one form or another since antiquity. Thus, although CP~CD may seem a bit strange, it actually embraces and entails both traditional and widely held contemporary ideas.

So what obstacles confront those of us who choose to pursue CP~CD? Possibly the most difficult one to overcome is the polarized either/or thinking that tends to completely dominate both philosophical and scientific communities, and for that matter, every other sphere of life. The dominance of either/or thinking is no accident. It is a self-sustaining, habitual, even addictive mind-set that strongly resists change. We do not expect that either/or thinking will go down quietly. Such fundamental change has tangible and perhaps not particularly palatable consequences. It may mean, for example, that scientists have to alter the way they do business. It may also bring into question some deeply held assumptions. Yet, as history clearly demonstrates, questioning and even abandoning assumptions is often the first step to insight, discovery, and invention.

One issue that CP~CD forces us to rethink is the nature of causality itself. Why? In trying to understand ourselves and the world we live in, we can no longer assume that one cause produces one effect, or that a small or large cause always leads to a small or large effect. Living things seldom, if ever, work according to such strict linear causality. In living things, apparently at all levels, a single cause can produce multiple effects, and multiple causes can produce the same effect. The slightest cause can produce a huge effect, and a large cause can produce little or no effect at all. In self-organizing coordination dynamics, the collective informs the parts while the parts, interacting by means of any viable medium, inform the collective. The complementary nature not only allows for, but embraces such circular, or as we prefer to call it, *reciprocal* causality. CP~CD is equipped to handle both the more familiar linear causality and the less familiar, seemingly paradoxical kind of nonlinear causality.

It is important to reiterate that the advent of CP~CD doesn't mean that one can't continue to advance the philosophy of complementary pairs (CP) and the science of coordination dynamics (CD) as individual disciplines and pursuits. On the contrary, as with any complementary pair, both aspects (CP and CD) are equally valid in themselves. While one *can* study complementary pairs and coordination dynamics in their own right (as we have done largely in Movements 1 and 2), insight and understanding may be enhanced significantly by studying them together.

As a philosophy~science of the complementary nature, CP~CD leads us to a threshold that, once crossed, has the potential to open up whole new vistas for discovery and invention. Importantly, such opportunities are within everyone's grasp. At least in the initial stages of applying CP~CD for one's own purposes, one need only to understand (1) the basic concept and syntax of complementary pairs, (2) the tenets of the philosophy of complementary pairs, and (3) the basic concepts of coordination dynamics. These basic concepts have not been watered down or artificially simplified for the novice, only to be replaced later with other, more advanced and accurate ones. The basic ideas are as relevant to those doing contemporary experiments and theoretical modeling in coordination dynamics as they are to someone who has never even heard of coordination dynamics. And they can certainly be used as a scaffold upon which to construct future, as yet unseen developments. In this respect, we can expect great things and big surprises—*from you!*

At any rate, regardless of its potential applications, CP~CD should cause us all to pause and reflect on the age-old adage "Life is rarely black and white," perhaps more seriously than ever before. CP~CD offers a way to get at life's shades of gray and inspires the move to transcend the habitual either/or mind-set. Considering the many marvelous theories, discoveries, and inventions that have resulted from the reconciliation of contraries throughout history, imagine what might be gained if 8,000 years of complementary pairs were now brought under the aegis of principles of coordination dynamics and a novel, more flexible interpretation of the ubiquitous complementary pair?

PUTTING CP~CD TO WORK

... so that in principle, pure theory and pure experimentation are, and always have been, untenable constructs, and should now be abandoned in favor of the more realistic stance of reciprocity.

At Home in the Universe (1994)

John Wheeler
(1911–)

In this day and age, for CP~CD to flourish and grow, we need to be able to use it. How might CP~CD be put to work? Practical and economic reasons are not the main driving forces behind the pursuit of CP~CD, any more than they were for all those who have pondered nature's secrets in the past. Rather, a primary goal is to use CP~CD to advance both philosophy and science by means of their successful reconciliation. As a flexible, dynamic approach to nature, CP~CD operates in the same way as the very philosophy~science it seeks to comprehend. So one immediate use of CP~CD is to promote a much-needed dialogue between philosophers and scientists. Of course, if all CP~CD ever did was help lead to the end of the either/or or some unnecessary conflict, crisis, or war, we would consider it a successful undertaking.

CP~CD eschews any approach that overemphasizes one complementary aspect over another as not only biased, but incomplete. For example, CP~CD treats parts~whole as a complementary pair and shows *why* it is one. Thus, in Movement 2 we demonstrated explicitly that the diverging, local, competitive tendency of the parts to behave independently coexists with the converging, global, cooperative tendency of the parts to work as a whole. So it's individual *and* collective, local *and* global, divergence *and* convergence, competition *and* cooperation, etc., not one *versus* the other. Likewise, CP~CD says that all complementary pairs obey coordination dynamics, even the complementary pair CP~CD! According to coordination dynamics, CP~CD itself must be both dynamical and multifunctional.

Now what does all this mean in practice? What research strategies, if any, can be devised based on CP~CD? Once again, the syntax of complementary pairs helps on a very basic level. Given CP~CD, two main complementary strategies immediately suggest themselves:

Complementary pairs of coordination dynamics To study the complementary pairs of coordination dynamics in order to aid and advance the science of coordination dynamics (CP of CD).

Coordination dynamics of complementary pairs To use the concepts, methods, and tools of coordination dynamics to understand complementary pairs wherever they are found (CD of CP).

Considering the ubiquity of complementary pairs and the context-dependent yet universal tendencies of coordination dynamics, these two strategies represent a point of departure, a very basic level of CP~CD application that is conceptually simple but not simplistic. Indeed, these strategies already point to some clear-cut ways to proceed. In particular, they lend themselves in a natural, straightforward fashion to two corresponding plans of action. A more ambitious goal of CP~CD might be to try to find a relationship between the complementary pairs and their coordination dynamics in all fields. But for now, let's see how far we can get by just elaborating these two basic complementary strategies.

COMPLEMENTARY PAIRS OF COORDINATION DYNAMICS (CP OF CD)

The CP of CD strategy aims to advance coordination dynamics by applying its complementary pairs to new areas of study. There are at least two ways to do this. One, which we'll shortly come to, concerns extensions of the scientific field of coordination dynamics itself. The other pertains to the study~teaching of coordination dynamics and addresses students and teachers who wish to understand more about coordination dynamics as a field of study. CP of CD is already being used as a didactic tool to familiarize audiences with coordination dynamics and to help them comprehend its theoretical~experimental basis. So why is this crucial?

Coordination dynamics is difficult to absorb at first, and may even seem downright alien to the uninitiated. As you are all too aware by now, coordination dynamics possesses its own jargon, the mathematical and scientific language of informationally based self-organizing dynamical systems. This is where the complementary pairs of coordination dynamics might be of some assistance. That is, the main concepts of CD can also be expressed in the more accessible and generally more user-friendly language of CP of CD. An additional benefit of the CP of CD strategy lies in its associated philosophical stance, namely the philosophy of complementary pairs, which counteracts the two-valued but one-way hypnotic spell of the either/or, dualist, and neutral monist mind-sets. Coordination dynamics bristles with apparent dichotomies and paradoxes at every turn. Yet, as we are not afraid of repeating, if one instead adopts a CP~CD mind-set, the dichotomies begin to vanish even as the dynamics are exposed. In CP~CD, apparent dichotomies are none other than complementary pairs: the complementary pairs of coordination dynamics.

We also expect that CP of CD will prove helpful to those who are already familiar with coordination dynamics, including those who work with it professionally. In this case, CP of CD amounts to a bit of an "upgrade" of coordination dynamics that highlights the key role played by complementary pairs. Henceforth, one can add the CP of CD strategy to established methods, as an aid in the design and implementation of new research projects. Moreover, the complementary pairs of coordination dynamics might greatly facilitate the interpretation and even the communication of the results of such research. We shall have to wait and see.

To summarize, complementary pairs are involved in all that one does when one practices the theoretical~empirical paradigm of coordination dynamics. The CP of CD strategy not only implies that principles of coordination dynamics can be cast as a set of complementary pairs, but reminds us that, historically speaking, those involved in the development and elaboration of coordination dynamics—whether they realized it or not—have been working with complementary pairs all along. Through its complementary pairs, CP of CD offers unique access to the concepts of coordination dynamics. By that we mean that someone outside the field should be able to comprehend the main ideas of coordination dynamics without the *immediate* necessity of mathematics or any advanced scientific~philosophical knowledge.

Of course, as people advance deeper into the complementary nature of their subjects, more advanced conceptual tools will likely be required. Nevertheless, the CP of CD strategy provides ways~means through which one can get to the heart of coordination dynamics via a different path than purely professional academic study and pedagogy, namely the examination and consideration of its related complementary pairs. Armed with such new information, one can use it to explain the principles of coordination dynamics to others, in one's own style, in one's own words. And why might that be important?

Well, while you are hopefully inspired by the science of coordination dynamics, you may not be in a position to run down to the lab and busy yourself applying the mathematics of nonlinear dynamics to your personal areas of interest—at least not yet. On the other hand, the concepts and principles of coordination dynamics may be sufficiently thought-provoking to have piqued your interest enough to encourage you to try to convey them to others. But where do you begin? One way to get personally connected to coordination dynamics is via a language that is much more familiar to most of us: the ordinary language of complementary pairs. The ordinary language of complementary pairs seems like a good way to then communicate the main ideas to others. It seems like a good place to start.

THREE CP OF CD VIGNETTES

The following vignettes are intended as brief demonstrations of the CP of CD strategy in action. In all of them, the chief focus is on advancing coordination dynamics itself, or upon the further elucidation of a problem to which coordination dynamics has already been applied. As you read these vignettes, remember that the concepts and principles of coordination dynamics are already well worked out. So, even though the CP of CD strategy is a relatively new addition to the arsenal of coordination dynamics, these vignettes are by no means speculative. For an entry point into the supporting literature, refer to *Dynamic Patterns* and the bibliography at the end of this book.

CP of CD #1: Understanding Coordination Dynamics via Its Complementary Pairs

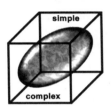

I do not see any way to avoid the problem of coordination and still understand the physical basis of life.

Howard Pattee (1926–), "Physical Theories of Biological Coordination" (1976)

The Simple~Complex and the Search for Laws of Coordination

In spite of, or perhaps because of the successes of modern molecular biology, the great unsolved problem remains: How are living things coordinated in space~time? The primary goal of coordination dynamics is to understand coordination in all its manifestations. Trying to comprehend the scope of coordination going on, say, in a single human nervous system, never mind the multiverse at large, is truly mind-boggling. Yet despite its complexities and complications, all of its uncountable structure~functions, aggregation~dispersions, persistence~changes, and excitation~inhibitions are coordinated at every conceivable level. How is this done? Some find this question unanswerable and have claimed quite strongly that ultimately, the human nervous system is far too complex to ever fully comprehend itself.

In the face of such bewildering complexity and complication, the goal of science is to somehow distill the essential principles and mechanisms upon which coordination might be based. Nobel laureate Murray Gell-Mann is credited with saying, "Surface complexity arises out of deep simplicity." Similarly, one of Albert Einstein's three main operating principles in life was, "Out of clutter, find simplicity."

Both $F = ma$ and $E = mc^2$ are great examples of underlying simplicity in the face of nature's apparent complexity. They express and promote humanity's hope that eventually all the secrets of the universe will be discovered. Thus enlightened, we will then be masters of the universe. This pervasive modus proceeds as Socrates and Plato might have envisioned it. That is, strip away irrelevant complexity or illusory perceptions to reveal the pure, naked, simple truth. Or, in another, more Hegelian direction, by following and transcending each new level of thesis~antithesis synthesis, humanity will move inexorably toward a zenith, a synthetic summit of the mountain of knowledge. From the view at the top, all foggy complexities will vanish. Thus accomplished, humanity is expected to enter a new golden age.

Though these noble pursuits toward a pure enlightened simplicity, whether they be reductive or synthetic, are lovely to think about, they are intrinsically flawed. Complexity is never far away from simplicity. To deny this is to delude oneself. For example, for Einstein to wrest the simple from the complex in his theory of relativity, he needed to employ a hyperbolic, non-Euclidean geometry devised by the Russian mathematician Minkowsky. Non-Euclidean geometries flew in the face of 2,000 years of mathematical tradition grounded in simple, solid regular forms, where never a rough edge was to be seen. In this critical case, the "pure and simple" was inappropriate. Reality was really much more (not less) complex than anticipated. Consider science's attempt to arrive at a final simplest piece of matter. All the discoveries of particle physics—including those of Gel Mann—rest on the totally unpredicted discovery that the atom was far more (not less) complex than anyone had ever conceived it to be. The old conception had lasted unscathed for over 2,000 years.

Add to this our old friend quantum mechanics, in which it was found that light wasn't just more complex than previously known, it was positively schizophrenic. Poor old light couldn't decide if it wanted to be a wave or a particle, and seemed to be able to change its mind! Further, quantum mechanics has uncertainty *embedded* in it—nothing too simple about that. And finally, the smiling Nobel laureate Richard Feynman comes to mind: He claimed that even though quantum mechanics was the most successful theory ever invented, it was so weird that *nobody* really understood it. So where did that "pure enlightened simplicity" go? Quantum mechanics and its interpretations are a lot of things, and almost none of them are simple.

In the science of coordination dynamics, complementary aspects are equally valid. Thus, not only does surface complexity arise from deep simplicity, but the opposite is true too: Surface simplicity also emerges out of deep complexity. This the humble centipede knows very well, otherwise it wouldn't be able to put one foot in front of the other. It seems possible that the eminent biologist E. O. Wilson does not get it quite right when he declares in his book *Consilience* that "complexity is what interests scientists in the end, not simplicity." As shown in Movement 2, *both* simple and complex behavior may emerge from the same self-organizing coordination dynamics. Surface (deep) complexity and deep (surface) simplicity are all possible outcomes of CP~CD. By now, this should come as no surprise. In the complementary nature, simple~complex is a complementary pair. In the complementary nature, each complementary pair is simple~complex.

Returning to our story, a primary goal of coordination dynamics is to identify laws and mechanisms of coordinated behavior both within and across levels of description. CP of CD helps us realize that complexity arising from simplicity and simplicity arising out of complexity are equally valid, complementary aspects. Assuming one of them is the more fundamental is unnecessarily limiting and almost certainly is begging for trouble. As much as we might like to become enlightened exclusively via a final reduction to simplicity, this is very unlikely indeed. CP~CD not only embraces the simple~complex, but provides a language (i.e., dynamics), a syntax (i.e., the "~"), and a methodology to understand it.

Functional Information~Self-Organization

Immanuel Kant
(1724–1804)

An organized being is then not a mere machine, for that has merely *moving* power, but it possesses in itself *formative* power of a self-propagating kind which it communicates to its materials though they have it not of themselves; it organizes them, in fact, and this cannot be explained by the mere mechanical faculty of motion.
Critique of Judgment (1790)

Isaac Newton
(1642–1727)

God who gave Animals self-motion beyond our understanding is without doubt able to implant other principles of motion in bodies which we may understand as little. Some would readily grant this may be a Spiritual one; yet a mechanical one might be shown.
letter to Oldenburg (1675), from James Gleick, *Isaac Newton* (2003)

What then of goal-directedness, of intentionality, of what Immanuel Kant called "formative power"? Is this not what distinguishes the living from the dead, the animate and the inanimate, the machines from the organisms? Are the organic and the mechanical ever to be successfully reconciled? Shall the two never meet? Testimony to the power of the CP of CD strategy is that in its most mature and general form, coordination dynamics itself is comprised of a complementary pair. One of its complementary aspects concerns self-organization. It is about the cooperative~competitive, individual~collective, macro~micro processes that give rise to the spontaneous formation of patterns and pattern change in complex systems at many levels, from genes to mind. The other complementary aspect of coordination dynamics deals with how functionally meaningful information is created de novo in such complex cognitive systems, and how it modulates and is modulated by the intrinsic dynamics of spontaneous, pattern-forming processes.

When it comes to living things especially, this informational aspect and the influence it exerts in the form of parameters of the coordination dynamics is crucial. Self-organizing dynamics creates and constrains functionally meaningful information, and functionally meaningful information constrains self-organizing dynamics. *Understanding life and its origins requires both.* Thus, from the most general frame of reference, coordination dynamics itself is best thought of as a complementary pair, self-organization~functional information.

The perspective of CP of CD becomes crucial because we are living in a world where enormous emphasis is placed on centralized directing agencies, two prime examples being most federal governments and the "genetic instructions" of DNA. Other voices, such as that of physician and complexity theorist Stuart Kauffman in his book *Investigations*, stress that life has an unalienable "wholeness" and that the central dogma of central directing agencies are *not* necessary to life. Kauffman prefers to think of life as being generated and controlled by "collective autocatalysis," which is a well-known form of self-organization that stretches back to Turing's early work on morphogenesis. From a TCN perspective, life is neither predominantly an informationally prescribed and controlled phenomenon nor predominantly a self-organized, autocatalytic phenomenon. From a TCN perspective, these two fundamental processes are inextricable and *both* necessary for life— they are a complementary pair. Thus, one can appreciate how the development of an animal's form depends not only on its complement of genes but on turning on and off patterns of gene expression. The role of these so-called genetic switches is to transform existing patterns of gene activity into new patterns of gene activity. Switches themselves are a result of self-organizing dynamics that reduce the astronomically large number of potential DNA sequences in such switches to a finite number. Cell types and tissues emerge as a result of switches that guide, and are guided by, self-organized spatial~temporal patterns of proteins and other molecules that give cells and tissues their unique properties. Development thus relies upon both genetic switches (a specific kind of functional information) and self-organized pattern-forming processes.

Here one can appreciate the power of the CP of CD strategy: Contained within this complementary pair is the basic essence of coordination dynamics, functional information~self-organization. Vis à vis Kant and Newton, "agency" as a steering factor arises from self-organization and derives its meaning within the context of the very self-organizing processes it may be said to steer.

Coordination Variable~Control Parameter: Dynamic Patterns

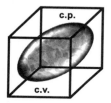

Finding the coordination variables that uniquely capture dynamic patterns of behavior in real systems on a chosen level of description and mapping them onto dynamical systems are at the core of understanding coordination. In coordination dynamics, control parameters are the complementary aspect of coordination variables. Both naturally occurring environmental conditions and intrinsic, endogenous factors may qualify as control parameters. Remember that a parameter is a control parameter if, as it changes smoothly and continuously, it causes abrupt, qualitative changes in coordinative behavior (continuous~discrete). Under such conditions, when a control parameter approaches a critical threshold, instability occurs, leading to the formation of new, different patterns (stability~instability).

The mathematics of nonlinear dynamics, or so-called "dynamical systems approaches," is not sufficient in its own right to provide a deep understanding of coordinated behavior. In every case, the mathematical terminology (attractors, stability, bifurcations, fluctuations, etc.) has to be filled with content. This means one has to do the difficult work of identifying meaningful pattern variables and control parameters. The payoff from discovering coordination variables and control parameters is nevertheless quite high. They enable one to obtain the dynamical laws, i.e., the equations of motion that describe how patterns of coordination persist~change on a given level of description—their so-called *pattern dynamics*. Coordination variables and control parameters are coexistent, complementary and coimplicative: You do not have one without the other. Moreover, they can be interchangeable depending on level of description.

For example, in studies of phase transitions in the brain, the frequency of neural driving (roughly the number of stimuli per second) is a control parameter that influences the coordination among brain areas. Changing this driving frequency continuously or even by a small amount may produce abrupt changes in the brain's dynamic behavior at the cellular level. Within a given brain area, however, the frequency of neural oscillation is also a coordination variable, reflecting the synchronized activity of thousands of neurons. So a control parameter can be a coordination variable on a different level, and vice versa. As with all complementary pairs, the distinction between complementary aspects is both changeless and mutable, in this case depending on the context and level of description (within~between).

Qualitative~Quantitative: Windows into Pattern Dynamics

Let's run a bit further with this central complementary pair, coordination variable~control parameter. In our little lesson on coordination dynamics in Movement 2, we said that the first step in finding laws of coordination is to find *relevant* coordination variables and control parameters. In a typical complex system, many things may be changing all at once, and it can be very hard to distinguish the variables that matter from the ones that don't. Qualitative change provides a clear distinction between one coordinative pattern and another. Thus, qualitative change is a big help in identifying relevant coordination variables.

This is not only because qualitative changes are easier to spot and thus attract our attention. Potential coordination variables that change abruptly are likely to be the most informative, for both the system itself and the person who wants to understand the system's dynamical behavior. The reason, as we've mentioned once or twice, is that such abrupt changes are due to dynamic instability, the main generic mechanism behind spontaneous self-organized pattern formation and change in natural systems. Thus, the most relevant coordination variables~control parameters are those that entail abrupt and qualitative changes in coordination. Notice that insight is not necessarily gained by peeling down the system to its barest form, or integrating it into some all-encompassing whole. Instead, we use a CP of CD method: We seek qualitative~quantitative changes in coordination variable~control parameters by observing stable~unstable coordination patterns.

As stressed earlier, using qualitative change as a point of entry into identifying the coordination dynamics in no way implies that coordination dynamics isn't a quantitative science. CP of CD doesn't favor qualitative over quantitative changes any more than it favors quantitative changes over qualitative ones: Qualitative~quantitative is a complementary pair. As complex systems approach critical regions of instability, they are predicted to and actually do become more variable. This is the phenomenon of "enhancement of fluctuations." Furthermore, near places of qualitative change, the system is far more sensitive to small perturbations —meaning it takes longer to recover from them—the phenomenon of "critical slowing down." In coordination dynamics, quantitative measures are routinely employed to test such theoretically motivated and predicted effects.

In an enterprise committed to uncovering the laws and mechanisms underlying coordination, CP of CD has at least two advantages. On the one hand, the underlying processes and mechanisms governing the stability, flexibility, and selection of coordination patterns become accessible near critical points where qualitative change occurs. On the other, complementary pairs such as control parameters~coordination variables, qualitative~quantitative, and stability~instability may be used strategically to uncover the cooperative~competitive nature of coordination dynamics.

Micro~Macro: The Concept of Levels in Coordination Dynamics

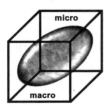

One of the attributes of coordination dynamics is that the task of identifying dynamic pattern variables and their pattern dynamics may be conducted at any level of description and for any activity one chooses. In the lingo of science, this is referred to as "scale- or level-independence." Obviously, if coordination dynamics works at many or most levels of description, it becomes imperative to be able to rein in its scale-independence on some particular level of interest. Now, deciding upon a level of description on which to employ coordination dynamics is at the chooser's prerogative, and is a key step in the creative process. There is no formula or recipe for this.

Practically speaking, in complex systems it is often very difficult to isolate individual components and to study their dynamics independently of the overall context. The fact is that when trying to understand interactions within and between levels, one always needs to pay attention not only to scale, but to context. An individual fish swimming in a school may be viewed as an isolated, individual entity. But when this individual fish is part of a school, which is, after all, a coordination pattern, it does not function entirely as an independent entity. The same might be said of the notion of "brain areas." Even though we have partitioned the brain into areas or regions, one may wonder whether evolution gives a hoot about our partitionings.

On a broader scale, a horse may be independent of its rider, but in a way it can be thought of as a different horse when a rider is on its back. Most horse~riders are keenly aware of this. This concept takes us back to our discussion of entailment. In the complex systems of coordination dynamics, there are no purely context-independent parts from which to derive a context-independent coordinative whole, even though we often try and occasionally succeed to analyze them as such. Again, in coordination dynamics, as in the brain itself, "the parts are not one nor are they separate" (said by Ashvaghosha in Joseph Campbell's book *The Hero with a Thousand Faces*).

In coordination dynamics, no level is more or less fundamental than any other. Higher and lower, bottom-up and top-down, macro and micro levels are complementary. If my atoms go away so do my organs, so do *I*. If *I* go away, so do my organs, so do my atoms. No individual fish, no collective school. No independent water molecules, no collective wave. No specialized neural regions, no integration between them. In coordination dynamics, any notion of defining absolute macro or micro levels is eschewed. The earth is a very macroscopic object relative to a human being and yet is a very microscopic object relative to the sun; the cell studied by a physiologist is a very macroscopic object relative to the molecular composition of the cell studied by the molecular biologist, and so on.

A complete description of context-dependent coordination dynamics on *any* chosen level of description always requires three different kinds of information: (1) control parameters and boundary conditions that establish a context of constraint for particular coordinative phenomena (collective behavior) to arise; (2) context-dependent coordination variables and their dynamics; and (3) individual context-dependent coordinating elements and their dynamics. With all three kinds of information available, plus a little ingenuity and a bit of luck, the behavior of the whole may be seen to emerge from the nonlinear interactions among the coordinating elements. Just as important, the behavior of the coordinating elements may be seen to be constrained by the behavior of the whole. In the science of coordination, both "top-down" reduction and "bottom-up" construction are not only possible but necessary (construction~reduction). Emergentism~reductionism is a key strategic complementary pair of coordination dynamics.

To summarize, adoption of the CP of CD strategy provides a different angle on the science of coordination and a useful way to think about principles and processes. Each guiding principle of coordination dynamics can be cast as a complementary pair. And these complementary pairs of coordination dynamics can be used both as an approach and as tools to advance understanding of complex~simple systems and complex~simple behavior. The philosophy of complementary pairs helps us remember that the complementary aspects found in coordination dynamics, like coordination variable~control parameter, stability~instability and qualitative~quantitative, are best treated as reconciled complementary pairs. Such thinking has helped elucidate one of the chief complementary pairs of coordination dynamics: dynamic patterns~pattern dynamics. Thereby we obtain a glimpse into how the syntax of complementary pairs and the CP of CD strategy may begin to bear fruit and a hint also, perhaps, of its largely unexplored potential.

CP of CD #2: Using Complementary Pairs to Advance the Science of Learning

Tom always started by trying to find out a horse's "history," as he liked to call it. Had he been ridden yet? Were there any special problems? There always were, but more often than not it was the horse who told you, not the owner.

Nicholas Evans, *The Horse Whisperer* (1995)

—Twenty-seven and twenty-seven, she said.
—*What?*
—Bottles.
—*What's in them?*
—Porter.
—*Fifty-four.*
Are you a genius, maybe? she asked.

Roddy Doyle, *A Star Called Henry* (1999)

How do human beings learn? How does a novice become an expert? How do complex systems and organizations adapt and change in a world that never stands still? Solving the great puzzles of learning and adaptation is a top priority of parents, teachers, institutions, and governments the world over. What insights might coordination dynamics reveal about learners and learning and how might the CP of CD strategy advance a new learning paradigm? According to coordination dynamics, one of the keys to understanding learning lies in identifying the *intrinsic dynamics* of individuals, groups, and organizations. The term "intrinsic dynamics" refers to patterns, preferences, and predispositions a system already possesses *before* it is confronted with something new to learn, some new information. Intrinsic dynamics exist at all levels. Each network in a complex system of networks (e.g., genetic, metabolic, neural, social, etc.) has its own unique dynamics, its own intrinsic patterns of activity. Intrinsic dynamics are important because they place constraints on what can be learned, and more generally on what can be changed and stabilized in memory. From this perspective, the transition from novice to expert entails the modification of a system's intrinsic dynamics by *functional information*. Functional information may take many forms. At the level of human behavior it may take the form of environmental constraints, the task to be learned, the motivation of the learner, and so forth. Functional information and intrinsic dynamics are complementary aspects of a complementary pair central to the science of learning.

Most would agree that the organism is not a blank slate. But just what does not being a blank slate mean? According to coordination dynamics, it means that each individual enters a given learning situation with a history, a preexisting repertoire consisting at the very least of spontaneous, self-organized tendencies. In practice, such preexisting biases, whether innate or the result of previous practices and experiences, have a potentially huge, but largely untapped effect on what can be learned, how rapidly we learn it, and ultimately how skilled we become.

Taking the horse whisperer's advice, it is just as important for a horse to tell the trainer what the trainer can do as it is for the trainer to tell the horse what the trainer wants it to do. If the trainer doesn't listen to the horse, teaching it new tricks is going to be a long and arduous process. Likewise, though the street urchin in Roddy Doyle's novel *A Star Called Henry* had never been taught arithmetic, he was surprisingly proficient at addition when the context was right: when the teacher's question contacted his past experiences as a youngster serving porter in a Dublin pub.

Even J. B. Watson, the father of behaviorism, recognized how crucial it was to be aware of and to understand an animal's intrinsic dynamics—what he termed "structural peculiarities"—before trying to teach him anything:

This gives us a key to what all animals of a particular species naturally do—i.e., the acts which they perform without training, tuition, or social contact with their fellow animals. It teaches the psychologist, too, the way to go about the animal's education—i.e., gives him a notion of the problems which the structural peculiarities of the animal will permit him to learn.... We must know the avenues through which we may appeal to him.

Four years after Watson wrote this in an article in *Harper's Magazine* in 1909, he published "Psychology as the Behaviorist Views It," in which he launched behaviorism, an extreme mutually exclusive either/or theory that held that an animal can be trained to associate any stimulus with any response. As a result, "the avenues through which we may appeal to him" fell silent. Only many years later did ethologists such as Konrad Lorenz and Tito Tinbergen—both Nobel laureates—draw attention again to the significance of instinct, innate patterns of behavior that can be imprinted by appropriate stimuli. But how do stimuli and innate patterns interact with one another? This is a good question, but a hard one to answer. Coordination dynamics says that the complementarity of functional information and intrinsic dynamics is the key to solving it.

Scientists in different fields have borrowed the CD concept of intrinsic dynamics and adapted it for their own purposes. For example, the social psychologists Robin Vallacher and Andrezj Nowak talk about "the intrinsic dynamics of social and psychological processes" and how "it is hard to imagine how any *situational factor* or *stimulus* could influence interpersonal thought and behavior independently of the person's goals, concerns and other internal mechanisms." The late developmental psychologist Esther Thelen and her colleague Linda Smith have also adopted the term "intrinsic dynamics" in their studies of how infants learn to reach:

First is the question of the infant's *intrinsic dynamics*. In the dynamic system's view infants *discover* reaching from an ongoing background of other non-reaching postures and movements. In other words, before reaching begins the system has a landscape with preferred attractor valleys that may be more or less deep, that reflect both the infant's history and his or her potential for acquiring new forms. This landscape constitutes the infant's *intrinsic dynamics* [italics theirs].

Similarly, although they do not use the term "intrinsic dynamics," other theorists such as Olaf Sporns and the Nobel laureate Gerald Edelman also stress that learning and development proceed in the context of preexisting capacities and predispositions. And earlier on we mentioned the work by Rafael Yuste and colleagues demonstrating very clear reverberating patterns of intrinsic activity in the brain.

The concept of intrinsic dynamics has clearly caught on in one form or another. The problem is that ways to evaluate this preexisting repertoire prior to learning are lacking or, as more often the case, totally ignored. For example, most theories of human memory do not usually concern themselves with how knowledge is actually acquired. Instead, focus is placed on how it is recognized or recalled (recall~recognition) once it is acquired. Coordination dynamics, however, does not disregard this very important point. It shows that how one *acquires* information is a powerful determiner of how well one remembers it (acquisition~retention).

Because discovering the nature of preferences and predispositions is so difficult and time-consuming, the science of learning has tried to sidestep the problem of intrinsic dynamics by using tasks that are as novel as possible. That is, investigators set up a learning situation that is as unrelated as possible to any existing knowledge that the learner might possess. As two well-known experts on skill acquisition, Richard Schmidt and Tim Lee, remark:

> Discovering the nature of existing capabilities has been difficult, and many experimenters therefore have attempted to avoid the issue by developing novel tasks that were as *dissimilar* as possible to any existing skills that a subject might possess.

Ironically, this ostrich-like strategy tends to limit, even prevent the elucidation of the very features of learning that are actually shared between individuals and across tasks, and that facilitate transfer and generalization of learning! The usual procedure is to study a group of people performing a novel task, record the errors they make, and then calculate an average of how well they perform. How the average changes with practice is then plotted as a "learning curve." Calculating an average essentially means that every learner's intrinsic dynamics is treated as if it were more or less the same. In contrast, coordination dynamics promotes the study of the individual learner, searching for learning processes that are common across individuals and examining how these processes affect the level of performance attained. It seeks to reveal rather than obscure the way an individual learns and the styles of learning that individuals share.

New methods are sorely needed to identify a learner's preexisting capabilities before learning actually begins. Only then can we begin to tailor or engineer new experiences (training regimens, teaching practices, rehab protocols, etc.) to the individual capabilities of the learner. Methods to identify the intrinsic dynamics are likely to be significant *any* time it is important to know something about a system's initial state, its preferences and biases. Knowledge of preexisting preferences and biases allows for any new information to be set on an individualized basis.

Such "new information" may take many forms, such as a new task to be learned, the administration of a drug, the design of a rehabilitation program, or the adoption of an alternative management strategy. Why do some people respond to a drug favorably and others not? If you were sick, wouldn't you rather receive therapy based on your own intrinsic dynamics, than a therapy based on the averaged behavior of some group of (usually) unrelated and unknown others? For that matter, why do some people get sick when exposed to a pathogen, while others remain healthy? Coordination dynamics says that the answer depends on how new information cooperates~competes with individual preexisting dispositions and tendencies. Because of their intrinsic dynamics, some individuals are more or less susceptible to outside influences than others. Effective learning, effective therapy, and effective training happen when new information is tailored to the intrinsic dynamics of the learner, the patient, or the trainee.

Let's examine a fairly typical situation. Imagine that very little if anything is known about a test subject's individual potentials and predispositions prior to exposure to some learning task. Now, as educators, doctors, or managers, we might be trying to induce beneficial changes in a way that is quite arbitrary and unconnected to the system's intrinsic dynamics. What happens? Results of the learning trials will predictably be mixed. Very much will depend upon what the individual learner (child, patient, client, horse, etc.) brings into the situation—his or her individual "signature." If the desired information to be learned doesn't match an individual's intrinsic dynamics—that is, the susceptibilities already present in the learner—coordination dynamics predicts that learning will be slow and inefficient. The reason is that the new information to be learned and the intrinsic dynamics *compete* with one another. It is just this competition that has been shown to lead to instabilities and phase transitions in learning.

Now imagine the alternative situation. Imagine that a great deal of effort is made to understand a learner's dispositions before introducing new information, to ascertain their intrinsic dynamics, as in fact some rare teachers and managers have always done. Now, the new material to be learned is structured such that it *does* resonate with the learner's existing intrinsic dynamics. If this undertaking is successful, new information, as we say in the scientific jargon of coordination dynamics, is in the same "pattern space" as the relevant coordination variables that characterize the learner's existing abilities *before* they try to learn something new. In such a situation, learning is greatly facilitated because the new material relates to the learner's existing information base or knowledge structure, like the "genius" in Roddy Doyle's novel. In this case, learning takes the form of a *cooperative* process. Newly learned patterns stabilize when the new information and the old information (as intrinsic dynamics) *cooperate*. As we said before, but it's worth repeating, how new information cooperates~competes with the learner's preexisting repertoire predicts the outcome of the learning process.

If one wants to enhance learning, coordination dynamics says one should structure the learning environment in such a way that it resonates with the intrinsic dynamics of the learner. What a learner already "knows"—in the broadest sense of that word—and how a learner tends to behave at a given point in time both significantly influence the nature of the learning process and, ultimately, what is learned. Although this might *sound* obvious, like a rehash of some conventional wisdom that has long since been woven into the core of professional learning methodology, it isn't as obvious as it seems. Incorporating it into policy and putting it into practice remain major challenges. In other fields though, such as medicine, treatments based on predispositions that arise, say, from an individual's unique genotype~phenotype is a very hot topic, and very much in the offing.

Research initiated and conducted by Scott Kelso and Pier-Giorgio Zanone, a psychologist trained in the Geneva school of Jean Piaget, has shown that learning can take on two main forms depending on the relationship between the new information to be learned and the learner's preexisting repertoire. In one, the "shift route," learning takes the form of a smooth and gradual shift in behavior toward the newly learned pattern. In the other, learning involves abrupt discontinuous changes, eureka-like phase transitions. And what do you know? The two routes to learning form a complementary pair: gradual~abrupt learning. Zanone and Kelso also found that when new information was stabilized in memory as a result of learning, so also were patterns *related* to the new material. Learning something new, in other words, not only involves both smooth shifts and phase transitions, it also alters the entire landscape of the learner's coordination dynamics. And *that* landscape, of course, acts as the learner's intrinsic dynamics for future learning situations.

This intriguing and provocative discovery has a number of potentially eye-opening implications, affecting issues such as how attitudes and biases form and change, and how knowledge is transferred from one domain to another. Now we get a feel for why people often have such a difficult time attempting to learn concepts that are antithetical to previously assimilated ideologies. A phase transition is required! According to coordination dynamics, for that to happen, new information must compete in the same pattern space as the old. Finding that space of relevant pattern variables, especially its representation in the individual brain, is a difficult task. Here again the study of qualitative change—*dynamic instability*—is and has been one of the keys to progress.

Is there any evidence that destabilization of brain activity patterns accompanies changes in learning? According to coordination dynamics, destabilization of brain activity happens when neural populations in the cerebral cortex are forced to reorganize their spatial~temporal behavior due to changes in nonspecific control parameters, such as fluctuating concentrations of neurotransmitters and neuromodulators. In the functional brain imaging work of Kelly Jantzen, Fred Steinberg, and Scott Kelso at Florida Atlantic University's Center for Complex Systems and Brain Sciences, such reorganization has been shown to take several forms. Activity in a given area of the brain crucial to performing the task may drop in amplitude as a result of learning, implying less energy and a certain economy of effort. Other regions important early in learning may drop out altogether later on when the pattern has become well-established. In relearning, for example after a stroke, new areas of the brain may actually be recruited to perform a task that normally does not need them. Context again plays a key role, establishing a kind of network memory in the brain that influences future behavior.

Global macroscopic brain activity changes that arise during learning are known to depend on local changes in the strength of synaptic connections (global activity~local synaptic strength). During learning, changes in synaptic connectivity occur. After learning, the task is no longer novel. This is reflected in decreased concentrations of neurotransmitters and ipso facto by decreased excitability and amplitude of ongoing neural activity.

Although much remains to be understood about the learning process and the brain mechanisms that underlie it, there is little doubt that the payoff from a deeper understanding of how human beings learn will be immense. Benjamin Disraeli's remark in 1874, "Upon the education of the people of the country the fate of the country depends," is as salient today as it ever was. On offer here are some summary principles that might be useful to all agencies of change, whether they be teachers~learners, doctors~patients, managers~employees, policy-makers~taxpayers, training officers~trainees. Associated complementary pairs together with their novel syntax and interpretation will hopefully stimulate new modes of thought and perhaps new solutions into how problems of learning and education may be solved. Nuances of several of these propositions are still under ongoing experimental scrutiny. As with the rest of CP~CD, there is still much work to be done.

The CP of CD Learning Paradigm

Leonardo da Vinci
(1452–1519)

An arch is two weaknesses which together make a strength.
The Notebooks of Leonardo da Vinci, trans. J. P. Richter (1888)

The following preliminary list, terse though it is, amounts to at least a partial re-evaluation of the social, psychological, and neuronal basis of learning. Although learning principles are encapsulated in complementary pairs, the learning process itself is described by coordination dynamics. A serious commitment toward assessment of individual preexisting capabilities as constraints on the learning process and the need to structure the learning environment in light of them has potentially significant consequences for education and educational policy, as well as many other fields. However, as a practical means to guide and improve learning, therapy, etc., this strategy remains largely untapped.

1. Novelty~Experience The individual is not a tabula rasa. To assume the opposite is to place an unreasonable bias on novelty over experience. Four months after conception, for example, the six-inch human embryo can suck its thumb, grasp with its hands, and kick with its feet; shortly thereafter, it can recognize its mother's voice. More generally, every individual enters a new learning situation with an existing set of innate~acquired capabilities, their own make-up, repertoire, dynamic landscape, signature, or fingerprint unique to them.

2. Individual Inference~Group Inference How or why people change the way they do will remain forever hidden in the absence of knowledge about individual susceptibilities and predispositions. Such an individualized approach holds great promise in education and also in medicine, allowing the clinician to base a patient's treatment and therapy on that person's unique "signature," or set of susceptibilities and predispositions. The coordination dynamics of learning, with its emphasis on identifying individual susceptibilities and predispositions, relies on identifying commonalities among the distinctive ways individuals learn, in addition to using conventional group averaging methods to assess how performance changes over time. Averaging often smears evidence of common behavior across individuals, missing essential details of how individuals change. On the other hand, anecdotal evidence about individual change is no substitute for the scientific method, i.e., identifying and measuring key variables, parameters, etc. The main strategy of coordination dynamics is to group people on the basis of the way they learn as individuals, not simply to average them together as one homogeneous population.

3. Intrinsic Dynamics~New Information In order to understand the nature of learning, the predispositions of the individual learner need to be identified. This does not mean that all predispositions have to be identified—*only the ones pertinent to the context or frame of what is to be learned.* The reason this is so important is that the preexisting knowledge and capabilities of the individual learning system influence the way new skills are learned and remembered. In the language of coordination dynamics, such predispositions and susceptibilities are collectively referred to as "intrinsic dynamics." Knowing the latter, new information can be structured in terms of the learner's intrinsic dynamics, thereby facilitating learning (training, therapy, treatment, drug response, organizational change, etc.).

4. Accommodation~Assimilation Learning, fundamentally, means the modification, expansion, and elaboration of preexisting capabilities and potential for change. It is not, or not only, a reinforcement-repetition-association process. In fact, CP~CD views reinforcement as the stabilization of functional information. Learning is a process in which new information becomes functional, serving to stabilize intrinsically unstable patterns of behavior. Just as the learning system must be able to accommodate new information, so also must it be able to assimilate such information into its preexisting repertoire or landscape, thereby modifying the learner's dynamics.

5. Cooperation~Competition During the learning process, new information cooperates~competes with the learner's current predispositions. Cooperative~competitive processes along with noise inherent in all complex systems determine the rate of learning: Learning tends to be fast when new information cooperates with the intrinsic dynamics, but tends to be slow and laborious when it competes. Transitions in learning occur when competition between new and old information is reduced, giving rise to pattern stabilization, a cooperative effect. Here again, we see how important it is to understand the intrinsic dynamics of the learner, that is, to have accurate probes/measures of preexisting biases, susceptibilities, and capabilities.

6. Gradual~Abrupt Change Learning isn't necessarily a smooth, gradual process. Nor does it necessarily consist of a series of quantum-like improvements in performance. Rather, depending on the degree of cooperation~competition between new and existing information (intrinsic dynamics), learning may involve smooth shifts in behavior or proceed via highly nonlinear, abrupt transitions (e.g., the eureka effect).

7. Stability~Instability Memory is a quantifiable network property of, e.g., genetic, metabolic, neural, and social systems that refers to the stabilization of learned functional information over time. In the brain, it is thought to be due to enhanced synaptic connectivity among neurons. An example taken from studies of neuroplasticity (an experimental neurophysiological model of learning) is long-term potentiation~long-term depression. Again, short-lived patterns of intrinsic activity in the brain may reflect ongoing circuit memory, which can be updated and stabilized by new inputs.

8. Local~Global Reorganization When the human brain is learning or relearning a skill, activity in local neural populations and the coordination among distant neural areas may undergo dramatic spatial~temporal reorganization. Moreover, measures of blood flow show that the individual brain, after it has learned, functions far more economically than one that has not (global~local; persistence~change; efficiency~cost).

9. Boundary~Domain The degree of functional~structural brain plasticity—that is, dynamic changes in the size and distribution of active regions in the cerebral cortex following learning (or following recovery of function after a stroke)—is remarkable, unexpected, and currently the subject of much serious investigation in both children and adults.

CP of CD #3: Understanding Rhythmical~Discrete Coordination Dynamics via Convergence~Divergence

Poetry is more a threshold than a path, one constantly approached and constantly departed from, at which reader and writer undergo in their different ways the experience of being at the same time summoned and released.

The Government of the Tongue (1989)

Seamus Heaney
(1939–)

Rhythms, as we noted in Movement 2, are ubiquitous in nature, from the vibrating string of the cosmos to the genes and proteins that regulate circadian clocks and beyond. In his remarkable book *Keeping Together in Time*, the eminent historian William McNeill shows how powerful a force rhythm is in forming social groups and holding them together. According to McNeill, sharing rhythms, as in dance and military drill, creates and sustains human communities, enabling them to cooperate and thereby enhance their survival. He calls this "muscular bonding," though the bond is based on the flow of information.

Many people believe that the way rhythmical activity is organized is very different from the way discrete kinds of behavior are organized, like reaching for a glass of beer, which seems to have a clear beginning and end. Some recent work on the coordination dynamics of rhythmical and discrete behavior indicates that this assumption is erroneous, or at least overly simplistic. The key to showing this is the relationship between convergence and divergence in the phase-space trajectories (please refer back to the tutorial on dynamics in Movement 2).

Remember that we showed that the integrative activity of the brain and behavior was due to coupling among individual coordinating elements and that these couplings could be both in-phase and antiphase? In 1984, the same year that Kelso's work on bistability and phase transitions appeared, the Japanese physicist Yoshiki Kuramoto demonstrated that *all* collectives of weakly coupled oscillators display in-phase and antiphase solutions! This is a generic, universal result. There is a catch, however: The stability of the solutions depends on the coupling strength of the elements involved. As we have shown in Movement 2, changes in coupling can induce instability and create new forms of coordination. We'll return to this point presently.

For now, we want to recognize that much of what human beings do on a daily basis is of a discrete episodic nature. That's hardly news. We get up in the morning, put on our clothes, brew a cup of coffee, and so on. Although this seems like one continuous stream of behavior, it can also be envisioned as a series of discrete, intentional actions. For such discrete behaviors, a mathematical translation of the behavior using phase equations of the Kuramoto or HKB type doesn't work in general. The question then arises, is the same metastable mechanism of integration~segregation that works for rhythmic coordination valid also for discrete behaviors? The answer is yes, but only if we treat this as a special case of a more general coordination dynamics that involves the complementary pair *convergence~divergence*.

Recent work by the physicist and theoretical neuroscientist Viktor Jirsa and Scott Kelso has demonstrated that all kinds of discrete behavior, from the false starts that sprinters make to the starter's gun to detailed navigational trajectories for how one goes from one place to another arise from the fundamentally convergent~divergent nature of coordination dynamics. This is shown in figure 3.1 (top). In the Jirsa and Kelso theory, a measure of convergence~divergence of trajectories in phase space is the distance $d(t)$. As the individual coordinating elements evolve in time, the convergent~divergent effects of their coupling tend to minimize~maximize the instantaneous distance d between them. For discrete *convergent* dynamics, the coupling creates a kind of a bottleneck in the phase space in which all the trajectories are bundled together (bottom part of figure 3.1). Thus, in the case of discrete movements of two limbs, for example, "convergence" causes the two limbs to coordinate their movements simultaneously and in parallel, as in the way a goalkeeper catches a ball.

The opposite is true for *divergent* dynamics. Here the couplings drive nearby trajectories away. In the case of discrete movements of two limbs, divergence causes one limb to be delayed relative to the other, producing sequential ordering in which one discrete action overlaps but follows the other, kind of like out-of-synch windshield wipers in your car. It is the difference in amplitudes of the coordinating components that provides the critical threshold for the convergence~divergence of trajectories.

What is the connection between the rhythmic and discrete? When the coupled coordination dynamics is on a limit cycle (figure 3.2), the greatest distance in the phase space corresponding to maximal divergence is antiphase motion (180° out of phase). Divergence means that the individual coordinative elements are maximally segregated. They can't get any further apart, or they will be converging again. Think again of windshield wipers in a city bus: together, a little different, a little more different, more, more ... now one is all the way down and the other all the way up ... one is down, the other is up, closer, closer, less different, a little different, together again. From this worldview, segregation is understood as a divergence of trajectories in the phase space of the underlying coordination dynamics. Alternatively, the shortest distance between coordinating elements in the phase

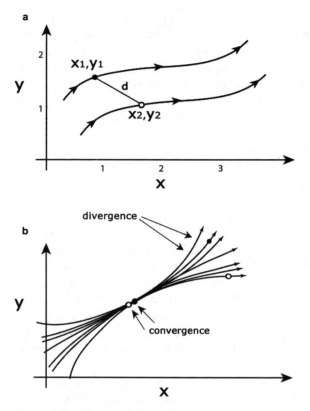

Figure 3.1 The Jirsa-Kelso excitator model of discrete coordination behavior. (a) The time-dependent distance, $d(t)$, in phase space provides a measure of similarity of the dynamics of two coordinating elements given by the coordinates (x_1, y_1) and (x_2, y_2). (b) As the coordinating elements evolve in time, the convergent effect of their coupling tends to minimize the instantaneous distance between them and create trajectory bundles of the kind illustrated.

space is in-phase motion. This is where maximal *convergence* of trajectories occurs, meaning that the individual coordinating elements are fully integrated.

From the foregoing analysis, we see that integration~segregation may actually be understood as a special case of convergence~divergence dynamics in the phase-space of coupled, individual coordinating elements. A key reminder here is that science always requires methods and means to measure things: The convergence~divergence complementary pair applies to both the continuous and the discrete cases of coordination and is measured by the same metric, namely, the distance in phase space. As Seamus Heaney might put it, convergence and divergence coex-

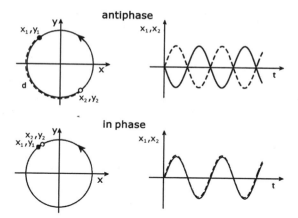

Figure 3.2 How the Jirsa-Kelso excitator model produces rhythmic coordination. The coordinating elements form limit cycles (seen as closed circles on the left). Divergence now results in antiphase motion (top row) because it maximizes the distance d along the closed circle. Convergence minimizes the distance d between the limit cycles and produces in-phase motion.

ist; trajectories can approach and depart, summon and release. There is a critical threshold that governs the transformation from one to the other.

A question that the curious might raise is how the brain might implement this threshold. Hearken back to our discussion of neurons as the basic cellular elements of information processing in the brain. Input to the neural population from, say, other neurons causes cellular activity to increase, usually in a linear fashion. At some point this activity saturates out, resulting in a so-called sigmoidal or S-shaped nonlinearity. The sigmoid represents the most natural form of coupling in biological systems. Multimodal responses such as the neuron's firing rate increasing and then decreasing are also known to occur. Mathematically and technically speaking, a truncation of the expansion of the sigmoid produces multimodal HKB-coupling. Ironically enough, analysis reveals that sigmoidal coupling leads only to convergence in the phase space, whereas the HKB coupling of coordination dynamics, due to its multimodal response to increasing input, displays both convergent and divergent behavior!

THE CP OF CD COLLECTION

From the last vignette, it appears that the CP of CD strategy is not only useful to express what is already known in the field of coordination dynamics; it can also be employed to advance other areas of research and development. One might wonder just how many complementary pairs are contained in coordination dynamics. This is a good CP of CD question. In an effort to answer it, we have begun consolidating all of the complementary pairs that seem to be inherent to coordination dynamics into an organized collection—the CP of CD collection.

The idea is to express the entire conceptual framework of coordination dynamics as a set of complementary pairs. At this early stage of development, though, it has to be said that the CP of CD collection is still very much an open collection. There are sure to be cases where complementary pairs should be included in the collection but aren't, as well as the inclusion of some that overlap with others in meaning. However, a little redundancy at this stage should be viewed as a help rather than a hindrance. On a positive note, all of the assembled pairs are based on established principles that are themselves grounded in experimental evidence and theoretical modeling.

CP~CD teaches us that in our reflections upon nature we should try to maintain a flexible outlook. Although staying flexible might introduce a degree of imperfection and uncertainty, it also leaves room for some surprises. Also, to be too driven by the pursuit of purity and perfection leads one inexorably down the path to polar extremes, which is exactly the kind of counterproductive either/or behavior we are trying to avoid. With further contemplation, work, and experience, the CP of CD collection will undoubtedly grow~shrink, ebb~flow. It will *evolve*. As it continues to evolve, hopefully it will become more and more useful.

In the meantime, one might wonder, Of what immediate use is the CP of CD collection? To begin with, a concise and organized overview of the range and scope of coordination dynamics principles might be quite valuable to anyone wanting to apply the CP of CD strategy. This initial collection also provides a sort of primary sequence of complementary pairs from which one might glean higher-order internal associations (archetypes, if you wish) that could lead to the discovery of new principles of coordination dynamics. More concretely, perhaps, thinking about complementary pairs can itself be a source of fertile ideas that one can incorporate into one's work and life. A couple of anecdotes might help to illustrate this point.

The morning after he'd given an invited lecture on the complementary nature in Vancouver, Kelso was approached by a well-known scientist who studies learning. The scientist asked, "What is the complementary aspect of learning?" Obviously, the person had been brewing over this question for a while. Kelso's response was, "Which complementary aspect would you like? Instinct? Innateness? Habit? Forgetting? There may be others." This answer was received by an "ah, ah" nod of the head. He got it.

At another recent scientific conference, Kelso was asked how he came up with the idea of studying syncopation (moving off the beat), a step that in turn led to the discovery of the transitions from syncopation to synchronization (moving on the beat) that occur when the beat gets faster. This paradigm has elucidated basic mechanisms of coordination stability and switching in both the brain and behavior. The answer lies in the mind-set of the complementary nature, thinking about things in terms of complementary pairs and their dynamics. Thus, Kelso's response was if there is in-phase, there must also be antiphase (in-phase~antiphase). If there is synchrony, there must also be syncopy (synchrony~syncopy). If things are together, they must also be apart (apartness~togetherness). And so on. The key point is that thinking this way led to novel experiments and theory: Thinking about complementary aspects introduces elements into the design of experiments that might otherwise have been ignored.

Consider the discoveries of the great physicists Paul Dirac and Wolfgang Pauli in light of complementary pairs. Is it so surprising that if there is matter there must also be antimatter (Dirac)? Is it so surprising that an electron can spin clockwise (spin up) and anticlockwise (spin down), a "two-valuedness" that led to Pauli's famous exclusion principle? One sees how thinking about one valid aspect of a complementary pair leads almost automatically to thinking about its partner, to such an extent that one is almost forced to think about their coordination dynamics. This is the complementary nature at work.

The CP of CD collection is already helping us envision how the complementary pairs of coordination dynamics are related to one another. Take how cooperation~competition is associated with stability~instability, for example. On first glance, one might think there is little or no connection between these two complementary pairs, but on further reflection of the coordination dynamics, of course there is. In self-organizing coordination dynamics, it is precisely the competition between opposing tendencies that creates instability and leads to pattern formation, a cooperative effect.

In any identified system and for any given situation, there is a lot of coordinating going on between and within the aspects one is considering, and complementarily, a lot of complementary pairs are interacting. The CP of CD collection provides a potential means to study this coordination, how the complementary pairs of coordination dynamics "dance" with one another. Does the CP of CD collection form an identifiable *taxonomy* of complementary pairs? These are puzzles and challenges for the future.

The Current CP of CD Collection

The current CP of CD collection furnishes a kind of consulting tool, an aid to study that can help one better understand the various concepts, principles, and phenomena of coordination dynamics. It also provides a nonmathematical, but no less valid access to coordination dynamics. Of course, one is always free to proceed by directly employing the scientific language of coordination dynamics and its mathematical underpinnings. Our point is that regardless of whether one takes one path or both, the complementary pairs of coordination dynamics are going to be right there in the game. Having the CP of CD collection in hand should enhance one's own comprehension of coordination dynamics as well as help explain coordination dynamics to others.

When considering how CP of CD might act as a tool to advance research~development, it is worth keeping in mind the possibility, even the strong likelihood, that some complementary pairs of coordination dynamics have yet to be discovered. As we present this brief "conceptual scaffold" for the science of coordination dynamics, a few further caveats are in order. First, not all the complementary pairs are unique to coordination dynamics, though many are. Second, notice that although each complementary pair can be expressed in its reversed or "complementary" direction, here only one ordering is presented.

agency~self-organization A distinguishing feature of coordination dynamics is that self-organizing processes are the source of agency and that agency is capable of steering the dynamics.

attraction~repulsion All nonlinear dynamical systems contain mathematically described objects called attractors and repellers. Coordination dynamics also describes behavior in terms of attractors and repellers. In its metastable regime, it also achieves *attraction and repulsion* with no attractors and no repellers—only *tendencies* for attraction and repulsion. In coordination dynamics, these tendencies arise when an attractive and a repelling fixed point "kiss" at a so-called saddle-node bifurcation, giving rise to the phenomenon of metastability. The latter has been called the Principle of Attraction Sans Attracteurs (the ASA principle).

between~within Coordination dynamics captures the coupling between individual elements and processes and also within individual elements and processes. To capture the latter, one most quantify the coupling between individual elements and processes on another level of description.

bifurcation~path The path a system follows can be smooth and linear, or, like a tree, it can have many branches. "Bifurcation" means the path splits into two as a result of the system crossing a *threshold*. This is sometimes called a "pitchfork bifurcation." Bifurcation can also occur in the reverse direction, in which case it's called a "reverse pitchfork bifurcation."

birth~death Experiments and theory in coordination dynamics show that under certain conditions, an existing stable fixed point can die at the same time as a new one is born. Dynamically, this means that an attractor turns into a repeller and a repeller turns into an attractor. This may sound strange, but it's true.

bistability~monostability When circumstances (e.g., in the form of control parameters) change continuously, monostability can give rise to bistability, and in general, multistability, and vice versa. In coordination dynamics, the latter regime gives rise to the Principle of Coexisting Equally Valid Alternatives (the CEVA principle). The mechanism of change is called a bifurcation by mathematicians and a nonequilibrium phase transition by physicists.

bottom-up~top-down Coordination dynamics reconciles purely top-down approaches to understanding and purely bottom-up approaches to understanding. Coordination dynamics stresses the importance of choosing a level of description or scale of observation. A complete account of the chosen level relies on looking one level up (to the boundary conditions, constraints, parameters, etc.) and one level down, to the individual components.

context-dependent laws~context-independent universality The laws of coordination dynamics are context-dependent. The same law may describe and explain how different kinds of things are coordinated, but may also be shaped by the things themselves. Context-dependent laws of coordination dynamics attest to the enormous diversity of nature. They are complementary to the context-independent "first principles" of physics that aim to unify nature. CP of CD says that the complementary nature cannot be understood without both.

control parameter~coordination variable Control parameters may be specific, as in stabilizing coordination states that would otherwise become unstable, or non-specific, as in moving a system through its coordination states. At places of qualitative change, control parameters reveal coordination variables and coordination variables reveal control parameters. In coordination dynamics, a control parameter at one level may be a coordination variable at another, and vice versa.

convergence~divergence In coordination dynamics, and in excitable media in general, the tendency of the flow of the dynamical system to converge coexists with the tendency of the flow to diverge. In the metastable regime of the coordination dynamics, these two opposing tendencies coexist, giving rise to the Principle of Coexisting Opponent Tendencies (the COT principle).

cooperation~competition In coordination dynamics, the relationship between cooperative and competitive processes determines the form self-organization takes, and hence the particular coordination patterns observed. In the metastable regime of the coordination dynamics, cooperation (the tendency for the parts to work together) and competition (the tendency for the parts to express their own individual character) coexist at the same time. This is another manifestation of the COT principle.

correlative inference~population inference The experimental designs used in coordination dynamics examine correlated changes in relevant variables as parameters are continuously varied in order to test key predictions underlying stability and change, e.g., critical slowing, fluctuation enhancement, etc. This complements conventional scientific experimental design, which randomizes the independent variable in order to draw inferences about the population.

coupling~components For coordination to occur and manifest the many forms it takes, coupling between components is necessary. In coordination dynamics both individual components and their couplings are context-sensitive. A component or coordinating element at one level may be a coupled dynamical system at another. Coordinating elements can be as large as the environment (organism~environment coupling) or as small as a molecule that binds to another (receptor~target). On any given level of description, coordinated patterns arise in a self-organized fashion as a result of nonlinear couplings among coordinating elements. In evolving living systems, the components themselves may carry some of the coupling or at least a remnant of previous interactions.

creation~annihilation (of information) In the metastable regime of coordination dynamics, functional information may be both created and destroyed by virtue of the system crossing a threshold. This is a basic selection, choice, or decision-making mechanism.

deterministic~stochastic In coordination dynamics, how a system behaves is based on deterministic and stochastic processes. All real systems have elements of both. Accident and necessity, choice and chance are inextricably connected.

discrete~continuous Discrete and continuous behaviors may arise not only as a result of activating different systems or mechanisms (the usual assumption) but as different parameterizations of the same underlying coordination dynamics (same~different).

dwell~escape In the metastable regime of the coordination dynamics, how long a system resides in the vicinity of a fixed point (its dwell time) and how quickly it escapes from this neighborhood (its escape velocity) are a function of how strongly the parts are coupled relative to how different the parts are from each other.

dynamic patterns~pattern dynamics In coordination dynamics, dynamic patterns are generated by self-organizing processes. These evolving patterns adapt, persist, and change according to context-dependent rules or laws, their pattern dynamics.

emergentism~reductionism Coordination dynamics sees no need to shift from an Age of Reductionism to an Age of Emergentism. Reductionism (e.g., breaking down into elementary parts) and emergentism (e.g., collective effects) are complementary strategies for understanding complex systems. The two may be reconciled by virtue of nonlinear interactions among components that may themselves carry some of the coupling.

fluctuations~states Fluctuations are a sign of dynamic instability and typically precede or anticipate switching among states depending on timescale relations. In coordination dynamics, fluctuations probe the stability of states and allow the system to discover and select new states. This is called the Principle of Selection via Instability (the SVI principle). SVI confers a kind of basic choice or decision-making capability on the system.

functional information~self-organization In coordination dynamics, spontaneous self-organizing processes create meaningful or "functional" information, and functional information guides (modifies, steers, directs, constrains, sets boundary conditions for) self-organizing processes. This crucial complementary pair locates coordination dynamics relative to other theories of self-organization. Each complementary aspect of this complementary pair constitutes a primary root of coordination dynamics. Both are crucial for the coordination of living things.

gradual~abrupt Due to its inherent nonlinearity, coordination dynamics may exhibit changes that are gradual (continuous) or abrupt (discrete) depending on where the system is located in its parameter space. In learning, the competitive~cooperative relationship between new information and the preexisting repertoire, landscape, or intrinsic dynamics determines whether change will be seen as a gradual adaptive shift or as an abrupt phase transition. These are the two routes to learning discovered by studies of coordination dynamics.

homogeneous~heterogeneous In coordination dynamics and living things in general, the individual coordinating elements may be all the same (e.g., as in the idealized case treated by the HKB model). More generally, they differ. In coordination dynamics the degree to which the individual coordinating elements differ is an important parameter~variable that affects the phenomena observed.

horizontal~vertical In coordination dynamics, interactions may occur side to side (within a level) and vertically (across levels in both directions). The resulting coordinative organization is heterarchical and coalitional, not only hierarchical.

individual~collective In coordination dynamics, a part is a whole and a whole is a part. Individual entities may retain their autonomy and independence within the collective at the same time as binding together to form a cohesive group. Individuals create the collective, and the collective, as it forms, affects how the individuals behave.

information~intrinsic dynamics Intrinsic dynamics refers to coordination tendencies and dispositions that exist as a kind of network memory at a given point in time. Once created via bifurcations and metastable dynamics, new information both modifies and is modified by the existing intrinsic dynamics. This principle applies at all levels and is especially relevant to situations where adaptation, learning, and change are at issue.

intention~dynamics In coordination dynamics, intention acts in the same space as the intrinsic dynamics, attracting the system toward an intended pattern. Intentions constrain and are constrained by the intrinsic dynamics. They may both stabilize and destabilize patterns of behavior.

integration~segregation In the metastable regime of the coordination dynamics, tendencies to integrate the parts coexist at the same time as tendencies for the parts to remain segregated, thereby retaining their individual autonomy. This has been called "the complementarity of the twenty-first century."

learning~memory In coordination dynamics, learning modifies the preexisting landscape or repertoire and the latter influences how new information is assimilated. Memory refers to the stabilization of functional information over time and can be stored both locally, e.g., in synaptic connections, and globally, as in a broadly distributed network.

linear~nonlinear Depending on surrounding circumstances, i.e., where the system is located in the space of its parameters, behavioral change may be smooth and linear or abrupt and nonlinear. Nonlinearity is a requirement for multistability and its biological manifestation, multifunctionality.

local~global In coordination dynamics, component elements such as specific areas of the brain may be coupled both locally (e.g., through neighborhood-based *intra*cortical connections) and globally (e.g., over large distances through *inter*cortical connections). Local~global and intra~inter complementary pairs are essential for information processing in complex systems.

macro~micro In coordination science, macro and micro are relative terms. What is micro at one level may be macro at another. This is called the Principle of Relative Levels (PRL).

metastability~information creation Self-organizing tendencies in the metastable regime of the coordination dynamics create~destroy functional information.

multifunctionality~functional equivalence Multifunctionality—the capacity for the same material structure to express multiple functions—and functional equivalence—the capacity for the same function to be realizable by multiple structures—are inherent aspects of coordination dynamics. The complementary pair multifunctionality~functional equivalence is manifest at all levels of biological organization and may be understood in terms of multistable and metastable coordination dynamics. The coexistence of multiple dynamic steady states and tendencies in coordination dynamics provides a scientific underpinning for the philosophy of complementary pairs.

multistability~metastability In coordination dynamics, as symmetry is broken and couplings are altered, multistability—in which several functional states may coexist for the same parameter values—gives way to metastability, in which only tendencies coexist. The generic mechanism is a saddle node or tangent bifurcation.

organism~environment Though by no means unique to coordination dynamics, this complementary pair is nevertheless central to it. In coordination dynamics, organisms do not exist independent of their environment and vice versa. One may be said to entail the other as an informationally coupled self-organizing dynamical system. Organism~environment is a general complementary pair of coordination dynamics which entails informationally coupled self-organizing dynamical systems at multiple levels, e.g., perception~action, stimulus~response, genotype~phenotype.

part~whole In coordination dynamics, the parts are not free of the context of the whole and vice versa. A whole is a part and a part is a whole.

perception~action In coordination dynamics, what an organism perceives is a function of how it acts, and how it acts is a function of what it perceives. Part and parcel of every action is perception, and part and parcel of every perception is action. The same applies to sensing and motion, which in living things are a complementary pair.

persistence~change *Plus ça change, plus c'est la même chose.* In coordination dynamics, whether a process is observed to persist or change depends on the timescale on which the process lives and the timescale of observation.

planning~execution Planning and execution, like mind and matter, sensory and motor, are but two complementary aspects of a single activity written in the language of biologically meaningful coordination variables and their dynamics.

preferences~exploration In the metastable regime of the coordination dynamics, a system exhibits tendencies or preferred locations in the phase space where it tends to reside, while also being able to explore the entire space of possibilities. How long the system dwells in a given preference before escaping to explore the phase space depends on how strong the parts are coupled relative to how different or heterogeneous the parts are. Dwell times and escape times are important empirical measures of coupling in systems that exhibit transient tendencies and dynamics.

qualitative~quantitative In coordination dynamics, quantitative consequences accompany and anticipate qualitative change (e.g., critical fluctuations, critical slowing down). Quantitative causes may or may not produce qualitative changes in behavior.

reaction~anticipation Whether a system reacts to or anticipates environmental information depends on parameters of the coordination dynamics (e.g., the rate of stimulation).

recruitment~annihilation In coordination dynamics, the ability to selectively recruit and annihilate degrees of freedom to accomplish a function is a crucial source of flexibility. These processes may go on at the same time as circumstances vary.

reduction~construction This complementary pair captures the strategic approach of coordination dynamics toward connecting levels of description. One of the mantras of coordination dynamics is, find the relevant pattern or coordination variables and their dynamics on a given level of description, then derive the latter from nonlinear interactions among components. This strategy allows one, scientifically speaking, to reduce down and construct up.

simple~complex In coordination dynamics, a complex system may crack itself into simple behavioral modes whose pattern dynamics may be rich. This complements the usual notion of nonlinear dynamics that simple nonlinear laws may give rise to surface complexity.

source~sink In the dissipative systems of coordination dynamics, the flow of energy from a source to a sink creates a cycle (Morowitz's theorem). The cycle is the archetype of all time-dependent behavior (Yates-Iberall conjecture).

space~time Coordination refers to functions that evolve in both space and time in the phase space of informationally relevant coordination variables.

stabilization~destabilization In coordination dynamics, functional information can both stabilize and destabilize patterns of behavioral coordination depending on context.

stability~instability In coordination dynamics, stability and instability are both indispensable attributes of pattern formation and change. One does not exist without the other. Selection among stable states occurs via instability (the SVI principle).

stable~unstable Fixed points of the coordination dynamics may be both stable and unstable. One can transform to the other. They can also collide or kiss and give rise to metastability.

states~tendencies In coordination dynamics, asymptotically stable states represent polarized ideal aspects. In between these asymptotic extremes lie metastable *tendencies* that are neither stable nor unstable states, but that possess remnants, ghosts, or traces of previously stable states. The coexistence of multiple tendencies and the convergent~divergent flow of their coordination dynamics undergird the philosophy of complementary pairs.

structure~function Structure and function are distinguished only by the multiple timescales on which they live. For example, in the developing organism, cells that fire together wire together, and cells that wire together fire together. In general, invariance of function under change of material structure (e.g., reconfiguration of connections among elements) is an intrinsic feature of coordination dynamics.

symbolic~dynamic After H. H. Pattee, symbolic, rate-independent descriptions and continuous dynamic descriptions are equally valid complementary aspects of complex systems. As a scientific strategy, coordination dynamics says exploit the dynamics to a maximum and trim the symbolic to a minimum.

symmetry~dynamics Symmetries refer to properties of a system that remain invariant under transformation (e.g., mirror, left-right symmetry; forward-backward time symmetry, etc.). In coordination dynamics, symmetries allow for the classification of *possible* coordination patterns or states. What you see in the real world, however, depends crucially on the coordination dynamics, that is, which patterns are more or less stable under current conditions.

symmetry~broken symmetry Symmetries allow for the classification of patterns in nature. Curie's principle, that symmetric causes produce symmetric effects, is not necessarily true in living things. Pattern diversity occurs when symmetries are broken. In coordination dynamics, symmetry breaking is a necessary condition for the emergence of metastable, converging~diverging tendencies. When differences between coordinative elements are eliminated, symmetry is restored or created.

togetherness~apartness Tendencies for togetherness coexist with tendencies for apartness. This is likely an inherent property of all complex organizations. For example, successful groups are loosely bound both by a commitment to a common goal and by the diverse needs and capabilities of their members. This is the essence of metastable coordination dynamics.

within~between As the science of coordination in living things, coordination dynamics seeks the laws, principles, and mechanisms underlying coordinated behavior in different kinds of system and at different levels of description. It aims to characterize the nature of the coordination within a part of the system (e.g., the firing of neurons in the brain), between different parts of the system (e.g., parts of the body, areas of the brain), and between different kinds of system.

COORDINATION DYNAMICS OF COMPLEMENTARY PAIRS (CD OF CP)

Now let's look at CP~CD the other way around. The aim of the CD of CP strategy is to understand the behavior of specific complementary pairs via the concepts, methods, and tools of coordination dynamics. Complementary pairs can but do not have to be drawn from the CP of CD collection. The key idea is that the complementary pair under consideration must entail coordination dynamics. If successful, the CD of CP strategy should elucidate this coordination dynamics as well as try to see how the pairs might fit into the bigger picture, which includes all the laws or rules of coordination. The CD of CP strategy thus provides a vehicle to study all complementary pairs and to see how they relate to each other. Coordination dynamics itself is not only chockfull of complementary pairs; it also shows how various complementary pairs form and change depending on context.

In principle, coordination dynamics should apply to any and all complementary pairs that people might choose to wonder about. Even though all CPs will likely share certain common dynamical features, this is not to say that all complementary pairs are governed by the same coordination dynamics. That would be too strong a claim and far too much to expect. Rather, the softer claim is that for every complementary pair, no matter the field of endeavor, there exists an underlying coordination dynamics waiting to be found. The hypothesis then is that the methods, concepts, and tools of coordination dynamics (including the CP of CD collection) will help reveal the dynamics of that CP and thereby lead to a deeper understanding of it. Down the road there may well be deeper associations between the coordination dynamics of particular instances and classes of complementary pairs, and between different fields of study. But this will have to wait until the CD of CP strategy is explored much further.

The CD of CP strategy always begins with the contemplation of one or more complementary pairs. The complementary pairs can be any that attract one's curiosity. They can range from, say, supply~demand in economics to foreground~background in perception and art. They can be ones found in the CP of CD collection and applied to some sphere of interest, or they can be drawn from the emerging Complementary Pair Dictionary, the beginnings of which will be presented shortly.

An immediately engaging possibility of CD of CP is that the scientific principles of coordination dynamics may be useful in everyday life. Using these principles alone, insights may be gained into complementary pairs without the immediate need to delve into the mathematical and technical details of coordination dynamics. For instance, in a usual setting an artist doesn't need to know all the detailed physics of weakly coupled, nonlinear oscillators and their capacity for spontaneous synchronization and self-organization. Nor for that matter does a butcher, baker, or candlestick maker, despite the fact that said coordination dynamics reveals the deep dynamical nature of complementary pairs! For many, whether scientist or nonscientist, soldier or sailor, pilgrim or poet, the CD of CP strategy can be put to good use in a metaphorical way. Only when one desires to move beyond metaphor to the mapping of particular phenomena onto mathematical dynamical systems is a fairly precise translation available. Metaphors in science are bound to be more useful when they are grounded in a testable theory.

The CD of CP strategy furnishes a couple of readily prescribed directions in which one might proceed. One way is to start with complementary pairs of interest, and use the principles of coordination dynamics (e.g., ASA, CEVA, SVI, COT) to advance understanding of those complementary pairs. Although this sounds a bit like the CP of CD strategy, remember that in that case the chief focus was on advancing the field of coordination dynamics itself. Remember also that CP of CD and CD of CP are themselves complementary undertakings. There is nothing to stop one from using the complementary pairs of coordination dynamics as stepping stones to discovering new complementary pairs as well as their dynamics in other fields, levels, and contexts. If the CP of CD collection, or "base set" as we sometimes call it, is helpful in getting a handle on a "new" or unknown complementary pair of interest, it should be used by all means! Anywhere there are complementary pairs, coordination dynamics can potentially be used to advance insight and understanding.

CD of CP assumes that all complementary pairs have some kind of coordination dynamics in common, and that coordination dynamics is common to all complementary pairs. Though there are probably thousands of complementary pairs one can think of (we have about 1,200 so far, and climbing), CP~CD assures us that any we are able to identify can ultimately be understood via coordination dynamics. Thus, commonalities between similar complementary pairs in different fields and levels and between different complementary pairs in similar fields and levels, and even between different complementary pairs in different fields and levels, can be expressed in a concise accessible form, namely *the language of coordination dynamics*. The language of coordination dynamics is pretty close to life itself. Its principles, which encompass multistable states, metastable tendencies and dispositions, spontaneity, attraction, crises, instability, transitions, synchronicity, coherence, and the like, lend themselves readily to organizing one's thinking and activities. That's because they constitute inherent dynamical aspects of brain~mind and brain~behavior.

The vignettes that follow are intended only as illustrations of the CD of CP strategy at work. Practically speaking, the idea is to take a complementary pair that is not part of the CP of CD collection and to see if it works according to some underlying coordination dynamics, thereby elucidating the dynamical nature of that complementary pair. Our choices of complementary pairs in what follows are pretty arbitrary, although all deal with issues of some contemporary interest. The intent is to provide a way to proceed whenever one is faced with contrarieties. At stake, of course, is how the language of coordination dynamics might be used to elucidate complementary pairs, and ultimately how it may be used to transcend different levels and fields.

THREE CD OF CP VIGNETTES

CD of CP #1: Coordination Dynamics Used to Understand Gene Regulation: The Promoter~Repressor Toggle

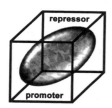

The business of genes is a hot topic at the moment, to say the least. As Eric Lander, one of the leaders of the human genome project said on TV recently, "now that we have the complete map of approximately thirty thousand human genes, what do we do with it?" How do these genes *work together* in normal function and disease? It's the old problem. Now that the genomic "Humpty Dumpty" has been partitioned into pieces, how do we put him back together again? This very problem is considered by Thomas Pynchon's character Mexico, in *Gravity's Rainbow*:

> "I don't want to get into a religious argument with you," absence of sleep has Mexico more cranky today than usual. "But I wonder if you people aren't a bit too—well, strong on the virtues of analysis. I mean, once you've taken it apart, fine, I'll be the first to applaud your industry. But other than a lot of bits and pieces lying around, what have *you* said?"

It's certainly a long, long way to go from the successful mapping of a genome composed of about 30,000 genes to understanding fully developed organisms. However, it now seems likely that understanding gene regulation—where genes make proteins that in turn influence gene expression to produce more protein or alter other genes—is going to be crucial. Given that mice and people have virtually the same set of genes and that our DNA is virtually identical to a chimp's, the development of an animal depends as much on turning genes on and off at the right time and place as on the genes themselves. Thus, it would be incredibly useful to have models of how gene regulatory networks work, in order to better intuit how biological processes can be manipulated at the molecular (DNA) level.

In this regard, there has been some rather startling progress recently in constructing synthetic gene networks. A main force behind this work is James Collins, a talented bioengineer who has turned his hand to analyzing and modeling a number of biological problems such as how animals change gaits or how human beings manage to remain upright. Collins's basic idea is that gene networks with virtually any desired property can be constructed from circuits composed of simple regulatory elements. One such undertaking is the so-called genetic toggle switch, a network in which each of two proteins regulates the synthesis of the other.

To get a better handle on this we need just a bit of background about the chief complementary pair involved: promoter~repressor. Notice that in this scenario, the complementary pair in question is not one of the complementary pairs of coordination dynamics. A "promoter" (more properly a "promoter region") refers to a particular segment of DNA where an RNA molecule binds and subsequently transcribes a gene into a messenger RNA molecule. In this context, we may speak of a promoter *driving* the transcription of a specific gene. As a result of this driving, a chemical sequence is begun in which a region of the gene is converted into amino acids, the protein building blocks of the organism. Gene expression in the

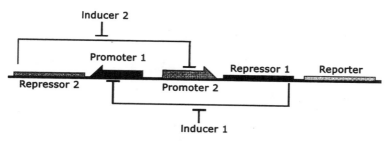

Figure 3.3 A genetic toggle switch. Repressor 1 inhibits transcription from Promoter 1 and is induced by Inducer 1. Repressor 2 inhibits transcription from Promoter 2 and is induced by Inducer 2 (adapted from Gardner, et al., 2000).

cell arises through transcriptional regulation, again in a dual fashion. Activation occurs through biochemical reactions that enhance polymerase binding at the promoter region. Repression occurs when polymerase binding is blocked at the promoter region (promoter~repressor, activation~supression).

So how does the toggle switch work, and what might its coordination dynamics be? The toggle switch is composed of two promoters and two repressors (figure 3.3). The idea is beautifully simple: Each promoter is inhibited by the repressor that's transcribed by the opposing promoter. Thus, protein A turns off the promoter for gene B and protein B turns off the promoter for gene A. Under certain conditions, control parameters called "inducers," such as temperature or a chemical substance, lead these systems to produce two stable steady states: one in which the concentration of one repressor is high and the other low, and one in which the concentration of one repressor is low while the other is high. Switching occurs when the rates of synthesis of the two repressors are no longer in balance (due to the inducer control parameter). Then only one stable steady state is possible.

Does this remind you of anything? It should, because the dynamics—bistability and qualitative pattern transitions that arise as a result of variation in control parameters—is nearly identical to the elementary coordination dynamics described in Movement 2. Figure 3.4 shows nicely how the toggle switch behaves. In figure 3.4(a), two stable states are separated (at a place called a *separatrix*) by an unstable steady state. Technically, this bistability depends on the cooperation between the two repressors. If the rates of synthesis of the two repressors are not in balance, only one stable steady state is possible. The geometrical~dynamical structure of the toggle switch creates two "basins of attraction." Any initial condition above the separatrix settles into state 1. If the toggle starts below the separatrix, it settles into state 2. Figure 3.4(b) shows the parameter space. As the rates of repressor synthesis (x- and y-axes) are increased, the size of the bistable region increases. The

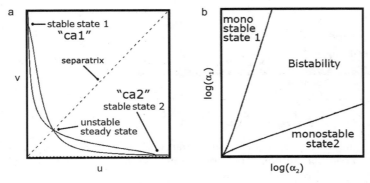

Figure 3.4 The coordination dynamics of gene regulation. (a) The phase portrait of the genetic toggle switch: u is the concentration of repressor 1 and v is the concentration of repressor 2. The curves are called "nullclines," which are defined as places where either the rate of change of u or the rate of change of v is zero, and reveal the flow of the dynamics. Here the nullclines intersect at three places: one unstable and two stable. (b) The parameter space shows the regions of bistability and monostability. α_1 and α_2 are the effective rates of repressor synthesis. The lines mark bifurcations between regions of monostability and bistability.

lines mark the range of control parameter values over which the transition from bistable to monostable behavior occur.

This is all terribly clever because it shows quite precisely how the network architecture is able to produce both stable and flexible behavior. Notice that the toggle switch is self-organized. That is, even though it is referred to as a toggle *switch*, no switch really exists. The promoter~repressor just behaves *as if* there is a switch. It is really a beautiful example of a nonlinear dynamical system at the molecular level. Switching occurs without switches when certain control parameters (the "inducers") cross a threshold value. Moreover, genetic noise likely plays a role in deciding between alternative states, such as choosing one developmental pathway over another (see fluctuations~states in the CP of CD collection).

It is not too difficult to imagine designing and constructing larger circuits of these self-assembled dynamical modules or "building blocks" to perform a broad range of biological functions, from "wet" nano robots delivering proteins into cells to controlling gene expression in disease. Our point here, however, is a bit different, namely to illustrate the potential of the CD of CP strategy. Beginning with a complementary pair (promoter~repressor) from a different field that is not part of the CP of CD collection, we (or rather Collins and colleagues) advanced our knowledge of it by identifying its coordination dynamics. In turn this contributes to the field of molecular biology a novel departure point for a potentially large amount of further research and development.

Of course, one could say that this is all a posteriori, and that Collins and his colleagues' work exists without any need for the philosophy~science of the complementary nature. We accept that. Yet this illustration shows how the CD of CP strategy might work in practice. The toggle switch represents a genuine technological advance. It embodies emerging notions of protein-DNA interactions (the epigenetic viewpoint) that are likely to be found to play a crucial role in the post-genomic era. It affords a quantitative description of gene regulation, and it has important applications in gene therapy and biotechnology. As you can see, it also incorporates many of the key notions of nonlinear coordination dynamics. And at its root is a real complementary pair—not just words or metaphor. Promoters and repressors are real proteins with real structure~functions. Last but not least, we can see that this complementary pair displays obvious and real multi-functional dynamics—real enough to point a finger at. The "toggle switch" is used to create real synthetic gene networks, right now.

CD of CP #2: Coordination Dynamics Used to Understand Buying~Selling in Economics

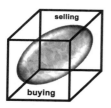

All kinds of factors affect the economy: prices, wages, interest rates, and profits; investment capital, distribution of wealth, and natural resources; and human factors such as age, birth and death rates, skills, habits, social customs, and the like. Some of these interdependent variables change slowly, such as the development of transportation systems, and others quite quickly, such as the cost of commodities like oil. Of course, most will agree that economic systems are constantly in flux, subject to different social and cultural influences. Think about what it takes to put that cup of coffee on your table every morning: the climate, the season, the bean, the picker, the farmer, the distributor, the ship—it's a huge list of interacting factors and interdependencies.

Traditional methods of analyzing the economy are associated with notions of linearity, stability, and static equilibria. As Kumaraswamy (Vela) Velupillai, the John E. Cairnes Professor of Economics at the National University of Ireland in Galway, notes, the "stability dogma" has a long history in economics and is entrenched in conservative policy circles. What was so bad about it? Velupillai quotes, as he says, Samuelson's candid confession:

What was so bad about the dogma ... that all economic business cycle models should have damped roots? Well, it slowed down our recognition of the importance of nonlinear auto-relaxation models of the van der Pol-Rayleigh type, with their characteristic amplitude features lacked by linear systems.

Visionaries like Velupillai have long argued in favor of models of economic dynamics with multiple unstable local equilibria and relaxation oscillations, but it is only fairly recently that mainstream economists have begun to take seriously the newer concepts and methods of analysis that surround complex systems and the science of coordination dynamics. Coordination dynamics, in addition to the more traditional notions of linearity, stability, and static equilibria, embraces nonlinearity, positive feedback, instability, nonequilibrium steady states, and of course *metastable tendencies*. One cannot fail to notice the relevant complementary pairs and their intimate relation with each other: positive feedback~negative feedback, stabilization~destabilization, linear~nonlinear, stability~instability, rest~cycle, birth~death. Many of the CPs discussed earlier in the context of human learning are also highly pertinent to understanding economic behavior and change in systems composed of truly complex, adaptive agents.

A prescient application coming from the field of econophysics (economics~physics—who would have thought *these* fields could be reconciled!) concerns the behavior of financial markets. Fundamentally, the market is about buying and selling, which is driven by demand. When does buying shares change to selling shares and vice versa (buying~selling)? What might the coordination dynamics of that be? Recent work published in the journal *Nature* by Gene Stanley and colleagues at Boston University shows that the financial market has *two phases*. One is an equilibrium phase, in which the market players are neither buying nor selling, and another is a nonequilibrium phase, in which they are selling about half the time and buying about half the time.

Are the bells ringing? They should be. These findings are identical to all that we know about phase transitions in coordination dynamics. Remember, phase transitions occur when a system undergoes a qualitative change in behavior at some critical value of a control parameter. And the change in behavior is quantified by the value of the order parameter or coordination variable.

Of course, in general, the control parameters and the coordination variables of complex systems are typically not known but have to be identified or (if worst comes to worst) guessed at. For the financial market, specifically all the transactions of the most actively traded stocks in the years 1994–1995, Stanley and colleagues identify a primary coordination variable to be the "distribution of demand." When the demand is near zero, the price of the stock is such that the probability of a buyer making a transaction is about the same as a seller making a transaction. There is a kind of residual noise in the system. Prices just fluctuate randomly around their equilibrium values, most likely because buyers and sellers are behaving idiosyncratically. There's no real information that would cause the stock price to go in a certain direction.

The action starts when there is a nonzero demand for the stock. Insider information? The latest quarterly figures? This is the nonequilibrium market phase. In this situation, the distribution of demand becomes bimodal: Now there is an excess of either buyers or sellers that causes the price of the stock to shift up or down in a less random way. Such demand drives the stock price to whatever the market's new evaluation of fair value is. The really interesting aspect of this is that the great and complex dynamical system we call the "stock market" is subject to multiple influences and consists of a very large number of human beings who talk to each other and are on the lookout for the latest information. And lo and behold, it exhibits phase transitions! This is a lovely example of a very basic human economic complementary pair, buying~selling, displaying bistability and bifurcations, key features of coordination dynamics.

Remember that we spoke in Movement 2 about the enhancement of fluctuations that typically accompanies nonequilibrium phase transitions? Might such fluctuation enhancement be connected to the volatile market movements that often precede major changes in the market? Pundits of economics have long studied the fluctuations in various stock indices as indicators of when the stock market is going to rise or fall, when investors should be bulls or bears. Our point is simply that trying to understand the coordination dynamics of the complementary pairs of economics could be quite, shall we say, *fruitful*. Indeed, there seems to be a growing if occasionally skeptical appreciation on the part of economic theorists of the need to study the emergence of dynamic patterns in the economy. Such study will require an understanding of the nonlinear interactions among intelligent sentient "agents" and their coupling to the aggregate, self-organizing economy (individual~collective).

Again, the words of the economic theorist Vela Velupillai seem to capture the spirit of the complementary nature in general and CD of CP in particular. Velupillai notes that "the citadel," a nickname for orthodox, general equilibrium theory, (1) is weak with regard to processes (e.g., ignores the typically adaptive behavior of economic agents); (2) is silent on increasing returns (e.g., the positive advantages such as monopoly that accrue to the first firms that enter the market with a new technology); and (3) can't handle disequilibria (hence no emergence). He asks whether a new "complexity vision" is needed, as some others have argued:

A more modest challenge would be the suggestion that the "complexity vision" does not challenge the foundations of the citadel but it offers complementary tools, methods and an alternative epistemology.... Such complementarities may well be the best case that can be made at this point for the development of the sciences of complexity.

In economics, the CPs seem to be staring you in the face. Some examples include supply~demand, increasing returns~diminishing returns, negative feedback~feedforward positive feedback, random events~determined events, equilibria~disequilibria, fixed point~multiple equilibria, induction~deduction, inhomogeneities~homogeneities, linearities~nonlinearities, microfoundations~macrofoundations, reductionism~emergentism. Economics seems ripe for the CD of CP approach. Indeed one might argue that, at its very core, this is pretty much what economic (coordination) dynamics is all about.

CD of CP #3: Coordination Dynamics Used to Advance Understanding of First~Third-Person Phenomenology

However honorable the motivating intentions and rigorous and intricate the research, practical, theoretical and verbal bridge-building compromises genuine reconciliation. Reconciliation free of colonization requires practices and strategies ... that do not turn their back on life itself and on the abiding stuff of animate life: meaning.

"Preserving Integrity against Colonization" (2002)

Maxine Sheets-
Johnstone
(1930–)

This final CD of CP scenario deals with the apparent incommensurability of descriptions of human experience (i.e., first-person accounts) and descriptions of the brain events that are sometimes correlated with them (i.e., third-person accounts). Contemporary neuroscience aims to correlate the biophysical, neurochemical, and neurophysiological levels of activity of a human brain with human behavior. While representing a staggering amount of research, not only have such correlations fallen far short of explanation, often these studies seem to be correlating apples and oranges. The two vocabularies—one external and objective, the other internal and subjective—appear independent and irreconcilable. How can two apparently different spheres, once torn asunder, ever be put back together again?

It will hardly come as a surprise that we think the greatest mistake was to ever separate them in the first place. CD of CP always seeks to reconcile what we experience as complementary aspects in terms of a meaningful and context-dependent coordination dynamics. At least for sentient life, external behavioral and internal physiological worlds are inextricable complementary aspects. To suppose that a separation exists between them and that these hypothetically separate domains then "interact" is to believe in sheer magic and mysticism. At best this supposition is dualistic, and like all dualistic interpretations is unnecessarily limiting.

Sometimes it takes a person outside a scientific field to communicate a difficult idea within that field clearly and effectively. In an effort to come to grips with the reconciliation of first~third-person accounts of phenomenal experience, we remind you of the philosopher Maxine Sheets-Johnstone's description of coordination dynamics (note the "complementary" language):

> The concepts of coordinated [sic] dynamics consistently tether experience and experimental evidence to theory and theory to experience and experimental evidence. The real-life coordinated actions of animals and the real-life coordinated actions of brain neurons are thus conceived as being cut from the same dynamic cloth. Integrity is in turn preserved because it is never threatened.

Sheets-Johnstone then goes on to exemplify how coordination dynamics is never "loose" and on its own but is actually firmly anchored in experience. She does this by recounting the discovery of self-organizing phase transitions in the voluntary movements of human beings:

> What Kelso was wondering at the time of his discovery was how to demonstrate self-organizing behavior, in particular how to provide experimental evidence that would anchor theory concerning self-organized dynamic patternings, such as those found in physics, to real-life happenings such as equine gait transitions (how a horse sustains a steady gait across a range of speeds, and how it switches to a new gait that is energy efficient and better fits the circumstances). In Kelso's own words:

> It is the winter of 1980 and I'm sitting at my desk in my solitary cubicle late at night. Suddenly from the dark recesses of my mind an image from an advertisement for the Yellow Pages crops up: "Let your fingers do the walking." To my amazement I was able to create a quadruped composed of the index and middle fingers of each hand. By alternating the fingers of my hands and synchronizing the middle fingers between the hands, I was able to generate a "gait" that shifted involuntarily to another gait when the overall motion of the fingers was speeded up.... On hindsight, the emergence of this idea was itself a kind of phase transition.

Sheets-Johnstone would have us pay special attention to the last statement of this passage. For her it is the *idea* of letting your "fingers do the walking" that constituted a spontaneous breakthrough into a new mode of thinking—in this case about spontaneous self-organizing processes. It was, in other words, an "ideational phase transition" that in a neurophysiological sense set in motion new coordinated patterns of brain activity that eventually evolve into a reconciliation of thinking and moving. For Sheets-Johnstone:

> Ideational phase transitions are as real as gait transitions, for example; constitution of one's own body and one's body in movement is as real as constitution of a melody or of the reflection of a bridge in the water below it. In short, coordination and constitution span mind

and body. While appearing to start at opposite methodological ends of the living spectrum, they have the conceptual breadth and staying power to span the whole.... *A genuine reconciliation of first- and third-person methodologies asks us to discover just such conceptual complementarities and to trace out in detail their common ground. It does not require bridge-building because the bridge is already there* [italics ours]. What it does require is listening and learning from each other, immersing ourselves in studies and concepts outside our disciplines, and expanding our understandings of life by solidifying and deepening the ties that bind us in a common quest for knowledge.

We could not say it much better than that. Nor could one better articulate the long-range agenda of CP~CD studies of the complementary nature. In fact, Sheets-Johnstone naturally adopts a style of description that resonates strongly with our philosophy of complementary pairs. Reconciliation of first~third-person accounts of natural phenomena promises much in helping us to understand ourselves, other creatures, and the world we live in. So what would come next? Sheets-Johnstone points to the great difficulty in understanding learning as a crucial area where more work is necessary. Recall, though, that CP of CD already opens up new ways to look at learning.

WAYS~MEANS OF DISCOVERING RELEVANT COMPLEMENTARY PAIRS

James Murray
(1837–1915)

The English language is not a square with definite sides containing its area; it is a circle ... nowhere bounded by any line called a circumference. It is a spot of colour on a damp surface, which shades away imperceptibly into the surrounding colourlessness.

from Bernadette Paton, *New-Word Lexicography and the OED* (1995)

In the preceding sections of Movement 3, we have spoken about two different but complementary means of putting CP~CD to work. On the one hand, CP of CD uses the complementary pairs of coordination dynamics, a collection of complementary pairs, to advance the field of coordination dynamics. The CP of CD collection is the result of a focused effort to survey the established concepts that define coordination dynamics, and to recast them as a set of complementary pairs.

On the other hand, the CD of CP strategy begins with focused contemplation of specific complementary pairs that currently reside outside the aegis of CD and proceeds by using the concepts, methods, and tools of coordination dynamics to advance understanding of those complementary pairs. When successful, this activity not only quickly leads to other complementary pairs associated with a particular context, but also links the field from which they are drawn to coordination dynamics.

One of the CD of CP scenarios followed the premise that promoter and repressor regions of the genome represent complementary aspects of a complementary pair promoter~repressor. We described how the switching dynamics of promoters and repressors parallels that found generally in coordination dynamics. Indeed, such dynamical behaviors may be quite well understood when contemplated within the overall theoretical~empirical framework of coordination dynamics. So at least for that particular scenario, it doesn't seem unreasonable to suggest that molecular biologists who actually study promoters and repressors might take the complementary pair promoter~repressor seriously. In such a fiercely competitive field, it might just give an individual~group a vital edge in its research~development.

In turn, this complementary pair immediately implies a nontrivial connection between molecular biology and coordination dynamics, and this in itself might be quite valuable. Part of everyday life, as the molecular biologist Richard Strohman remarks, involves the fact that all our cells are changing patterns of gene expression all the time in response to changing circumstances. In an influential article in *Nature Biotechnology* he writes, "We are trying to fit dynamic nonlinear change into a linear theory of the gene, and it will not fit." Strohman sees coordination dynamics as a candidate theory to fill the gap.

Recall from the history of ideas that when the reconciliation of a single complementary pair, like time~space or wave~particle or electricity~magnetism or mass~energy, has been successfully made, the ramifications have been staggering. Even a single productive reconciliation can lead to the desire for more: not only more relevant connections between molecular biology and coordination dynamics, but complementarily to the search for more complementary pairs relevant to molecular biology. And those complementary pairs will again carry with them the potential for coherent understanding via coordination dynamics.

Eventually, such resonance could well shim if not shake the foundations of both fields, such that the walls separating molecular biology, with its roots in biochemistry and genetics, and coordination dynamics, with its roots in self-organization and functional information, will come tumbling down. But notice, the aim here wouldn't be to reduce coordination dynamics to the physics of molecular biology any more than it would be to totally replace existing molecular biological methodology with that of coordination dynamics. CP~CD doesn't work that way, nor do we think, does the complementary nature.

The purpose of the foregoing discussion is not to ruminate upon points made previously, but to lead up to the following question: What, if any, protocol might one follow in order to choose salient complementary pairs? In almost every case, there are bound to be many to choose from. But which are the relevant ones, the ones that can lead to palpable advances via the CD of CP strategy? Is there a way to proceed systematically, to choose complementary pairs in such a way that one can avoid overlooking some complementary pairs, while choosing others redundantly? It seems clear that the CD of CP strategy would be more compelling if some ways were identified to go about this task. Let's assume that there isn't just one correct way to go about it; there are probably many different, equally valid ways. Here are four different but related methods:

Brainstorming CPs

The first method is to just think of complementary pairs off the top of one's head. While certainly a good place to start, in most cases this method isn't expected to be terribly comprehensive or systematic. It does offer a couple of pragmatic advantages, though. Brainstorming is free and accessible to everyone. All it requires is that one puts in the effort and takes time for the necessary contemplation. In fact, all the methods described here require one to *think* about complementary pairs. As some of the earlier anecdotes reveal, such thinking doesn't always come automatically to people. In our experience, it is fascinating not only how thinking about one complementary aspect evokes its partner, but also how contemplation of one complementary pair invariably leads to the contemplation of others. This is an interesting phenomenon in its own right that would seem to warrant further investigation. In particular, *why* it happens remains a bit of a mystery.

Brainstorming CPs implies that the process of thinking about and jotting down new possible complementary pairs when they pop into one's head is a worthy and potentially valuable activity. In terms of putting the CD of CP strategy to work, brainstorming is likely to be more useful to those with a lot of experience in their field of interest, who perhaps are able to pursue the coordination dynamics of those particular complementary pairs. The downside to brainstorming is well known to all: Sometimes the head doesn't want to play. Under a deadline, one may very much welcome other ways to access complementary pairs that don't completely rely on "epiphany-on-demand." On the other hand, this undertaking can be used to fair advantage: The next time someone asks you what you are doing just sitting there, tell them you are actually hard at work brainstorming for relevant complementary pairs.

Collecting CPs by Field of Interest

In this method, one studies a collection of complementary pairs compiled by oneself or somebody else for a particular field, level, or system of interest. Of course, a good example of this is the CP of CD collection itself. It was compiled for the field of interest called coordination dynamics, and hopefully will prove valuable to all those interested in the science and practice of coordination. Collections of complementary pairs central to other fields of interest are also likely to prove valuable especially as CP~CD evolves further.

Although no two experts~novices are expected to generate exactly the same CP collection for their fields, similarities are not unlikely. Moreover the *differences* between CP lists might prove to be as instructive as the *similarities*. At any rate, this method is attractive because it makes available a broad range of CPs in a particular field or endeavor to those interested in using the CD of CP strategy.

The idea of creatively compiling a collection of complementary pairs by field might very well lead to new directions, associations, and discoveries. Of the four methods outlined here (there may be more), this one is quite appealing, not only because it challenges one in the way a word puzzle does, but also because its subject matter is intellectually engaging to the person compiling the collection.

For instance, take art as a general field of interest. We challenge you to compile a collection of 50 to 100 complementary pairs that have to do with art; same for management, health sciences, environmental studies, law, medicine, education, engineering, or any other field of interest. At the end of the book we have compiled short CP collections for a number of fields—just to act as seeds to get you started. Note that the activity of collecting CPs by field may proceed without immediate regard for their ultimate order or importance, redundancy, cross-association, etc. That can all come later, and is likely to be facilitated by coordination dynamics and all its consequences and implications. But first one need only collect a number of CPs that one thinks are important, and write them down.

Notice that compiling complementary pairs by field of interest also relies a good deal on brainstorming. That said, the search for CPs using this method is bound to be more systematic since it will naturally follow the established lines and terminology of the field itself. Once again, a good example is the CP of CD collection, which follows the established lines of coordination dynamics. The whole endeavor, then, may be envisioned as a style of guided brainstorming. The Internet and its lightning-fast keyword search engines (like google.com) give one a great way to discover complementary pairs and possible complementary aspects cross-referenced by field. Try it. Type in the name of your subject (like geography, art history, economics) along with words like complementary, complementary pairs, complementarity, and see what comes up. You may be surprised at what you find!

With collections of complementary pairs by field, the study of complementary pairs can take on a more systematic, orderly, and comprehensive form. For instance, as a preemptive effort toward applying coordination dynamics to one's field of interest, one could begin by collecting as many complementary pairs as possible that seem indispensable to that field. An organizational phase would be expected to ensue, where the collected complementary pairs would be consolidated, checked for redundancies, and cross-referenced in some way pertinent to future studies and interests. Not only may such CP collections provide relevant complementary pairs to be used directly in CD of CP studies; the process may also be expedited by comparing them to the CP of CD base set, using that as an aid to align said field with the complementary pairs of coordination dynamics.

Application of the CP of CD Collection

In this next method, one studies the CP of CD collection itself and tries to see how some or all of those complementary pairs might be directly applicable to one's own field. For the time being at least, this method is more mature and systematic than the first two, and should become more important as the CP~CD paradigm takes hold. For example, one might wonder what role cooperation~competition plays in one's system of interest. Or one could reflect upon the complementary pair individual~collective, and try to determine which phenomena represent individual aspects, and which collective, in the context of one's own interests.

The immediate advantage of this method is that it lends itself to direct interpretation of the complementary pairs in terms of coordination dynamics. Of course there are sure to be other relevant complementary pairs not found in the CP of CD collection that might be missed if one only uses it to the exclusion of all else. In all likelihood new complementary pairs relevant to coordination dynamics will eventually be discovered in other fields. If that turns out to be the case, the process of compiling CP collections in other fields will serve to advance coordination dynamics!

Discovering CPs via a General CP Collection

The final approach to discovering relevant complementary pairs is somewhat akin to using a dictionary or thesaurus. In this method, one is provided with a general collection of CPs—a complementary pair dictionary, if you will. This collection of complementary pairs would be *much larger* than, though inclusive of, the current CP of CD collection and all the CP collections compiled for specific fields of interest combined. It would also likely be much larger than anything most individuals would be able to (or care to) produce on the spot via brainstorming. Such a complementary pair dictionary would be intended to be an aid, providing the thinker~doer with the advantages inherent in any such organized list.

For example, if one was inclined to believe that some aspect of one's life is a *complementary* aspect, one might want to look that aspect up in a complementary pair dictionary, and seek out its complementary aspects. The example referred to earlier—"why study syncopation?"—is a bit like that. Thinking of the complementary aspects to one's target of interest can open up possibilities. A complementary pair dictionary might also be useful for compiling collections by field of interest, and for brainstorming. Study of the dictionary could help one spot many complementary pairs that one might have thought of eventually via brainstorming or guided brainstorming, but far more expediently. It also might help people discover complementary pairs that they might never have considered, and create others.

Of course, finding relevant complementary pairs from a CP dictionary could be a bit overwhelming, like trying to study a dictionary in the hope of discovering relevant words to develop an idea, a poem, or a novel. A CP dictionary may be a useful reference tool to have at one's disposal, but it would surely be more helpful to readers who have an idea of what they are looking for. Still, as CP~CD continues to evolve, a general collection of complementary pairs, or CP dictionary, might well become an indispensable reference tool.

From Brainstorming to a Complementary Pair Dictionary

As for *brainstorming*, hopefully reading this book has piqued your interest in complementary pairs enough that you have already begun noticing when they pop up in different contexts in your daily life. If your brain~mind has begun to brainstorm for complementary pairs spontaneously, that is a good sign. Years ago, when ideas about self-organization were undergoing rapid development, an example frequently referred to was the beautiful hexagonal patterns that arise in fluid dynamics. Once one sees how this works, one starts to see hexagons everywhere—in clouds, beehives, and rock formations, like the Giant's Causeway near Bushmills in Northern Ireland. In a similar vein, perhaps the many examples given in this book from the history of philosophy~science will cause you to notice and think about the ubiquity and relevance of complementary pairs.

The complementary nature is a subject~object that seems to be heating up lately. For example, a recent research article by Stephen Grossberg of Boston University, one of the founders of the field called "neural networks," is entitled "The Complementary Brain." This paper was written quite independently of the present work. We include it here as a compelling contemporary example of another's effort to ground scientific research in the complementary nature and vice versa. And perhaps unsurprisingly, with a title so related to all that is found herein, one finds in Grossberg's paper a short list of complementary pairs, a fine example of a small CP collection (9 CPs) made by field of interest. Note that the title of this table also comes from Grossberg's paper.

Some Complementary Pairs of Brain Processes

Boundary Surface	Boundary Motion
'What' learning and matching	'Where' learning and matching
Attentive learning	Orienting search
Object tracking	Optic flow navigation
Color	Luminance
Vergence	Spherical angle
Motor expectation	Volitional speed
Sensory cortical representation	Learned motivational feedback
Working memory order	Working memory rate

Thus, the present authors are not the only ones who currently believe that thinking about the complementary nature and brainstorming for complementary pairs and complementary aspects are worthwhile undertakings. Take a look at the conclusion of Grossberg's paper. Does it sound familiar?

Thus, just as in the organization of the physical world with which it interacts, it is proposed that the *brain is organized to obey principles of complementarity*, uncertainty, and symmetry-breaking. In fact, it can be argued that known *complementary properties* exist because of the need to process *complementary* types of information in the environment. The processes that form perceptual boundaries and surfaces provide a particularly clear example of this hypothesis. The *'complementary brain'* may thus perhaps best be understood through analyses of the cycles of perception, cognition, emotion, and action whereby the brain is intimately linked to its physical environment through a continuously operating feedback cycle. One useful goal of future research may be to study more directly how *complementary aspects* of the physical world are translated into *complementary brain* designs for coping with this world. [italics ours]

"To study more directly how complementary aspects of the physical world are translated into complementary brain designs for coping with the world." Indeed. This is one of the guiding themes in TCN. Grossberg's call-to-arms is already well under way! And how might complementary pairs be "translated" into complementary brain designs? Let's recast the question a bit: How can the language of complementary pairs be translated into the language of complementary brain designs? We have two answers. For one thing, complementary brain designs *are* complementary aspects of the physical world. But taking the question the way it was intended, then our answer is: *Two ways*—CP of CD and CD of CP.

It is clear that as soon as one begins to brainstorm, think about, get the gist of, clue into, attend to, or focus upon complementary pairs, one almost immediately begins to make associations between different complementary pairs. As one thinks of some complementary pair of consequence, it leads to another, and another. Thus, brainstorming for complementary pairs evolves from a more general and sporadic whimsy to a more specific motivated hunt, as is clear in the case of Grossberg's research. Though we consider this a real step forward, it is just the beginning. As one collects complementary pairs, new connections naturally emerge that one can hardly resist pursuing. It seems that the very act of brainstorming very quickly leads to guided brainstorming by field of interest.

Our list of complementary pairs by fields of endeavor found in the back of the book provides a starting point, a point of departure. One may add Grossberg's complementary pairs in the field of cognitive neuroscience, and in so doing recognize his efforts. Grossberg is clearly aware of the complementary nature; he has got the idea we are trying to convey to you in this book. He and we are by no means alone. Complementary pairs as processes (e.g., integration~differentiation) are instantiated in the human brain and central to the brain's development. The pharmacologist Sungchul Ji of Rutgers University has even tied what he calls "complementarism" to the hemispheric specialization of the human brain. Moreover, certain complementary pairs seem to take place in particular locations within the brain, such as the hippocampus and the brainstem. Listen to the recent words of the eminent neurobiologist Sacha Nelson from Brandeis University:

> Superman and Lex Luthor or Yoda and Darth Vader, most physiological processes come in matched pairs.... Whether at the level of excitatory and inhibitory circuits, potentiation and depression of individual synapses, or phosphorylation and dephosphorylation of individual proteins, opposing processes are often balanced to maintain a setpoint homeostatically.

Of course, TCN has labored to show that opposing tendencies coexist on multiple levels and are based on the mechanisms and principles of multistable and metastable coordination dynamics. But that is not the key point here. Nelson is addressing new research findings demonstrating that long-term potentiation *and* long-term depression work together to alter the stability~plasticity of neural circuitry. This work reveals that the rewiring of brain circuits by learning and experience occurs "in a coordinated fashion" at multiple sites within a neural circuit. Nelson concludes with the hypothesis that "Multiple learning rules, implemented through different molecular mechanisms, are likely to coexist in all flexible neural circuits." Such mutifunctionality and coexistence of rules, states, and tendencies are intrinsic features of self-organizing coordination dynamics and the complementary nature.

If any further evidence were required that TCN thinking is on the rise, witness the research on the lowly fungus, which Jeremy Thorner and colleagues at UC Berkeley refer to as "the Jekyll and Hyde of the microbial world." Fungi, the most familiar of which is baker's yeast, happen to be a key experimental model system for working out how cells work at a molecular level and what happens when they don't. So what is the connection to Robert Louis Stevenson's Dr. Jekyll and Mr. Hyde? The fungus appears to be a classical bistable system, the coordination dynamics of which has yet to be worked out. Thorner's Dr. Jekyll version of baker's yeast has a spherical or ovoid form which proliferates by budding. The Mr. Hyde version consists of filaments of sticky elongated cells that can invade host tissue.

The two forms or morphologies constitute a complementary pair for the molecular biologist and the eukaryotic way of life. Like the proverbial characters in Stevenson's novel, all that needs to happen is that nature (or the experimenter) limits the nutrients. When that happens, the yeast undergoes a "dimorphic transition" (phase transition or bifurcation, in our lingo) from a spherical form to a thin, oblong form resembling a sausage string. The so-called switch appears to be controlled by at least three signaling inputs. Be that as it may, the two forms constitute a complementary pair. Both are stable for certain values of control parameters, only one when a threshold is crossed. *Sound familiar?*

There is likely an underlying dynamics to the switching process which, though physically realized by diverse molecular mechanisms, has yet to be explored. For example, is the dimorphic transition reversible? Yes, when rich growth conditions are restored. Is hysteresis present? No one knows. Symmetries and broken symmetries may also be involved. Theoretical~empirical characterization of such dynamical processes may help to unveil the complementary nature at the molecular level, in particular the self-organizing processes that enable cells to be multifunctional at the same time that they adapt uniquely and appropriately to specific and nonspecific environmental information.

Returning to the business at hand, our goal was to provide ways~means to discover relevant complementary pairs that may subsequently be studied using both the CP of CD and CD of CP strategies. Supplementing the CP of CD collection are what we have called brainstorming methods and short representative CP collections by fields of interest, the latter ranging from anatomy to sports. Although we are not experts in all these fields, the idea is to inspire you to improve and extend these collections. A more complete set of tools, however, might include a complementary pair dictionary, so let's talk about the prospects and challenges of creating one. Think of it. If a complementary pair dictionary were on tap, one would then have ample ways~means available to explore the science behind the philosophy and the philosophy behind the science of the complementary nature.

THE COMPLEMENTARY PAIR DICTIONARY

By a complementary pair dictionary, we mean a general comprehensive list, collection, and repository of complementary pairs. Such a list should be strategically organized, relatively easy to use, and as comprehensive as possible. The Complementary Pair Dictionary we are developing isn't conceived to be a list of word definitions. After all, the definitions of complementary aspects are the same as

nd in normal dictionaries. As it stands at the moment, the CP dictionary
f a long list of complementary pairs, along with a couple of necessary de-
ces. Although the potential exists for developing some kind of cross-
g scheme such as one finds in a thesaurus, that is still on the drawing board.
dictionary is a different kind of dictionary because in order to use it
one needs to keep in mind what is meant by a complementary pair,
ntary aspects, and just as crucial—coordination dynamics. The purpose
of the dictionary is to enable the user to locate complementary pairs and complementary aspects. Remember, complementary pairs aren't simply paired concatenations. That squiggle character (~) sitting between written complementary aspects isn't just a fancy hyphen. It stands both for the inextricability of complementary aspects and for the dynamical potential of complementary pairs: the context-dependent multistable and metastable coordination dynamics inherent in all complementary pairs. The squiggle (~) is telling you that those pairs are much *more* than the words composing their complementary aspects.

If the *Oxford English Dictionary* (OED) is anything to go by, producing comprehensive organized lists of words can take a long, long time. The OED project was originally estimated to take three years. It was apologetically extended to ten, and so on (and on). The 70-some-odd years it actually required to produce the finished OED took a lot of people by surprise. Taking this lesson to heart, let us say we have only *begun* this predictably lengthy and challenging project, and just like the earlier brave souls who tried to compile a comprehensive dictionary of the English language, we hope to inspire the public in the near future to help with this endeavor.

After a good deal of initial labor-intensive effort, it eventually became apparent to the original compilers of the OED that they needed help. So they hired more and more (and more!) people to work on it, but to little avail. The stroke of insight came from the realization that if the search for words and definitions was opened to the public, things might proceed much more rapidly. Ironically, the flood of public response did not shorten, but actually lengthened the duration of the OED project. However, the OED stands today as a tour de force. It seems the CP dictionary may be in a similar situation, in that there is potentially a mountain of possible complementary pairs that might be included in it. Taking the challenge to the public from the very beginning seems a wise course of action. Hopefully, efforts to compile a CP dictionary can proceed a bit more quickly with the advent of the Internet, email, smart databases, and data processing techniques.

The Complementary Pair Dictionary follows at the back of this book. There is also an official TCN website (thecomplementarynature.com), where, hopefully after some healthy study of the complementary pairs presented here, readers and others are invited to send in complementary pairs that are not yet included in the current version. To our knowledge, the Complementary Pair Dictionary is the first of its kind. Like the short collection of CPs by fields of interest that precedes it, it is a functioning draft, and is not even close to being finished. Yet already the dictionary contains over a thousand complementary pairs: The challenge now is to find more for inclusion in future versions. Of course, the idea of forwarding complementary pairs is a take~give situation. Once an adequate threshold of new entries has been reached, the growing CP Dictionary will be posted on the Net, *giving them back to you*. This will also be helpful to reduce the number of redundant contributions.

When studying the CP Dictionary, there are a couple of points to be aware of. First, note that complementary pairs are always written as a syntactical juxtaposition of two complementary aspects. By convention, the two aspects are referred to as ca1 and ca2. The written complementary aspects ca1 and ca2 (e.g., ca1 = body and ca2 = mind) are joined syntactically with the tilde or squiggle "~" character, like so: "ca1~ca2" (body~mind). The squiggle is used to remind us that though ca1 and ca2 are discernible, differentiable aspects, they are nonetheless inextricable—they coexist and are complementary.

The most straightforward and practical description of the CP Dictionary at present is that it is a comprehensive alphabetized list of complementary pairs. The first entry of a given ca1~ca2 pair is based on which complementary aspect appears first in the alphabet (e.g., body~mind). The second entry in the CP Dictionary, obviously further down the list, is that same complementary pair written the other way around (i.e., mind~body). Without this convention, it would be rather difficult to locate half of the complementary aspects, thus defeating the purpose of having such a dictionary in the first place. The problem is simply and effectively solved by entering each CP twice, one for each complementary aspect.

But this isn't all there is to say about the written order of complementary pairs. Recall that in Movement 1 we deferred on the issue of whether we thought complementary pairs ultimately had a different meaning if the order of their complementary aspects was reversed. At that point, we said for the time being one should just assume that, for example, individual~collective equals collective~individual. That is, both refer to exactly the same complementary pair, and both are written in a way that indicates their complementary nature. Make no mistake about that. However, considering the inherent ordering process that takes place in reading~writing (e.g., left to right in English), one could imagine that the two written forms may be used to emphasize one complementary aspect relative to another.

Continuing our example, one could use **individual**~collective to refer to an individual who also happens to be a collective. This might be used to emphasize that individuals such as human beings also happen to be members of the collective species *Homo sapiens*. Or that in an economy, individual freedom coexists with

THE COMPLEMENTARY PAIR DICTIONARY

collective planning. But we can also play the same game, just the other way around. A **collective**~individual could be used to emphasize the coordinated behavior of a collective, as one observes in a school of fish, a flock of geese, an audience clapping, a mob engaging in a riot, etc. In all these examples, a collective acts *as one*. In short, as a window into the complementary nature, individual~collective and collective~individual describe exactly the same complementary pair. In addition, from a pragmatic, literary standpoint, **individual**~collective and **collective**~individual are both perfectly valid ways of writing this complementary pair. The liberty to write the same CP two ways can be used as a literary device to emphasize two of the more straightforward manifestations of this complementary pair.

For both of the reasons just explained, each complementary pair is entered twice in the CP Dictionary, one for each of its possible written orders. We apologize for belaboring this point. But our goal is for no one to waste time wondering why complementary pairs appear *twice* in the CP Dictionary! Below is a small complementary pair collection just to provide the gist of the concept. The prototype awaits your perusal in the back of the book. Notice in the list below that the CP "absorption~reflection" (in bold) appears a second time further down the list as "reflection~absorption." We can't resist making one more point regarding duplication: Double (multiple) entries of a word in a word reference book aren't nearly as uncommon as they may seem. Remember that in a normal thesaurus, the synonym of any given thesaurus entry is also a thesaurus entry itself, which refers back to the original entry—*as a synonym*. The complementary nature strikes again!

a~z
absorption~reflection
aggregation~dispersion
brain~neuron
ca1~ca2
ca2~ca1
cell~organ
collective~individual
dark~light
dispersion~aggregation
DNA~protein
emergence~reduction
environment~organism
gravitation~radiation
green~red
individual~collective
light~dark
many~one
nature~nurture
neuron~brain
nurture~nature
one~many
organ~cell
organism~environment
packing~unpacking
part~whole
particle~wave
person~society
protein~DNA
qualitative~quantitative
quantitative~qualitative
radiation~gravitation

red~green
reduction~emergence
reflection~absorption
society~person
unpacking~packing

wave~particle
whole~part
.
.
z~a

Basic Layout of the Complementary Pair Dictionary

At this point in time, it is unknown how many complementary pairs might exist. The very roughest, possibly absurd estimate would be that every word in a language has a complementary aspect. In this case, the number of entries would approximately equal the number of words in the language, since it takes two words to tango, and each complementary pair must appear twice. While this possibility may seem far-fetched, it is good to use as a benchmark, as a hypothetical bound. The remaining possibilities are that there are many *more* complementary pairs than words in a normal dictionary, and that there are many *fewer* complementary pairs than words in a normal dictionary. In the case that there are many more CPs, the explanation must at least include the phenomenon that one complementary aspect can be a member of more than one complementary pair (multifunctionality), and that some complementary aspects aren't single words, but are combinations of words, as modified words, phrases, sentences. It may even be that entire themes can be shown to be complementary aspects.

On the other hand, it might be found that only a much smaller subset of entries in a normal dictionary can really be thought to be complementary aspects. If such is the case, then any given complementary aspect would actually have many fewer meanings (i.e., synonymic redundancy). One thing is certain: The present prototype version already contains over a thousand CPs, with no end in sight. Can the many CPs listed in the CP Dictionary be classified into families? The jury is out on that question. Were such families actually identified, the basis CPs for them might well be revealing of possible archetypal forms of the complementary nature. An informal observation relates to this: When we have asked students of both genders across a broad range of ages and cultural and ethnic backgrounds to provide a short list of complementary pairs that come immediately to mind, they often come up with the same ones. This high degree of redundancy, if true, and the form it takes might suggest a kind of archetypal memory shared by us all.

To be honest, we are not sure where the CP Dictionary project is going to take us. Regardless of where it goes, to our knowledge it is one of a kind. A normal dictionary gives us word definitions. A thesaurus gives us words with similar meaning, and often also provides (usually fewer) antonymic possibilities (synonym~antonym). The CP Dictionary gives us words that are related to each other via the meaning of the squiggle (~). All CPs by definition share the definition of a complementary pair. Hence, each entry in the CP Dictionary comprises two complementary aspects *and* the ability to move between them—a coordination dynamics. All the entries in the CP Dictionary carry with them the general philosophical~scientific meanings of complementary pairs and their implicate dynamics.

Right in the back of your copy of TCN a short prototype of the Complementary Pair Dictionary is provided to be studied at your leisure. Will you discover missing complementary pairs? Yes. Would we like to know what those missing CPs are? Yes. Please let us know about them at our website dedicated to TCN, ⟨www.thecomplementarynature.com⟩. Together with the CP collections by fields of endeavor, the CP of CD collection, whose CPs are also present in the CP Dictionary in bolded font, and the core principles of coordination dynamics, this book contains the necessary information and reference tools necessary for you to begin working with CP~CD, using the CP of CD and CD of CP strategies. Now we encourage you to get out a notebook or go to your computer (*or both* . . .) and *get to work!*

A PHILOSOPHY~SCIENCE OF THE COMPLEMENTARY NATURE

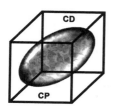

CP~CD Is More than and Different from "CP of CD" + "CD of CP"

The complementary pair CP~CD implies that coordination dynamics is relevant wherever complementary pairs are found, and complementary pairs are relevant wherever coordination dynamics is found. Pause for a moment and think about what this implies. In light of our historical perspective of Movement 1, it implies that every philosopher and scientist, from Lao Tzu to Aristotle to Aquinas, from Descartes to Kant to Bohr and Einstein, has been pointing at something like coordination dynamics all along. CD~CP means that complementary pairs are not only philosophical constructs, not just mental metaphors that we use to explain our existence, but are also phenomena in their own right. Actually, it also implies that mental metaphors aren't "just" mental metaphors either!

In the philosophy~science CP~CD, one uses the complementary pairs found in coordination dynamics as a way to understand and advance coordination dynamics itself—the CP of CD strategy. We also use coordination dynamics to advance our understanding of individual complementary pairs and complementary pair collections—the CD of CP strategy. These strategies can both be employed as ways~means of investigating other fields that, on first blush at least, have nothing whatsoever to do with coordination dynamics. Finally, through CP~CD and its two most obvious spinoffs, CP of CD and CD of CP, the possibility exists of relating disparate ideas, processes, fields, and systems in a systematic fashion. Those agents~agencies wishing to find a common language across fields and disciplines might just want to give CP~CD a little attention.

As with any complementary pair one might wish to consider, both the concepts of complementary pairs and coordination dynamics have their distinct advantages and disadvantages. This comes as no surprise. Complementary pairs have been on the scene a very long time—a whole lot longer than the science of coordination dynamics. They are old and familiar. They have been the subject of endless discussion and study throughout the ages. Human beings are at least many thousand years more familiar with complementary pairs than they are with the scientific language of coordination dynamics (notwithstanding the CP of CD collection!). For all of our familiarity with contraries, though, there has yet to be any satisfactory comprehensive philosophical~scientific interpretation of them. An exaggeration? Not really. The mind~body problem is a nice representative case of a deadlocked debate that has stretched over 300 years and is still raging on many fronts.

Even though coordination dynamics is much newer than the 8,000-odd years of debate over contraries, it is still a quite well-developed scientific discipline. Coordination dynamics has solid roots in physics, mathematics, psychology, and brain research. It has been the basis of many hundreds of research articles, lectures, didactics, and a whole lot of research~development effort. Its basic principles have been tested again and again and appear to grow stronger with every challenge. Of course, like any other scientific theory and research program, coordination dynamics is not etched in stone, but subject to modification and change, even replacement by some better, still more encompassing framework. When that happens, we'll be the first to applaud.

In Movement 3, we have described what we think are the ramifications of coordination dynamics entailed by complementary pairs and of complementary pairs entailed by coordination dynamics. We have unfurled CP~CD, our philosophy~science of the complementary nature. As a first step in the development of this high-level reconciliation, we introduced two strategies to put CP~CD to work. It is important to remember that though it is illuminating, CP of CD and CD of CP still *emphasize* CP~CD from the frame of reference of each of its complementary aspects, in each case using one to understand and advance the other. These two strategies, regardless of their utility and promise, will never by themselves exhaust the full potential of CP~CD. Why? The definition of complementary pairs reminds us that for all that can be gained by these two strategies, CP~CD is much more and much different than the sum of these two strategies.

An intriguing hypothesis that follows from TCN is that CP~CD, like all complementary pairs, has itself the potential for a dynamical metastable regime where new functional information may be created and destroyed. We propose, at least as a possibility, that the creation~destruction of functional information can be understood in the simplest of human terms: It is the level on which new~old ideas about the complementary nature are created~discovered. From a neuroscientific angle, it seems entirely plausible that CP~CD might apply to the very ways~means that humans create~discover theory~practices in the first place.

Stop for a moment and try to imagine a philosophy~science that not only provides an explanation for its intended subject~object, but also provides an explanation for its ability to explain its subject~object! Far-fetched? Perhaps and perhaps not. By itself, coordination dynamics has already been quite successful in explaining certain aspects of the structural~functional, spatial~temporal dynamics of the human brain and behavior. And now complementary pairs, which have played a central role in human reckoning as far back in history as humans can reckon, may have found a plausible and useful scientific grounding at long last. It is difficult, perplexing, and very exciting to see where CP~CD goes from here. One thing we know, though: If it fulfills its primary objective, it will lead us inexorably toward a better understanding of the complementary nature.

WHAT IS THE COMPLEMENTARY NATURE?

Good question. In Movement 1 we defined the complementary nature like so:

The complementary nature A set of mutually dependent principles responsible for the genesis, existence, and evolution of the universe relating to or suggestive of complementing, completing, or perfecting relationships and being complemented in return.

Simply put, the complementary nature is the name we have given to nature itself, while at the same time voicing our belief that nature itself entails complementarity in its manifold expression. It is our thesis that the complementary nature is experienced by human beings as complementary pairs. Now, hearkening back for a moment to the history section in Movement 1, some of the most influential thinkers in history have labored to reconcile complementary pairs. Their attempts at reconciliation of complementary aspects reveal that these thinkers had a deep intuition about the complementary nature. For Thomas Aquinas, it was faith~reason; for Immanuel Kant it was rationalism~empiricism; for Mohandas Gandhi, it was freedom~constraint and lawfulness~disobedience. For Jesus it was friend~enemy and love~hate; for Shakyamuni Buddha it was happiness~suffering; for Aristotle, it was form~matter, and so on.

In the last few hundred years, though, the complementary nature has become increasingly important to human beings, as they have developed more and more sophisticated methods to study and manipulate nature in order to understand it. When James Maxwell made the stunning discovery that electricity and magnetism were inextricable complementary aspects, he was demonstrating the complementary nature. Likewise, when Albert Einstein showed that time and space were inextricably connected as complementary aspects, he was demonstrating the complementary nature.

Bohr's Copenhagen interpretation of quantum mechanics was an incredible insight into the complementary nature, because in his attempt to explain the wave~particle nature of light, Bohr not only appreciated their complementary nature, but expressed it as such. He actually called it by name. Moreover, he realized that complementarity was not only important for explaining quantum mechanics, but that it was likely to be a principle that had far-reaching implications in its own right. Bohr had a deep intuition for the complementary nature.

Electricity~magnetism, time~space, energy~matter, and wave~partcle are four very striking examples of the complementary nature, because they are physical examples and have had profound scientific and technological implications. They are complementary pairs par excellence, because they can be demonstrated as such. Things don't get more "natural" than electricity, magnetism, time, space, energy and matter, waves and particles. Of course, each of these complementary pairs also provides strong metaphorical images. In the hands of Einstein, for example, matter is a knot of energy.

Even the person on the street who doesn't understand the physics and mathematics of relativity can imagine all too well that there is unbelievable energy in a small amount of matter called plutonium. It is not too hard to imagine an atomic bomb exploding. How is it that one can appreciate the concept of energy~matter even if one doesn't happen to understand atomic physics? A possible answer is that human beings understand complementary pairs in a native sense. They get the gist of a complicated complementary pair because, educated or not, they understand the complementary nature. And why might that be? We propose it is because human beings and human brains are themselves *manifestations* of the complementary nature. And why is that? Because human beings and human brains work according to the principles of coordination dynamics, and because coordination dynamics explains the coexistence of complementary aspects and the way they transform from one to another.

Human nature is inextricable from the complementary nature. We human beings are physical entities of the multiverse, just like planets, stones, and rain. We have a physical form, possess physical capacities, and are constrained by physical limitations. Moreover, we are mental beings of the multiverse, with an amazing capacity to adapt and learn, to think, to intend, to act, to love, and to hate. We have a sense of our physical nature and also a sense of self, of being conscious and aware. We understand complementary pairs because we *are* complementary pairs, and entail them in our brain~bodies. Our entire human spectrum of being~becoming is permeated with them. Human existence is so inundated with complementary pairs that we can get the gist of very difficult concepts if we can digest them as complementary pairs.

Human beings can be thought of quite literally as the complementary nature *observing itself*. This indicates that nature must entail some kind of non-trivial self-reference, what Douglas Hofstadter, author of *Gödel, Escher, Bach*, likes to call "strange loops." In the case of human beings, complex nonlinear self-organizing systems of energy~matter have managed to evolve to the point of organizing a sense of self~other. Sentience and self-reference have been making trouble for philosophers for centuries. How does the brain do it? How does self-awareness work, and where does it come from?

Luckily, CP~CD, with its enfolded nonlinear dynamics, provides feasible ways~means out of the apparently incessant, infinite loop of self-reference. It offers a preliminary but principled account of the origins of human agency based on self-organizing processes. It provides ways~means to comprehend how such unimaginable complexity can produce the routine day-to-day coordinated movements that people and animals make. CP~CD obeys the complementary nature that it seeks to explain. Thus, its ways~means are neither purely reductionistic nor purely emergent, neither purely individualistic nor purely collectivist, neither purely materialistic nor purely idealistic, neither purely empirical nor purely theoretical. And yet it is all of these aspects, and none of them. It is everything they are when they are added together, and much, much more.

Our planet earth and its biosphere are also an expression of the complementary nature, as is our "scenario multiverse." The macrocosm~microcosm is the complementary nature, the mechanics of quantum particles~waves and the larger cosmos, like stars, galaxies, nebulae, black holes, and *binary* pulsars (!). The complementary nature is every conceivable level of every conceivable complementary aspect, of every single complementary pair. In CP~CD, levels of the complementary nature are woven together. They are coordinated via an informationally meaningful, self-organizing coordination dynamics, a web~weaver. In the complementary nature there is no level that is most fundamental. All levels of the complementary nature are fundamental in their associated contexts, and may just as easily be irrelevant in others.

The philosophy~science of the complementary nature called CP~CD is also an interpretational extension of coordination dynamics, in the same way that complementarity is an interpretational extension of quantum mechanics. Coordination dynamics has already begun to establish itself in diverse fields and on different levels. Coordination dynamics appears in various guises. People talk about neural coordination dynamics, cognitive coordination dynamics, social coordination dynamics, and so forth. The extension of coordination dynamics that is CP~CD should enhance the potential of establishing meaningful dialogues between disparate fields and levels of inquiry. CP~CD may be a valuable paradigm not only because its hypotheses are testable but also because it gives us both a conceptual language and the methods and tools with which to point at the complementary nature. Of course, CP~CD is still in its early days. As people apply it to their own fields and levels of interest, we shall perhaps together eventually get more in touch with our own nature . . . *the complementary nature.*

END~BEGINNING

What we call the beginning is often the end
And to make an end is to make a beginning
The end is where we start from.

"Little Gidding," from *Four Quartets* (1943)

T. S. Eliot
(1888–1965)

The epitaph on Karl Marx's gravestone says, "The philosophers have only interpreted the world in various ways; the point is to change it." We have come to the end of this preliminary excursion into the unusual but exciting subjects of the complementary nature, complementary pairs, coordination dynamics, CP~CD, and the CP of CD and CD of CP strategies. But the ending of this book is also a beginning. The truth is we are still getting used to the language and philosophy~ science of the complementary nature ourselves. The concepts and ideas in this book constitute an effort to see whether there is a way to accomplish reconciliations between separate fields, systems, and levels while still respecting them as individual entities. The full consequences of the TCN paradigm, of course, will depend on whether we can do something apart~together about these ideas.

On the other hand, perhaps you should forget all about this CP~CD business, put the book on the shelf, and carry on believing that everything in life is always about the mighty either/or. Our hope is that after reading *The Complementary Nature* this might be easier said than done. So we leave the very complementary last words to the poet David Budbill:

The Three Goals

The first goal is to see the thing itself
in and for itself,
to see it simply and clearly for what it is.
No symbolism please.

The second goal is to see each individual thing
as unified, as one, with all the other ten thousand things.
In this regard, a little wine helps a lot.

The third goal is to grasp the first and the second goals,
to see the universal and the particular, simultaneously.
Regarding this one, call me when you get it.

(P.S. from the authors: No, email *us* when you get it ...)

COMPLEMENTARY PAIR COLLECTIONS BY FIELDS OF ENDEAVOR

While many, even most, complementary pairs essential to a particular field of endeavor are immediately familiar to those individuals interested in that field, it is nevertheless still surprisingly difficult to find organized collections of complementary pairs by fields of endeavor. What one usually finds is a more detailed account of the complementary nature of a particular complementary pair. As we discussed in Movement 3, the potential of such lists becomes much greater when formulated and applied in the context of CP~CD. What follows are complementary pair collections by 20 fields of endeavor. The collections are not meant to be comprehensive, but are rather presented as points of departure for the fields covered. It is our hope that these collections will not only be used and extended, but will also inspire similar collections to be generated for other fields of endeavor.

anatomy
afferent~efferent
analysis~synthesis
anatomy~physiology
appendages~trunk
axon~dendrite
cell~organ
connection~joint
cranial~caudal
dendritic~somatic
difference~similarity
dorsal~ventral
form~function
function~structure
head~tail
inferior~superior
internal~external
organ~organism
part~whole
position~size

art
absorption~reflection
abstract~realistic
adjacency~opposition
aesthetics~meaning
art~science
artist~artwork
artist~medium
classical~modern
color~complement
complementarity~contradiction
creation~destruction
dark~light
expression~meaning
expressionism~realism
foreground~background
inner~outer
maker~onlooker
medium~technique
original~replication

painting~painter
sensation~perception
talent~training
technique~inspiration

cultural
acceptable~unacceptable
adult~minor
arts~sciences
child~parent
city~country
civilized~barbaric
dependent~independent
familiarity~strangeness
family~society
father~mother
freedom~slavery
government~religion
hunter~gatherer
individual~society
innovation~stagnation
language~culture
modern~traditional
morality~immorality
music~dance
nationalism~globalism
native~foreigner
old~young
permissible~taboo
poverty~wealth
public~private
together~apart
unite~divide
woman~man

economics
analysis~prediction
boom~bust
buy~sell
capital~interest
central planning~competition
coordination~conflict
cycle~rest
employment~unemployment
equilibrium~disequilibrium
feedback~feedforward
free trade~protectionism

increasing returns~diminishing returns
inhomogeneities~homogeneities
labor~capital
labor~management
loan~interest
macroeconomics~microeconomics
prosperity~poverty
reaction~anticipation
reductionism~emergentism
risk~insurance
save~spend
stabilization~destabilization
supply~demand
taxation~spending

education
abrupt change~gradual change
accommodation~assimilation
classical~modern
collective inference~individual inference
competition~cooperation
conservative~liberal
context~information
desegregation~segregation
ethical~unethical
evaluation~performance
experience~novelty
forget~remember
information~data
innate~learned
instability~stability
intrinsic dynamics~new information
knowledge~ignorance
memorization~conceptualization
progressive~traditional
public education~private education
standard~innovation
student~teacher

entertainment
acceptance~rejection
acting~actor
advertisement~sponsorship
agent~talent
amateur~professional
applause~silence
attractive~repulsive

COMPLEMENTARY PAIR COLLECTIONS BY FIELDS OF ENDEAVOR

audience~entertainer
comedy~tragedy
cops~robbers
director~producer
fame~obscurity
fiction~reality
good guy~bad guy
leading~supporting
player~performance
popularity~ratings
privacy~publicity
prop~set
sad~funny

mathematics
1~0
addition~subtraction
analysis~topology
analytical~geometrical
axiom~proof
calculation~estimation
computation~visualization
deduction~induction
differentiation~integration
discrete~continuous
element~set
equivalence~inequality
explicit~implicit
finite~infinite
fractional~whole
imaginary~real
irrational~rational
linear~nonlinear
negative~positive
problem~solution

medical and health sciences
anesthesia~surgery
benefit~cost
circulation~stagnation
competence~malpractice
curative~palliative
decay~healing
desired effect~side effect
diagnosis~symptoms
disease~health
doctor~nurse
doctor~patient
drug~placebo
effective treatment~iatrogenesis
efficacy~side effects
epidemic~epidemiology
healthy~unhealthy
invasive~noninvasive
pediatrics~geriatrics
prevention~cure
public~private
sick~well

military
allegiance~disloyalty
ally~enemy
ammunition~gun
army~soldier
attack~retreat
battle~casualties
central~distributed
certainty~uncertainty
commander~troops
death~life
defensive~offensive
discipline~training
intelligence~spying
loser~winner
order~disorder
peace~war
prerogative~rank
prison~freedom
projectile~target
strength~weakness
vanquished~victorious

molecular biology
A~T
C~G
cDNA~DNA
clone~parent
coding DNA~junk DNA
continuous~discrete
DNA~protein
DNA~RNA
expression~translation
genetics~epigenetics
genotype~phenotype

gradualism~punctuated equilibrium
host~virus
inducer~inhibitor
instability~stability
intron~exon
metabolism~replication
mutation~selection
pleiotropic~polymorphic
promotor~repressor
retrovirus~virus

mythology
beauty~ugliness
child~mother
damsel~prince
death~life
deity~worshipper
father~mother
good~evil
heaven~underworld
hero~heroine
hero~quest
hero~villain
love~hate
magic~wizard
moon~sun
mortal~immortal
oracle~prophecy
peace~war
rule~trickster
story~storyteller
strong~weak

navigation and driving
acceleration~deceleration
anticipation~reaction
avoiding~targeting
brake~accelerate
calculation~estimation
collision~obstruction
controlled~uncontrolled
destination~departure
distance~speed
down~up
east~west
fuel~vehicle
leaving~staying

left~right
lost~found
navigator~pilot
north~south
on course~off course
optic flow~heading
training~trainee

neuroscience
action potential~graded potential
afferent~efferent
agonist~antagonist
axon~dendrite
brain~mind
brain~neuron
central~peripheral
chemical~electrical
dorsal stream~ventral stream
excitation~inhibition
glia~neuron
hearing~sound
left~right
light~vision
Na^+~K^+
potentiation~depression
presynaptic~postsynaptic
sensation~perception
sensory~motor
serial~parallel
smell~taste
spike~threshold
subthreshold~superthreshold

pharmacology
addiction~withdrawal
binding~transduction
bonding~breaking
central~peripheral
chemical~electrical
dependence~tolerance
dose~response
effect~side effect
efficacy~potency
efficacy~toxicity
excitation~inhibition
extracellular~intracellular
habituation~sensitivity

ligand~receptor
nicotinic~muscarinic
nonspecific~specific
parasympathetic~sympathetic
receptor~transmitter
release~uptake

philosophy
aesthetic~ethical
being~becoming
being~nothingness
collective~individual
construction~deconstruction
democracy~tyranny
democrat~republican
dualism~monism
empiricism~rationalism
existence~essence
existentialism~Marxism
faith~reason
finite~infinite
form~content
holism~reductionism
individual~collective
materialism~idealism
mental~physical
mind~body
monism~dualism
noumena~phenomena
novelty~experience
physical~spiritual
public~private
question~answer
sympathy~antipathy
truth~falsehood

physics
absorption~reflection
amplitude~frequency
animate~inanimate
certainty~uncertainty
change~persistence
coherence~incoherence
compression~tension
determinism~uncertainty
electricity~magnetism
energy~matter

entropy~self-organization
equilibrium~nonequilibrium
experiment~theory
gravitation~radiation
homogeneous~heterogeneous
law~regularity
linearity~nonlinearity
local~global
macrocosm~microcosm
matter~antimatter
measurement~error
observer~frame of reference
particle~wave
position~momentum
probability~mechanism
quantum mechanics~relativity
time~energy
time~space

politics
ally~enemy
balances~checks
boundary~domain
campaign~candidate
capitalism~socialism
centralization~decentralization
conservative~liberal
constituency~government
enforcement~law
fundamentalism~secularism
individual~collective
intelligent design~evolution
liberal~conservative
order~disorder
peace~war
privacy~security
representation~taxation
responsibilities~rights
stability~instability
state~church

psychology
abnormal~normal
abstinence~indulgence
abuse~care
acceptance~rejection
action~perception

cognitive~motor
computational~dynamic
conscious~unconscious
conservative~liberal
dream~sleep
dynamic~static
emotion~cognition
expert~novice
external~internal
extraversion~introversion
fantasy~reality
feeling~thinking
individual~social
intuition~sensation
irrational~rational
learned~innate
learning~forgetting
massed~distributed
mental~physical
mind~body
object~subject
passive~aggressive
pattern completion~pattern separation
plasticity~stability
proactive~retroactive
self~other
specific~general
stimulus~response

religion
angel~demon
atheist~agnostic
atonement~sin
Buddha~mana
clergy~congregation
confess~repent
conservative~liberal
creation~evolution
ecumenical~secular
eternal~transient
faith~despair
faith~reason
finite~infinite
fitna~jihad
fundamentalism~unity
God~Devil
God~human

good~evil
grief~joy
heaven~hell
holy~pagan
immortal~mortal
life~death
love~hate
love~justice
marriage~celibacy
permissible~taboo
physical~spiritual
religion~science
sin~virtue
transcendent~material
unity~divisiveness
yin~yang

sports
amateur~professional
attack~defend
best~worst
cheating~honesty
coach~player
competition~cooperation
contestants~match
defense~offense
experience~novelty
expert~novice
fans~players
game~gear
game~rules
individual~team
lose~win
luck~skill
performance~practice
players~team
professional~amateur
rules~referee
strategy~implementation
strength~stamina
talent~training
work~play

THE COMPLEMENTARY PAIR DICTIONARY PROTOTYPE

All CPs appear twice, one for each direction. Complementary pairs from the CP of CD collection are in bold.

0~1
1~0
1~2
2~1
A~T
abandon~claim
abandoned~occupied
abase~elevate
abate~intensify
abbreviate~lengthen
abdicate~reign
abdication~accession
abhor~admire
aboriginal~alien
abort~execute
abortion~birth
abound~lack
abridge~expand
abrogate~establish
abrupt~gradual
absence~presence
absent~present
absolute~limited
absolute~relative
absolution~blame
absolve~condemn
absorb~exude
absorb~reflect
absorbent~waterproof
absorption~reflection
abstain~partake

abstemious~acquisitive
abstinence~indulgence
abstract~concrete
absurd~reasonable
abundance~scarcity
abuse~care
abyss~mountain
accede~reject
accelerate~decelerate
acceleration~deceleration
accept~reject
acceptable~unacceptable
acceptance~rejection
accessibility~inaccessibility
accessible~inaccessible
accession~abdication
accident~substance
accidental~planned
acclaim~criticism
acclaimed~acclamation
acclamation~acclaimed
accommodating~disobliging
accommodation~assimilation
accompanied~alone
accomplishment~failure
accord~discord
accordingly~conversely
accretion~loss
accrue~dwindle
accumulate~scatter
accumulation~distribution

accuracy~error
accuracy~inaccuracy
accuracy~precision
accusation~breach
accusation~confession
accuse~confess
accustomed~unfamiliar
acid~base
acme~depth
acquaintance~stranger
acquiesce~resist
acquire~lose
acquire~purchase
acquisition~retention
acquisitive~abstemious
acquit~indict
action~act
action~actor
action~consequence
action~inaction
action~perception
action~reaction
action potential~graded potential
activation~inactivation
active~dormant
active~passive
activity~inactivity
activity~quiescence
actor~play
actual~possible
actual~potential
actual~theoretical
actual~virtual
actuality~model
actuality~percept
acute~obtuse
adamant~flexible
adamant~yielding
adaptable~inflexible
adaptation~extinction
adaptation~mutation
adaptation~survival
addict~addiction
addict~withdraw
addiction~addict
addiction~withdrawal
addition~subtraction

adenine~thymine
adept~inept
adequacy~inadequacy
adequate~inadequate
adhere~bond
adhere~repel
adherent~opponent
adhesive~solvent
adjacency~separation
adjourn~convene
adjust~leave alone
adjust~parameter
ad-lib~scripted
admire~abhor
admissible~inadmissible
admit~deny
admit~exclude
admonish~commend
ado~calm
adorn~strip
adorning~stripping
adrift~moored
adult~child
advance~retreat
advantage~disadvantage
advent~end
adventurous~unadventurous
adversary~ally
adverse~favorable
advertise~disclaim
advertise~product
advisable~inadvisable
advocacy~opposition
affect~mood
affected~unaffected
affection~loathing
affectionate~unaffectionate
afferent~efferent
affiliate~disassociate
affinity~aversion
affinity~strength
affirm~refute
affirmation~negation
affirmative~negative
affix~detach
afflict~relieve
affliction~relief

THE COMPLEMENTARY PAIR DICTIONARY PROTOTYPE

affluence~poverty
affordance~effectivity
afraid~unafraid
after~before
age~youth
ageless~transient
agency~self-organization
agent~control
agent~counteragent
aggravate~alleviate
aggravate~improve
aggravation~amelioration
aggregation~dispersion
aggressive~passive
agility~clumsiness
aging~presence
agitate~calm
agitation~pacification
agonist~antagonist
agonist~receptor
agony~ecstasy
agony~pain
agree~disagree
agreeable~disagreeable
agreement~disagreement
agricultural~industrial
ahead~behind
aid~obstruct
ailing~healthy
aim~target
aiming~target
akin~unrelated
alertness~lethargy
alien~aboriginal
alien~citizen
aligned~opposed
alike~different
alive~dead
all~none
allegiance~disloyalty
alleviate~aggravate
alliance~separation
allocate~withhold
allow~forbid
ally~adversary
alone~accompanied
altercation~amends

altering~retaining
altitude~depth
altruism~selfishness
always~never
amateur~professional
ambiguity~clarity
ambiguous~clear
ambition~apathy
ambivalent~definite
amelioration~aggravation
amends~altercation
ample~meager
amplify~reduce
amplitude~frequency
anabolism~catabolism
analog~digital
analysis~synthesis
analytic~holistic
analytical~geometrical
anatomical~physiological
anatomy~physiology
ancient~modern
angel~devil
animal~machine
animals~plants
animate~deaden
animate~inanimate
annihilation~creation
annihilation~recruitment
anode~cathode
answer~question
antagonist~agonist
antibody~antigen
anticipation~reaction
anticipatory~reactive
antidote~poison
antigen~antibody
antiparticle~particle
antiphase~in-phase
antithesis~thesis
antonym~synonym
apart~together
apartness~togetherness
apathy~ambition
a posteriori~a priori
appetite~satiation
applied~basic

a priori~a posteriori
arousal~awareness
arrive~leave
arrow~target
art~science
artist~artwork
artist~medium
artwork~artist
ascend~descend
ask~tell
asleep~awake
aspirant~mentor
assimilation~accommodation
associate~remember
association~cognition
association~disassociation
assumption~condition
assurance~doubt
asymmetry~symmetry
atheist~deist
atman~brahman
atom~void
attend~ignore
attention~inattention
attentive~distracted
attentive learning~orienting search
attraction~repulsion
attractor~repeller
audience~performance
audience~performer
audience~storyteller
augmenting~diminishing
authority~freedom
authority~novelty
automatic~controlled
autopoiesis~program
autumn~spring
avocation~occupation
avoid~meet
awake~asleep
aware~unaware
awareness~arousal
awareness~viability
back~front
background~foreground
backward~forward
bad~good

barbaric~civilized
barren~reproductive
base~acid
basic~applied
beautiful~ugly
beauty~ugliness
because~why
becoming~being
before~after
begin~end
beginning~ending
behavior~mind
behind~ahead
being~becoming
being~nothingness
belief~fact
belief~proof
believer~nonbeliever
believer~skeptic
best~worst
between~within
between subjects~within subjects
bifurcation~bistability
bifurcation~path
big~small
biology~physics
birth~abortion
birth~death
bistability~bifurcation
bistability~monostability
bit~computer
bit~it
black~white
blame~absolution
blame~forgiveness
bland~spicy
blue~orange
blue~red
blunt~sharp
body~mind
boil~freeze
bond~adhere
bond~connection
bonding~separating
book~writer
bore~excite
boring~entertaining

THE COMPLEMENTARY PAIR DICTIONARY PROTOTYPE

both/and~either/or
bottom~top
bottom up~top down
bound~free
boundary~domain
boundary motion~boundary surface
boundary surface~boundary motion
bourgeoisie~proletariat
boy~girl
brahman~atman
brain~mind
breach~accusation
break~fix
breaking~bonding
breaking~making
Buddha~mana
build~destroy
buy~sell
buying~selling
calculus~geometry
calm~ado
calm~agitate
camera~photography
campaign~candidate
candidate~campaign
care~abuse
carry through~abort
casual~disciplined
catabolism~anabolism
catch~throw
catching~throwing
cathode~anode
causality~synchronicity
cause~effect
caution~confidence
CD~CP
ceiling~floor
celestial mechanics~terrestrial mechanics
cell~organ
center~surround
central~peripheral
certainty~uncertainty
certitude~lie
CEVA~COT
challenge~retreat
chance~choice
change~form

change~persistence
chaos~order
chase~pulse
child~adult
child~parent
choice~chance
citizen~alien
city~country
civilized~barbaric
civilized~primitive
claim~abandon
claim~disclaim
clarity~ambiguity
clarity~nebulosity
classical~modern
clean~dirty
clear~ambiguous
clear~fuzzy
clear~nebulous
client~lawyer
climbing down~climbing up
climbing up~climbing down
clone~original
close~far
close~open
closing~opening
closure~openness
clothing~dressing
clumsiness~agility
cluster~disperse (randomize)
code~junk
cognition~association
cold~hot
collect~distribute
collecting~dispersing
collective~individual
collective inference~individual inference
collective variable~control parameter
collision~precession
color~luminance
color~wavelength
come~go
comedy~tragedy
coming~going
commend~admonish
communication~isolation
competence~incompetence

competence~performance
competition~cooperation
complain~compliment
complementarity~unity
complete~incomplete
completeness~consistency
complex~simple
complexity~simplicity
complicate~simplify
compliment~complain
composition~improvisation
composition~material
compositional~propositional
compression~tension
computer~bit
concave~convex
concentrate~relax
concentration~relaxation
concept~conception
conception~concept
concrete~abstract
concrete~ephemeral
condemn~absolve
condemned~executioner
condensing~sectioning
condition~assumption
conduction~insulation
conductor~insulator
confess~accuse
confession~accusation
confidence~caution
conflict~resolution
connected~disconnected
connection~bond
conscious~unconscious
consciousness~unconsciousness
consequence~action
conservative~liberal
consistency~completeness
constant~intermittent
constant~variable
constraint~freedom
construction~reduction
constructionist~instructionist
consume~eliminate
consumption~production
container~contents

content~form
content~frame
context~scenario
context~text
context-dependent laws~context-independent laws
context-independent laws~context-dependent laws
contextual~universal
contingency~necessity
continuous~discrete
contract~expand
contraries~dialectic
contrast~compare
contrast~edge
contrast~haze
control~agent
control~laws
controlled~automatic
controlled~wild
control parameter~coordination variable
control parameter~order parameter
convene~adjourn
conventional~unconventional
convergence~divergence
conversely~accordingly
convex~concave
cooking~ingredient
cool~warm
cooling~warming
cooperate~oppose
cooperation~competition
cooperativity~freedom
coordinate~dimension
coordinated~uncoordinated
coordination~coupling
copy~innovation
corner~edge
corporation~university
correct~incorrect
cortical~subcortical
COT~CEVA
counteragent~agent
country~city
couple~uncouple
coupling~coordination
coupling~element

THE COMPLEMENTARY PAIR DICTIONARY PROTOTYPE

covert~overt
CP~CD
CPU~peripheral
create~destroy
creation~annihilation
creation~destruction
creation~discovery
creation~evolution
creativity~technique
crisp~vague
criticism~acclaim
cruelty~kindness
cry~laugh
cues~optic flow
culture~nature
culture~population
curative~palliative
cure~heal
cure~prevention
curiosity~question
current~resistance
curved~straight
cycle~rest
cytosine~guanine
damned~saved
dark~light
dated~modern
daughter~son
dawn~dusk
day~night
dead~alive
deaden~animate
death~birth
death~life
debate~issue
decelerate~accelerate
decipher~encrypt
declarative~nondeclarative
declarative~procedural memory
decrease~increase
decreasing~increasing
deduction~induction
deep~shallow
defense~offense
deficiency~surplus
deficit~spending
definite~ambivalent

definite~indefinite
deist~atheist
demand~supply
democracy~dictatorship
deny~admit
dependent~independent
depth~acme
depth~altitude
depth~shallowness
descending~ascending
design~implementation
desire~need
desire~purpose
destabilization~stabilization
destroy~build
destroy~create
destruction~creation
detach~affix
determined~free
determined~random
determinism~free will
deterministic~stochastic
diachronic~synchronic
dialectic~contraries
dialogue~lecture
diastole~systole
dictatorship~democracy
die~live
different~alike
different~similar
differential~existential
differentiate~integrate
differentiation~integration
difficult~easy
diffusion~reaction
digital~analog
dimension~coordinate
diminishing~augmenting
diminishing~growing
diminishing returns~increasing returns
direct~indirect
directedness~self-organization
dirty~clean
disadvantage~advantage
disagree~agree
disagreeable~agreeable
disagreement~agreement

disassociate~associate
disassociation~association
discipline~disobedience
discipline~talent
disciplined~casual
disciplined~nondisciplined
disclaim~advertise
disclaim~claim
disconnected~connected
discord~accord
discounted~valued
discover~explore
discover~invent
discovering~inventing
discovery~exploration
discovery~implication
discovery~invention
discrete~continuous
disdain~recognize
disease~health
disinhibit~inhibit
disinterested~interested
disobedience~discipline
disobliging~accommodating
disorder~order
disperse~cluster
dispersed~massed
disregard~notice
disrespect~respect
dissimilar~similar
dissimilarity~similarity
dissipate~gather
dissipate~recycle
dissipating~recycling
dissipation~feedback
dissipation~oscillation
distension~pressure
distracted~attentive
distraction~vigilance
distribute~collect
distributed~focal
distributed~massed
distribution~accumulation
divergence~convergence
diversity~unity
divide~link
divide~unite
dividing~uniting
divination~mensuration
division~multiplication
DNA~enzyme
do~don't
do~prepare
do~say
doctor~patient
doing~saying
domain~boundary
dominate~yield
dominating~yielding
don't~do
dormant~active
dorsal~ventral
dose~response
doubt~assurance
doubt~trust
down~up
down-going~up-going
downgrade~upgrade
downturn~upturn
downward~upward
drain~fill
dream~myth
dreaming~awake
dressing~clothing
dressing~undressing
driving~vehicle
drown~swim
drunk~sober
dry~wet
dualism~monism
dull~sharp
dumb~smart
dusk~dawn
duty~honor
dwell~escape
dwell~roam
dwindle~accrue
dying~living
dynamic~static
dynamic~symbol
dynamical~informational
dynamical~intentional
dynamic patterns~pattern dynamics
dynamics~geometry

THE COMPLEMENTARY PAIR DICTIONARY PROTOTYPE

dynamics~information
dynamics~symmetry
dysfunction~function
early~late
earth~Gaia
east~west
easy~difficult
easy~hard
eating~food
ebullience~malaise
ecstasy~agony
edge~contrast
edge~corner
effect~cause
effectivity~affordance
efferent~afferent
efficacy~potency
efficacy~toxicity
efficiency~style
either~neither
either/or~both/and
electric~magnetic
electricity~magnetism
electron~positron
element~coupling
element~set
element~system
elevate~abase
eliminate~consume
embryo~womb
emergentism~reductionism
emotion~intellect
emotion~reason
empirical~formal
empirical~theoretical
empiricism~rationalism
employee~employer
employer~employee
employment~unemployment
emptiness~fullness
empty~full
encrypt~decipher
end~advent
end~begin
ending~beginning
endogenous~exogenous
enemy~friend

energy~dynamics
energy~matter
energy~time
enforcement~law
enlightenment~ignorance
enlightenment~*mokyo*
enquiry~findings
en-soi~pour-soi
entertain~entertainer
entertainer~entertain
entertaining~boring
enthusiasm~trepidation
environment~heredity
environment~organism
enzyme~DNA
ephemeral~concrete
epistemology~ethics
equilibrium~nonequilibrium
error~accuracy
error~trial
escape~dwell
esse~posse
essence~name
essence~possibility
essentialism~progressivism
establish~abrogate
established~new
estimate~measure
estimation~measurement
estimation~mensuration
eternal~transient
ethics~epistemology
Euclidean~non-Euclidean
even~odd
even~uneven
event~law
evoked~spontaneous
evolution~Intelligent Design
exam~examinee
examinee~exam
examinee~examiner
exception~rule
excess~moderation
excitation~inhibition
excitatory~inhibitory
excite~bore
exclude~admit

execute~abort
execution~planning
execution~program
executioner~condemned
exhalation~inhalation
existence~nonexistence
existential~differential
exogenous~endogenous
expand~abridge
expand~contract
expected~unexpected
experience~novelty
experienced~inexperienced
experiment~theory
explicate~implicate
explicit~implicit
explode~implode
exploration~discovery
exploration~preferences
explore~discover
explosion~implosion
extension~flexion
extensor~flexor
external~internal
extinction~adaptation
extinction~survival
extracellular~intracellular
extravagance~frugality
extrinsic~intrinsic
extroversion~introversion
extrovert~introvert
exude~absorb
fabricate~harvest
facilitation~inhibition
fact~belief
fact~fiction
failure~accomplishment
failure~success
faith~reason
fake~genuine
fall~rise
fall~spring
false~true
falsify~verify
familiar~strange
family~kin
family~society

famous~infamous
famous~unknown
far~close
far~near
fast~slow
faster~slower
father~mother
favorable~adverse
fear~hope
fear~threat
fearful~fearless
fearless~fearful
feedback~dissipation
feedback~feedforward
feedforward~feedback
feeling~thinking
feeling~touch
female~male
fertile~impotent
few~many
fiction~fact
fiction~reality
field~particle
figure~ground
fill~drain
finding~searching
findings~enquiry
finish~start
finite~infinite
fire~water
first~last
fix~break
fixed~moving
flexible~adamant
flexible~inflexible
flexible~rigid
flexion~extension
flexor~extensor
floor~ceiling
flow~map
flow~resist
flow~stasis
fluctuations~states
fluidity~viscosity
focal~distributed
focus~wander
folding~stretching

follower~leader
food~chef
food~eating
food~shelter
forbid~allow
foreground~background
forget~remember
forgiveness~blame
form~change
form~content
form~ideal
form~substance
formal~empirical
forward~backward
forward~reverse
forward modeling~inverse modeling
found~lost
fractal~integral
fragile~robust
frame~content
frame of reference~observer
free~bound
free~determined
freedom~authority
freedom~constraint
freedom~cooperativity
free will~determinism
freeze~boil
frequency~amplitude
friend~enemy
friend~foe
friend~lover
from~to
front~back
front~rear
frugality~extravagance
full~empty
fullness~emptiness
function~dysfunction
function~structure
functional information~self-organization
future~past
fuzzy~clear
Gaia~earth
gain~lose
gaining~losing
gamble~risk

game~players
gate~portal
gather~dissipate
gather~distribute
gathering~distributing
general~particular
general~specific
genotype~phenotype
genuine~fake
geometrical~analytical
geometry~calculus
geometry~dynamics
girl~boy
give~receive
give~take
global~local
go~come
go~stop
goal~task
going~coming
going~stopping
good~bad
government~politicians
graded potential~action potential
gradual~abrupt
gradualism~saltationism
grail~seeker
grasp~release
gravity~mass
gravity~radiation
green~red
ground~figure
growing~diminishing
guanine~cytosine
guest~host
guilt~innocence
habit~instinct
happiness~sadness
happiness~suffering
happy~sad
hard~difficult
hard~soft
hardware~software
harvest~fabricate
has~hasn't
hasn't~has
haze~contrast

heads~tails
heal~cure
health~disease
health~illness
healthy~ailing
healthy~sick
hearing~sound
hearing~speech
heaven~hell
heavy~light
hell~heaven
help~hinder
heredity~environment
hero~villain
hers~his
heterogeneous~homogeneous
hide~show
high~low
hill~valley
hinder~help
his~hers
history~future
holism~reductionism
holistic~analytic
homogeneity~inhomogeneity
homogeneous~heterogeneous
honor~duty
hope~fear
horizontal~vertical
host~guest
host~vector
hot~cold
housing~support
how~what
humble~proud
hunger~satiety
hunted~hunter
hunter~hunted
hurry~linger
husband~wife
hypothesis~observation
I~thou
ice~steam
idea~percept
ideal~form
idealism~materialism
ignorance~knowledge

ignore~attend
ignored~respected
illness~health
illness~wellness
illogical~logical
immanence~transcendence
immature~mature
immortal~mortal
immortality~mortality
impermeability~permeability
implementation~design
implementation~strategy
implication~discovery
implicit~explicit
implode~explode
implosion~explosion
impossibility~possibility
impossible~possible
impotent~fertile
impotent~potent
improve~aggravate
improvisation~composition
impulse~threshold
in~out
inaccessible~accessible
inaccuracy~accuracy
inaction~action
inactivate~activate
inactive~active
inactivity~activity
inadequacy~adequacy
inadequate~adequate
inadmissible~admissible
inadvisable~advisable
inanimate~animate
inattention~attention
incident~reflection
incidental~intentional
incomplete~complete
incorrect~correct
increase~decrease
increasing~decreasing
increasing returns~diminishing returns
indefinite~definite
independent~dependent
indict~acquit
indirect~direct

THE COMPLEMENTARY PAIR DICTIONARY PROTOTYPE

individual~collective
individual inference~collective inference
induction~deduction
indulgence~abstinence
industrial~agricultural
inept~adept
inexperienced~experienced
infamous~famous
infection~vector
inferior~superior
infinite~finite
inflexible~adaptable
inflexible~flexible
information~dynamics
information~energy
information~intrinsic dynamics
information~meaning
informational~dynamical
information creation~metastability
ingredients~cooking
inhalation~exhalation
inhibit~disinhibit
inhibit~facilitate
inhibited~uninhibited
inhibition~excitation
inhibition~facilitation
inhibitory~excitatory
inhomogeneity~homogeneity
injustice~justice
innate~learned
innocence~guilt
innovation~copy
innovation~stagnation
inorganic~organic
in-phase~antiphase
input~output
inside~outside
inside out~outside in
insider~outsider
instability~stability
instinct~habit
instruction~selection
instructionist~constructionist
instrument~musician
insufficiency~sufficiency
insulation~conduction
insulator~conductor

integer~real
integral~fractal
integrate~differentiate
integration~differentiation
integration~segregation
intellect~emotion
intellect~wisdom
Intelligent Design~evolution
intelligent design~physical law
intensify~abate
intentional~dynamical
intentional~incidental
intentional~serendipitous
intentional~unintentional
interested~disinterested
intermittent~constant
internal~external
interpretation~prediction
intracellular~extracellular
intrinsic~extrinsic
intrinsic dynamics~information
introversion~extroversion
introvert~extrovert
intuition~logic
invent~discovery
inventing~discovering
invention~discovery
invincibility~vulnerability
invisibility~visibility
invisible~visible
inward~outward
ions~salt
irrational~rational
irregular~periodic
irrelevance~relevance
isolation~communication
issue~debate
it~bit
it~me
jittery~smooth
join~quit
judge~disregard
junk~code
justice~injustice
kensho~koan
key~lock
kin~family

kind~mean
kindness~cruelty
kinetic energy~potential energy
knotted~untied
knowledge~ignorance
known~unknown
koan~*kensho*
labor~leisure
labor~rest
lack~abound
lack~plenitude
lamination~layer
landing~take-off
language~grammar
large~small
last~first
late~early
laugh~cry
law~enforcement
law~event
law~order
laws~control
laws~measurement
laws~rule
lawyer~client
layer~lamination
leader~follower
learned~innate
learning~memory
leave~arrive
leave~stay
leave alone~adjust
lecture~dialogue
left~right
leisure~labor
lengthen~abbreviate
lessee~lessor
lessor~lessee
lethargy~alertness
liberal~conservative
liberty~obligation
lie~certitude
lie~truth
life~death
ligand~receptor
light~dark
light~heavy

light~shadow
light~translucent
limited~absolute
linear~nonlinear
linger~hurry
link~divide
link~Web page
linked~unrelated
liquid~vessel
liquefy~solidify
listener~storyteller
live~die
living~dying
loathing~affection
local~global
lock~key
logic~intuition
logical~illogical
long~short
long-term memory~short-term memory
loosen~tighten
lose~acquire
lose~gain
lose~win
losing~winning
loss~accretion
loss~profit
lost~found
loud~quiet
love~hate
lover~friend
low~high
LTD~LTP
LTP~LTD
luminance~color
lyrics~melody
machine~animal
macro~micro
macrocosmic~microcosmic
macroscopic~microscopic
magic~magician
magician~magic
magnetic~electric
magnetism~electricity
making~breaking
malaise~ebullience
male~female

THE COMPLEMENTARY PAIR DICTIONARY PROTOTYPE

man~woman
mana~Buddha
manager~laborer
manifestation~source
many~few
many~one
map~flow
mass~gravity
massed~dispersed
massed~distributed
master~servant
mastering~slaving
material~composition
materialism~idealism
matter~energy
matter~mind
mature~immature
me~it
meager~ample
mean~kind
mean~nice
meaning~information
meaningful~meaningless
meaningless~meaningful
means~way
measure~estimate
measurement~estimation
measurement~laws
mechanicalism~organicism
mechanism~phenomenon
mechanism~vitalism
meditation~zazen
medium~artist
medium~style
meet~avoid
melody~lyrics
membrane~permeability
memory~learning
mensuration~divination
mensuration~estimation
mental~physical
mentor~aspirant
metabolism~replication
metamorphosis~transition
metastability~information creation
metastability~metastasis
metastability~multistability

metastasis~metastability
meter~rhythm
micro~macro
microcosmic~macrocosmic
microscopic~macroscopic
mild~severe
mind~behavior
mind~body
mind~brain
mind~matter
mirror~reflection
mission~objective
modality~relation
model~phenomenon
moderation~excess
modern~ancient
modern~classical
modern~dated
mokyo~enlightenment
momentum~position
monism~dualism
monitoring~qualification
monogenic~polygenic
monomer~polymer
monostability~bistability
mood~affect
moored~adrift
morphogenesis~morphogens
morphogens~morphogenesis
mortal~immortal
mortality~immortality
mother~father
motor~sensory
mountain~abyss
movement~stillness
moving~fixed
multifunctionality~equivalence
multiplication~division
multiplicity~singularity
multistability~metastability
muscle~tendon
musician~instrument
mutation~adaptation
myth~dream
nadir~zenith
naive~wise
name~essence

name~thing
narrow~wide
nation~people
nature~culture
navigator~pilot
near~far
nebulosity~clarity
necessity~contingency
need~desire
negation~affirmation
negation~reality
negative~affirmative
negative~positive
negativity~positivity
negotiation~value
neither~either
neither~nor
neuron~synapse
neurotransmitter~neuroreceptor
never~always
new~established
new~old
nice~mean
night~day
no~yes
nobody~somebody
noise~signal
noisy~quiet
nondisciplined~disciplined
none~all
nonequilibrium~equilibrium
non-Euclidean~Euclidean
nonexistence~existence
nonlinear~linear
nonspecific~specific
nor~neither
normal~parallel
normal~peculiar
normal~weird
north~south
nothing~something
nothingness~being
notice~disregard
noumenon~phenomenon
noun~verb
novel~traditional
novelty~authority

novelty~experience
novelty~repetition
now~then
object~subject
objective~mission
objective~subjective
objective~trajectory
objectivity~subjectivity
obligation~liberty
obscure~recognized
observation~hypothesis
observed~observer
observer~frame of reference
observer~observed
obstruct~aid
obtuse~acute
occupation~avocation
occupied~abandoned
odd~even
off~on
offense~defense
old~new
old~young
on~off
one~many
one~zero
onset~termination
opaqueness~transparency
open~close
opening~closing
opponent~adherent
oppose~cooperate
opposed~aligned
opposition~advocacy
optic flow~cues
optimistic~pessimistic
orange~blue
order~chaos
order~disorder
order~law
order parameter~control parameter
organ~cell
organic~inorganic
organicism~mechanicalism
organism~environment
organization~self-organization
organize~disorganize

THE COMPLEMENTARY PAIR DICTIONARY PROTOTYPE

orienting search~attentive learning
original~clone
original~replication
oscillation~dissipation
oscillation~rhythm
other~self
out~in
output~input
outside~inside
outside in~inside out
outsider~insider
outward~inward
over~under
overt~covert
pacification~agitation
pack~unpack
packing~unpacking
page~link
pain~agony
pain~pleasure
paint~painter
painter~paint
painter~painting
painting~painter
palliative~curative
panspermia~seed
parallel~normal
parallel~serial
parameter~adjust
parasympathetic~sympathetic
parent~child
part~whole
partake~abstain
particle~antiparticle
particle~field
particle~wave
particular~general
passion~reason
passive~active
passive~aggressive
past~future
path~bifurcation
patient~doctor
patient~provider
pattern~transience
pattern dynamics~dynamic patterns
peace~war

peculiar~normal
people~nation
percept~actuality
percept~idea
perception~action
performance~audience
performance~competence
performer~audience
periodic~irregular
periodic~quasiperiodic
peripheral~central
peripheral~CPU
perish~survive
permeability~impermeability
permeability~membrane
permeate~resist
permission~prohibition
persistence~change
perspective~observer
pessimistic~optimistic
phase locking~phase scattering
phase scattering~phase locking
phasic~tonic
phasing~wrapping
phenomenon~model
phenomenon~noumenon
phenotype~genotype
philosophy~religion
philosophy~science
photographer~photography
photography~camera
photography~photographer
photon~ray
physical~biological
physical~mental
physical~spiritual
physical~virtual
physical law~intelligent design
physiological~anatomical
physiology~anatomy
pilot~vehicle
planned~accidental
planning~execution
plant~seed
planter~seed
plants~animals
plasticity~stability

play~actor
play~work
pleasure~pain
pleiotropic~polymorphic
plenitude~lack
plot~story
plural~singular
plurality~unity
poison~antidote
polygenic~monogenic
polymer~monomer
polymorphic~pleiotropic
poor~rich
portal~gate
position~momentum
positive~negative
positivity~negativity
positron~electron
posse~esse
possibility~essence
possibility~impossibility
possible~actual
possible~impossible
postsynaptic~presynaptic
potassium~sodium
potency~efficacy
potent~impotent
potentiation~depression
pour-soi~en-soi
poverty~affluence
power~stamina
powerful~weak
pranayama~yoga
precession~collision
precision~accuracy
predator~prey
prediction~analysis
prediction~interpretation
preferences~exploration
prepare~do
prerogative~responsibility
presence~absence
present~absent
pressure~distension
presynaptic~postsynaptic
prevention~cure
prey~predator

primary~secondary
primitive~civilized
privacy~publicity
private~public
procedural~declarative
process~function
process~product
process~structure
product~advertise
product~process
production~consumption
professional~amateur
profit~loss
prognostication~reflection
program~autopoiesis
program~execution
progress~regress
progress~tradition
progressivism~essentialism
prohibition~permission
proletariat~bourgeoisie
promoter~repressor
proof~belief
propositional~compositional
proud~humble
prove~refute
provider~patient
public~private
publicity~privacy
pull~push
pulling~pushing
pulse~chase
pulse~flow
punishment~reward
puppeteer~puppet
puraka~rechaka
purchase~acquire
purpose~desire
push~pull
pushing~pulling
qualification~monitoring
qualitative~quantitative
quality~quantity
quantitative~qualitative
quantity~quality
quantum mechanics~relativity
quasiperiodic~periodic

question~answer
question~curiosity
quiet~loud
quiet~noisy
radiation~gravity
random~determined
rational~irrational
rationalism~empiricism
ray~photon
reaction~action
reaction~anticipation
reaction~diffusion
reactive~anticipatory
read~write
reading~writing
real~imaginary
real~integer
reality~fiction
reality~negation
rear~front
reason~emotion
reason~faith
reason~passion
reasonable~absurd
rebellion~tyranny
recall~recognition
receive~give
receiver~transmitter
receptor~agonist
receptor~ligand
receptor~transmitter
rechaka~puraka
recognition~disdain
recognition~recall
recognized~obscure
recruitment~annihilation
recycle~dissipate
recycle~synthesize
recycling~dissipating
red~blue
red~green
reduce~amplify
reduction~construction
reductionism~emergentism
reductionism~holism
reflect~absorb
reflection~absorption

reflection~incident
reflection~mirror
reflection~prognostication
reflex~synergy
refute~affirm
refute~prove
regress~progress
reign~abdicate
reject~accede
reject~accept
rejection~acceptance
relation~modality
relative~absolute
relativity~quantum mechanics
relax~concentrate
relaxation~concentration
relaxation~tension
release~grasp
release~uptake
relevance~irrelevance
relief~affliction
relief~stress
relieve~afflict
religion~philosophy
religion~science
religious~sacrilegious
remember~associate
remember~forget
repel~adhere
repeller~attractor
repetition~novelty
replication~metabolism
replication~original
repressor~promoter
reproductive~barren
repulsion~attraction
res cogitans~res extensa
research~teaching
res extensa~res cogitans
resist~acquiesce
resist~flow
resist~permeate
resist~submit
resistance~current
resolution~conflict
respected~ignored
response~dose

response~stimulus
responsibility~prerogative
responsibility~right
rest~cycle
rest~labor
retaining~altering
retention~acquisition
retreat~advance
retreat~challenge
reverence~ridicule
revolt~moderation
reward~punishment
rhythm~meter
rhythm~oscillation
rhythm~timing
rich~poor
ridicule~reverence
right~left
right~responsibility
right~wrong
rigid~flexible
rise~fall
risk~gamble
risk~security
roam~dwell
robust~fragile
rough~smooth
rule~exception
rule~laws
ruled~ruling
ruler~subject
ruling~ruled
rural~urban
sad~happy
sadness~happiness
saint~sinner
salt~ions
saltationism~gradualism
same~different
satiation~appetite
satiety~hunger
saved~damned
say~do
saying~doing
scale~size
scarcity~abundance
scatter~accumulate

scenario~context
scene~set
science~art
science~philosophy
science~religion
science~technology
script~actor
scripted~ad-lib
searching~finding
secondary~primary
sectioning~condensing
security~risk
seed~panspermia
seed~plant
seed~planter
seed~womb
seeker~grail
segregation~integration
selection~instruction
self~other
selfishness~altruism
self-organization~agency
self-organization~directedness
self-organization~functional information
self-organization~organization
selling~buying
semantics~syntax
sensitivity~habituation
sensory~motor
separating~bonding
separation~adjacency
separation~alliance
serendipitous~intentional
serial~parallel
servant~master
set~element
severe~mild
shade~sunlight
shadow~light
shallow~deep
shallowness~depth
sharp~dull
shelter~food
Shiva~Vishnu
short~long
short~tall

THE COMPLEMENTARY PAIR DICTIONARY PROTOTYPE

short-term memory~long-term memory
show~hide
sick~healthy
sickness~health
side-to-side~up-down
signal~noise
similar~dissimilar
similarity~dissimilarity
simple~complex
simplicity~complexity
simplify~complicate
sin~virtue
singular~plural
singularity~multiplicity
sink~source
sinner~saint
size~scale
skeptic~believer
slow~fast
slower~faster
small~big
small~large
smart~dumb
smell~taste
smooth~jittery
smooth~rough
sober~drunk
society~family
sodium~potassium
soft~hard
software~hardware
solid~porous
solidarity~solitary
solidify~liquefy
solitary~solidarity
solvent~adhesive
somebody~nobody
something~nothing
son~daughter
song~musician
source~manifestation
source~sink
south~north
space~time
spatial~temporal
speaker~subject
specialized~versatile

specific~general
specific~nonspecific
speech~hearing
spending~deficit
spherical angle~vergence
spicy~bland
spiritual~physical
spontaneous~evoked
spring~autumn
spring~fall
stability~instability
stability~plasticity
stabilization~destabilization
stagnation~innovation
stamina~power
start~finish
stasis~flow
state~transition
states~fluctuations
states~tendencies
static~dynamic
static~fluid
stay~leave
staying~leaving
steady~unsteady
steam~ice
stimulus~response
stochastic~deterministic
stop~go
stopping~going
story~plot
storyteller~audience
straight~curved
strange~familiar
stranger~acquaintance
strategy~implementation
strength~affinity
strength~weakness
stress~relief
stretching~folding
string~weaver
strip~adorn
stripping~adorning
strong~weak
structure~function
structure~process
student~teacher

style~efficiency
style~medium
style~technique
subcortical~cortical
subject~object
subject~ruler
subject~speaker
subjective~objective
subjectivity~objectivity
submit~resist
substance~accident
substance~form
subthreshold~threshold
subtraction~addition
success~failure
suffering~happiness
sufficiency~insufficiency
summer~winter
sunlight~shade
superior~inferior
supply~demand
support~housing
surplus~deficiency
surround~center
survival~adaptation
survival~extinction
survive~perish
swim~drown
symbionts~symbiosis
symbiosis~symbionts
symbol~dynamic
symmetry~asymmetry
symmetry~broken symmetry
symmetry~dynamics
symmetry breaking~symmetry making
symmetry making~symmetry breaking
sympathetic~parasympathetic
synapse~neuron
synchronic~diachronic
synchronicity~causality
synchrony~syncopy
syncopy~synchrony
synergy~reflex
synonym~antonym
syntax~semantics
synthesis~analysis
synthesis~recovery

synthesize~recycle
system~elements
systole~diastole
tails~heads
take~give
take-off~landing
talent~discipline
tall~short
tame~wild
target~aim
target~aiming
target~arrow
target~trajectory
task~aim
taste~goal
teacher~student
teaching~research
technique~creativity
technique~style
technology~science
tell~ask
temporal~spatial
tendencies~states
tendon~muscle
tension~compression
tension~relaxation
termination~onset
terra~ocean
terrestrial mechanics~celestial mechanics
text~context
then~now
theoretical~actual
theoretical~empirical
theory~experiment
thesis~antithesis
thick~thin
thin~thick
thing~name
thinking~feeling
thou~I
threat~fear
threshold~impulse
threshold~subthreshold
throw~catch
throwing~catching
thymine~adenine
tie~untie

THE COMPLEMENTARY PAIR DICTIONARY PROTOTYPE

tighten~loosen
time~energy
time~space
timing~rhythm
to~from
together~apart
togetherness~apartness
tolerance~intolerance
tolerant~intolerant
tonic~phasic
top~bottom
top down~bottom up
touch~feeling
toxicity~efficacy
traditional~novel
tragedy~comedy
trajectory~objective
trajectory~target
transcendence~immanence
transient~ageless
transient~eternal
transition~metamorphosis
transition~state
translucent~light
transmitter~receiver
transmitter~receptor
transparency~opaqueness
trepidation~enthusiasm
trial~error
true~false
trust~doubt
truth~lie
truthful~untruthful
tying~untying
tyranny~rebellion
ugliness~beauty
ugly~beautiful
unacceptable~acceptable
unadventurous~adventurous
unaffected~affected
unaffectionate~affectionate
unafraid~afraid
unaware~aware
uncertainty~certainty
unconscious~conscious
unconsciousness~consciousness
unconventional~conventional

uncoordinated~coordinated
uncouple~couple
under~over
undressing~dressing
unemployment~employment
uneven~even
unexpected~expected
unfamiliar~accustomed
uninhibited~inhibited
unintentional~intentional
unite~divide
uniting~dividing
unity~complementarity
unity~diversity
unity~plurality
universal~contextual
universal~variable
university~corporation
unknown~famous
unknown~known
unpack~pack
unpacking~packing
unrelated~akin
unrelated~linked
unsteady~steady
untie~tie
untruthful~truthful
untying~tying
unusual~usual
unwilling~willing
unwinding~winding
unzip~zip
up~down
up-down~side-to-side
up-going~down-going
upgrade~downgrade
uptake~release
upturn~downturn
urban~rural
usual~unusual
vague~crisp
valley~hill
value~negotiation
valued~discounted
vanquished~victor
variable~constant
variable~universal

vector~host
vector~infection
vehicle~driving
vehicle~pilot
ventral~dorsal
verb~noun
vergence~spherical angle
verify~falsify
versatile~specialized
vertical~horizontal
vessel~liquid
viability~awareness
victor~vanquished
vigilance~distraction
villain~hero
violet~yellow
virtual~actual
virtual~physical
virtue~sin
viscosity~fluidity
viscous~dilute
Vishnu~Shiva
visibility~invisibility
visible~invisible
vitalism~mechanism
void~atom
vulnerability~invincibility
wander~focus
wane~wax
war~peace
warm~cool
warming~cooling
water~fire
waterproof~absorbent
wave~particle
wax~wane
way~means
weak~powerful
weak~strong
weakness~strength
weaver~string
weaver~weaving
weaving~weaver
Web page~link
website~WWW
weird~normal
wellness~illness

west~east
wet~dry
what~how
what~where
where~what
white~black
whole~part
why~because
wide~narrow
wife~husband
wild~tame
willing~unwilling
win~lose
winding~unwinding
winter~summer
wisdom~intellect
wise~naive
withdrawal~addiction
within~between
within subjects~between subjects
woman~man
womb~embryo
womb~seed
work~play
worst~best
wrapping~phasing
write~read
writer~book
writing~reading
wrong~right
WWW~website
yang~yin
yellow~violet
yes~no
yield~dominate
yielding~adamant
yielding~dominating
yin~yang
yoga~*pranayama*
young~old
youth~age
zazen~meditation
zenith~nadir
zero~one
zip~unzip

BIBLIOGRAPHY

PREFACE

Kelso, J. A. S., Delcolle, J., and Schöner, G. (1990). Action-perception as a pattern formation process. In M. Jeannerod (ed.), *Attention and Performance XIII*. Hillsdale, NJ: Erlbaum.

Kelso, J. A. S. (2000). Principles of dynamic pattern formation and change for a science of human behavior. In L. R. Bergman, R. B. Cairns, L.-G. Nilsson, and L. Nystedt, *Developmental Science and the Holistic Approach*. Mahwah, NJ: Erlbaum.

PRELUDE

Anonymous (1958). *The Complete Letters of Vincent van Gogh, vol. III, Arles, 21 (February 1888–8 May 1889)*. London: Thames and Hudson.

Bohr, N. (1958). *Atomic Physics and Human Knowledge*. New York: John Wiley.

Emerson, R. W. (1983). *First Series: History*. In *Ralph Waldo Emerson: Essays and Lectures*. New York: Literary Classics of the United States.

Fitzgerald, F. S. (1936). *The Crack-Up. Esquire*, 5 (February 1936), 41, 64.

His Holiness XIV Dalai Lama. (1995). *The Dalai Lama's Book of Wisdom*. London: Thorsans.

Kelso, J. A. S. (2002). Design for living. Editorial, *Sun Sentinel*, January 2. Can be found at http://www.thecomplementarynature.com.

MOVEMENT 1

Aquinas, T. (1947). *Summa Theologica*. New York: Benziger Bros.

Audi, R. (ed.) (1999). *The Cambridge Dictionary of Philosophy*. Cambridge: Cambridge University Press.

Ayer, A. J., and O'Grady, J. (eds.) (1992). *A Dictionary of Philosophical Quotations*. Oxford, UK, and Cambridge, MA: Blackwell.

Blake, W. (1975). *The Marriage of Heaven and Hell*. Paris: Oxford University Press in association with Trianon Press.

Bohr, N. (1998). *The Philosophical Writings of Niels Bohr: Causality and Complementarity*. Woodbridge, CT: Oxbow Press.

Brumbaugh, R. S. (1961). *Plato on the One: The Hypotheses in the Parmenides*. New Haven: Yale University Press.

Capra, F. (2000). *The Tao of Physics*. Boston: Shambala.

Cooper, D. E. (1996). *World Philosophies: An Historical Introduction*. Oxford, UK, and Cambridge, MA: Blackwell.

Crick, F. H. C. (1966). *Of Molecules and Men*. Seattle: Washington University Press.

Diogenes Laertius (1979–1980). *Lives of the Eminent Philosophers*. R. D. Hicks (ed.). Cambridge, MA: Harvard University Press.

Einstein, A. (1950). *Out of My Later Years*. New York: Philosophical Library.

Emerson, R. W. (1983). *Second Series*. In *Ralph Waldo Emerson: Essays and Lectures* (1844). New York: Literary Classics of the United States.

Feng, G., and English, J. (1973). *Tao Te Ching*. Hampshire: Wildwood House.

Fox-Keller, E. (2000). *The Century of the Gene*. Cambridge, MA: Harvard University Press.

Frank, P. (1957). *Philosophy of Science*. Englewood Cliffs, NJ: Prentice-Hall.

Gibran, K. (1998). *The Prophet*. Penguin Arkana.

Goldman, E. (1969). *Anarchism and Other Essays*. New York: Dover.

Gould, S. J. (2003). *The Hedgehog, the Fox, and the Magister's Pox*. New York: Harmony Books.

Hawking, S. W. (1988, 1996). *A Brief History of Time*. New York: Bantam Books.

Hegel, G. W. F. (1956). *The Philosophy of History*. J. Sirbee (trans.). New York: Dover.

Hodges, A. (1983). *Alan Turing: The Enigma*. New York: Simon & Schuster.

Jung, C. G., Adler, G., and Hull, R. F. C. (1959). *Aion*. In *Collected Works of C. G. Jung*, vol. 9, part 2, R. F. C. Hull (trans.). New York: Bollingen Foundation.

Kant, I. (1998). *Critique of Pure Reason*. P. Guyer (ed.), P. Guyer and A. Wood (trans.). Cambridge: Cambridge University Press.

Lewontin, R. C. (2000). *The Triple Helix*. Cambridge, MA: Harvard University Press.

Lovelock, J. (1979). *Gaia: A New Look at Life on Earth*. Oxford: Oxford University Press.

Mackay, A. L. (1991). *A Dictionary of Scientific Quotations*. Oxford, UK, and Cambridge, MA: Blackwell.

Marx, K. (1964). *Economic and Philosophic Manuscripts of 1844.* D. J. Struik (ed.), M. Milligan (trans.). New York: International Publishers.

Maxwell, J. C. (1890). On physical lines of force. In *The Scientific Papers of James Clerk Maxwell,* I. W. D. Niven (ed.). Cambridge: Cambridge University Press.

Minkowski, H. (1956). *The World of Mathematics.* J. R. Newman (ed.). New York: Simon and Schuster.

Misner, C. W., Thorne, K. S., and Wheeler, J. A. (1973). *Gravitation.* San Francisco: W. H. Freeman.

Newton, Isaac (1947). *Isaac Newton's Mathematical Principles of Natural Philosophy and His System of the World.* A. Motte (trans.). Berkeley: University of California Press.

Nietzsche, F. (1924). *Beyond Good and Evil.* H. Zimmern (trans.). New York: Macmillan. Available online at Project Gutenberg.

Pais, A. (1982). *Subtle Is the Lord.* Oxford: Oxford University Press.

Pais, A. (1991). *Niels Bohr's Times.* Oxford: Clarendon Press.

Palmer, M. (1991). *Elements of Taoism.* Longmead, UK: Element Books.

Pascal, B. (1995). *Pensées.* A. J. Krailsheimer (trans.). London: Penguin.

Pattee, H. H. (1982). The need for complementarity in models of cognitive behavior. In W. B. Weimer and D. S. Palermo (eds.), *Cognition and the Symbolic Processes.* Hillsdale, NJ: Erlbaum.

Plato (1972). *Phaedo.* R. Hackforth (trans. and ed.). Cambridge: Cambridge University Press.

Poincaré, H. (1909). *Science and Method.* F. Maitland (trans.). London: Nelson.

Read, H., Fordham, M., and Adler, G. (eds.) (1960). *The Collected Works of C. G. Jung.* New York: Pantheon Books.

Robbins, T. (2000). *Fierce Invalids Home from Hot Climates.* New York: Bantam Books.

Rohman, C. (1999). *A World of Ideas: A Dictionary of Important Theories, Concepts, Beliefs and Thinkers.* New York: Ballantine Books.

Rössler, O. (1998). *Endophysics: The World as an Interface.* Singapore: World Scientific.

Russell, B. (1945). *A History of Western Philosophy.* New York: Simon & Schuster.

Spinoza, B. (1960). The Origin and Nature of the Emotions. In *Spinoza's Ethics.* Dent: Everyman.

Strohman, R. C. (1997). The coming Kuhnian revolution in biology. *Nature Biotechnology* 15, 194–200.

Strohman, R. C. (2002). Maneuvering in the complex path from genotype to phenotype. *Science* 296, 701–703.

Tirtha, S. B. K. (1965). *Vedic Mathematics or Sixteen Simple Mathematical Formulae from the Vedas*. Hindu Vishvavidyalaya, Sanskrit Publication Board, Banaras Hindu University, Banaras.

Watson, J. D. (1980). *The Double Helix*. New York: Norton.

Watson, J. D., and Crick, F. H. C. (1953). Genetical implications of the structure of Deoxyribonucleic Acid. *Nature,* May 30, 964–967.

Weinberg, S. (1992). *Dreams of a Final Theory*. New York: Vintage Books.

Wilson, E. O. (1998). *Consilience*. Thorndike, ME: Thorndike Press.

Zukav, G. (1980). *The Dancing Wu Li Masters*. New York: Bantam Books.

MOVEMENT 2

Akil, M., Kolachna, B. S., Rothmond, D. A., Hyde, T. M., Weinberger, D. R., and Kleinman, J. E. (2003). Catechol-O-Methyltransverase genotype and dopamine regulation in the human brain. *Journal of Neuroscience* 23, 2008–2013.

Arbib, M., Erdyi, P., and Szentagothai, J. (1997). *Neural Organization: Structure, Function and Dynamics*. Cambridge, MA: MIT Press.

Babloyantz, A. (ed.) (1991). *Self-Organization, Emerging Properties and Learning,* Series B, vol. 260, 41–62. New York: Plenum.

Basar, E. (2004). *Memory and Brain Dynamics*. Boca Raton: CRC Press.

Beckett, S. (1929). Dante...Bruno.Vico..Joyce. In *Our Exagmination Round His Factification for Incamination of Work in Progress*. Paris: Shakespeare and Company.

Beer, R. D. (1995). Computational and dynamical languages for autonomous agents. In R. F. Port and T. van Gelder (eds.), *Mind in Motion*. Cambridge, MA: MIT Press.

Blaedel, N. (1988). *Harmony and Unity: The Life of Niels Bohr*. Berlin: Springer-Verlag.

Blekhman, I. I. (1988). *Synchronization in Science and Technology*. New York: ASME Press.

Bressler, S. L., and Kelso, J. A. S. (2001). Cortical coordination dynamics and cognition. *Trends in Cognitive Sciences* 5, 26–36.

Buckley, P., and Peat, F. D. (1979). *Conversations in Physics and Biology*. Toronto: University of Toronto Press.

Buzsáki, G., and Draguhn, A. (2004). Neuronal oscillations in cortical networks. *Science* 304, 1926–1929.

Capra, F. (2000). *The Tao of Physics*. Boston: Shambhala.

Chemero, A. (2000). Anti-representationalism and the dynamical stance. *Philosophy of Science* 67, 625–647.

Churchland, P. S. (1986). *Neurophilosophy: Toward a Unified Science of the Mind-Brain.* Cambridge, MA: MIT Press.

Clark, A. (1997). *Being There: Putting Brain, Body and World Together Again.* Cambridge, MA: MIT Press.

Delbruck, M. (1986). *Mind from Matter.* Osney Mead: Blackwell Scientific.

Edelman, G. M. (2004). *Wider than the Sky.* New Haven: Yale University Press.

Edelman, G. M., and Tononi, G. (2000). *A Universe of Consciousness.* New York: Basic Books.

Eigen, M., and Winkler, R. (1983). *Laws of the Game.* New York: Harper & Row.

Elsasser, W. M. (1987). *Reflections of a Theory of Organisms.* Baltimore: Johns Hopkins University Press.

Faye, J., and Folse, H. J. (eds.) (1998). *The Philosophical Writings of Niels Bohr.* Woodbridge, CT: Oxbow Press.

Fingelkurts, An. A., and Fingelkurts, Al. A. (2004). Making complexity simpler: Multivariability and metastability in the brain. *International Journal of Neuroscience* 114, 843–862.

Freeman, W. J. (1992). Tutorial on neurobiology: From single neurons to brain chaos. *International Journal of Bifurcations and Chaos* 2, 451–482.

Friston, K. J. (1997). Transients, metastability and neural dynamics. *NeuroImage* 5, 164–171.

Friston, K. J. (2000). The labile brain: I. Neuronal transients and nonlinear coupling. *Philosophical Transactions of the Royal Society* 355, 215–236.

Gell-Mann, M. (1995). *The Quark and the Jaguar.* London: Abacus.

Goodwin, B. C. (1994). *How the Leopard Changed Its Spots.* New York: Charles Scribner.

Gould, S. J. (2003). *The Hedgehog, the Fox and the Magister's Pox.* New York: Harmony Books.

Greene, H. S. (2000). *Information Theory and Quantum Physics.* Heidelberg: Springer.

Grillner, S. (2003). The motor infrastructure: From ion channels to neuronal networks. *Nature Reviews Neuroscience* 4, 573–586.

Haken, H. (1984). *The Science of Structure: Synergetics.* New York: Van Nostrand Reinhold.

Haken, H. (1988). *Information and Self-Organization.* Berlin: Springer-Verlag.

Ho, Mae-Wan. (1993). *The Rainbow and the Worm.* Singapore: World Scientific.

Holst, E. von. (1973). *The Behavioral Physiology of Animals and Man.* Coral Gables, FL: University of Miami Press.

Hoppensteadt, F. C., and Izhikevich, E. M. (1997). *Weakly Connected Neural Networks.* New York: Springer.

Iberall, A. (1978). Physical basis for complex systems. *Collective Phenomena* 3, 9–24.

Ikegaya, Y., Aaron, G., Cossart, R., Aronaov, D., Lampl, I., Ferstner, D., and Yuste, R. (2004). Synfire chains and cortical songs: Temporal modules of cortical activity. *Science* 304, 559–564.

James, W. (1890). *The Principles of Psychology*. New York: Henry Holt.

Jirsa, V. K., and Kelso, J. A. S. (eds.) (2004). *Coordination Dynamics: Issues and Trends*, vol. 1 in Springer series in Understanding Complex Systems. Berlin-Heidelberg: Springer.

Jorgensen, E. M. (2004). Dopamine: Should I stay or should I go now? *Nature Neuroscience* 7, 1019–1021.

Katchalsky, A. K., Rowland, V., and Blumenthal, R. (1974). Dynamic patterns of brain cell assemblies. *Neurosciences Research Program Bulletin* 12.

Kauffman, S. A. (1993). *The Origins of Order*. New York: Oxford University Press.

Kauffman, S. A. (2000). *Investigations*. New York: Oxford University Press.

Kelso, J. A. S. (1991). Behavioral and neural pattern generation: The concept of Neurobehavioral Dynamical System (NBDS). In H. P. Koepchen and T. Huopaniemi (eds.), *Cardiorespiratory and Motor Coordination*. Berlin: Springer-Verlag.

Kelso, J. A. S. (1992). Coordination dynamics of human brain and behavior. *Springer Proc. in Physics* 69, 223–234.

Kelso, J. A. S. (1995). *Dynamic Patterns: The Self-Organization of Brain and Behavior*. Cambridge, MA: MIT Press.

Kelso, J. A. S., and Haken, H. (1995). New laws to be expected in the organism: Synergetics of brain and behavior. In M. Murphy and L. O'Neill (eds.), *What is Life? The Next 50 Years*. Cambridge: Cambridge University Press.

Kelso, J. A. S., Ding, M., and Schöner, G. (1992). Dynamic pattern formation: A primer. In A. B. Baskin and J. E. Mittenthal (eds.), *Principles of Organization in Organisms*. Reading, MA: Addison-Wesley. Reprinted in L. B. Smith and E. Thelen (eds.), *A Dynamic Systems Approach to Development*. Cambridge: MIT Press, 1993.

Kelso, J. A. S., Fuchs, A., and Jirsa, V. K. (1999). Traversing scales of brain and behavioral organization. I–III. In C. Uhl (ed.), *Analysis of Neurophysiological Brain Functioning*. Berlin: Springer-Verlag.

Kelso, J. A. S., Mandell, A. J., and Shlesinger, M. F. (eds.) (1988). *Dynamic Patterns in Complex Systems*. Singapore: World Scientific.

Kerner, B. S. (2004). *The Physics of Traffic*. Heidelberg: Springer.

Kopell, N. (1988). Toward a theory of modelling central pattern generators. In A. H. Cohen, S. Rossignol, and S. Grillner (eds.), *Neural Control of Rhythmic Movements in Vertebrates*. New York: Wiley.

Koestler, A. (1969). Beyond atomism and holism: The concept of the holon. In A. Koestler and J. R. Smythies (eds.), *Beyond Reductionism*. Boston: Beacon Press.

Kugler, P. N., and Turvey, M. T. (1987). *Information, Natural Law and the Self-Assembly of Rhythmic Movement*. Hillsdale, NJ: Erlbaum.

Kuhn, T. (1966). *The Structure of Scientific Revolutions*. 3rd ed. Chicago: University of Chicago Press.

Kuramoto, Y. (1984). *Chemical Oscillations, Waves, and Turbulence*. Berlin. Springer-Verlag.

Llinas, R. R. (2002). *I of the Vortex*. Cambridge, MA: MIT Press.

Mandelbrot, B. (1982). *The Fractal Geometry of Nature*. New York: W. H. Freeman.

Miller, A. (2002). Erotica, aesthetics and Schroedinger's wave equation. In G. Farmelo (ed.), *It Must Be Beautiful*. London: Granta Books.

Moore, W. (1989). *Schrödinger, Life and Thought*. Cambridge: Cambridge University Press.

Morowitz, H. J. (1987). *Energy Flow in Biology*. Woodbridge, CT: Oxbow Press.

Nadis, S. (2002). The sight of two brains talking. *Nature* 416, 364.

Nicolis, G., and Prigogine, I. (1989). *Exploring Complexity: An Introduction*. New York: W. H. Freeman.

Nunez, P. L. (1995). *Neocortical Dynamics and Human EEG Rhythms*. New York: Oxford University Press.

Oullier, O., de Guzman, G. C., Jantzen, K. J., Lagarde, J. F., and Kelso, J. A. S. (2005). Spontaneous interpersonal synchronization. In C. Peham, W. I. Schöllhorn, and W. Verwey (eds.), *European Workshop on Movement Sciences: Mechanics-Physiology-Psychology*. Cologne: Sportverlag.

Pais, A. (1991). *Niels Bohr's Times*. Oxford: Clarendon Press.

Pattee, H. H. (1976). Physical theories of biological coordination. In M. Grene and E. Mendelsohn (eds.), *Topics in the Philosophy of Biology*, vol. 27. Boston: Reidel.

Pauling, L. (1948). Molecular architecture and the processes of life. 21st Sir Jesse Boot Foundation Lecture, University of Nottingham.

Pikovsky, A., Rosenblum, M., and Kurths, J. (2001). *Synchronization: A Universal Concept in Nonlinear Sciences*. Cambridge: Cambridge University Press.

Port, R. F., and van Gelder, T. (eds.) (1995). *Mind as Motion*. Cambridge, MA: MIT Press.

Prigogine, I. (1980). *From Being to Becoming*. San Francisco: W. H. Freeman.

Rosen, R. (1991). *Life Itself*. New York: Columbia University Press.

Schöner, G., and Kelso, J. A. S. (1988). Dynamic pattern generation in behavioral and neural systems. *Science* 239, 1513–1520. Reprinted in K. L. Kelner and D. E. Koshland, Jr. (eds.), *Molecules to Models: Advances in Neuroscience*. Washington, DC: AAAS.

Sheets-Johnstone, M. (1999). *The Primacy of Movement*. Amsterdam: John Benjamins.

Sheets-Johnstone, M. (2004). Preserving integrity against colonization. In *Phenomenology and the Cognitive Sciences* 3.3, 249–261. Kluwer Academic Publishers.

Singer, W., and Gray, C. M. (1995). Visual feature integration and the temporal correlation hypothesis. *Annual Reviews of Neuroscience* 18, 555–586.

Stewart, I., and Golubitzky, M. (1992). *Fearful Symmetry*. Oxford: Blackwell.

Strogatz, S. H. (1994). *Nonlinear Dynamics and Chaos*. Reading, MA: Addison-Wesley.

Strohman, R. (2002). Maneuvering in the complex path from genotype to phneotype. *Science* 296, 701–703.

Tanimoto, H., Heisenberg, M., and Gerber, B. (2004). Event timing turns punishment to reward. *Nature* 430, 983.

Tass, P. A. (1999). *Phase Resetting in Medicine and Biology*. Berlin: Springer-Verlag.

Thompson, E., and Varela, F. J. (2001). Radical embodiment: Neural dynamics and consciousness. *Trends in Cognitive Sciences* 5, 418–425.

Truckses, D., Garrenton, L. S., and Thorner, J. (2004). Jekyll and Hyde in the microbial world. *Science* 306, 1509–1511.

Tschacher, W., and Dauwalder, J. P. (eds.) (2003). *The Dynamical Systems Approach to Cognition: Concepts and Empirical Paradigms Based on Self-Organization, Embodiment and Coordination Dynamics*. Singapore: World Scientific.

Turvey, M. T. (1990). Coordination. *American Psychologist* 45, 938–953.

Varela, F., Lachaux, J. P., Rodriguez, E., and Martenerie, J. (2001). The brainweb: Phase synchronization and large-scale integration. *Nature Reviews Neuroscience* 2, 229–239.

von der Malsberg, C. (1995). Binding in models of perception and brain function. *Current Opinion in Neurobiology* 5, 520–526.

Wheeler, J. A. (1990). Information, physics, quantum: The search for links. In W. H. Zurek (ed.), *Complexity, Entropy and the Physics of Information*, vol. 8. Reading, MA: Perseus Books.

Wheeler, J. A. (1994). *At Home in the Universe*. Woodbury, NY: AIP Press.

Wigner, E. P. (1979). *Symmetries and Reflections*. Woodbridge, CT: Oxbow Press.

Woolf, V. (1925). *Mrs. Dalloway*. London: Hogarth Press.

Yates, F. E. (1987). *Self-Organizing Systems*. New York: Plenum Press.

MOVEMENT 3

Budbill, D. (1999). *Moment to Moment: Poems of a Mountain Recluse*. Port Townsend, WA: Copper Canyon Press.

Campbell, J. (2004). *The Hero with a Thousand Faces*. Princeton, NJ: Princeton University Press.

De Broglie, L. (1941). *L'avenir de la science*. Paris: Plon.

Doyle, R. (1999). *A Star Called Henry*. New York: Penguin Putnam.

Eliot, T. S. (2001). *Four Quartets*. London: Faber and Faber.

Evans, N. (1995). *The Horse Whisperer*. New York: Dell.

Gardner, T. S., Cantor, C. R., and Collins, J. J. (2000). Construction of a genetic toggle switch in *Eschericia coli*. *Nature* 403, 339–342.

Grossberg, S. (2000). The complementary brain: A unifying view of brain specialization and modularity. *Trends in Cognitive Sciences* 4, 233–246.

Heaney, S. (1989). *The Government of the Tongue: Selected Prose 1978–1987*. New York: Farrar, Straus and Giroux.

James, W. (1890). *The Principles of Psychology*, vol. 1. New York: Dover Publications.

Jantzen, K. J., Steinberg, F. L., and Kelso, J. A. S. (2002). Practice-dependent modulation of neural activity during human sensorimotor coordination: A functional magnetic resonance imaging study. *Neuroscience Letters* 332, 205–209.

Ji, S. (1995). Complementarism: A biology-based philosophical framework to integrate Western science and Eastern Tao. *Proc. Int. Congress of Psychotherapy*, 518–548.

Kant, I. (1987). *Critique of Judgment*. W. S. Pluhar (trans.). Indianapolis: Hackett Publishing Company.

Kelso, J. A. S. (1981). Contrasting perspectives on order and regulation in movement. In A. Baddeley and J. Long (eds.), *Attention and Performance* 9. Hillsdale, NJ: Erlbaum.

Kelso, J. A. S. (1995). *Dynamic Patterns: The Self-Organization of Brain and Behavior*. Cambridge: MIT Press. For discussions and background on the concept of intrinsic dynamics, see especially chs. 5 & 6.

Kelso, J. A. S. (2000). Principles of dynamic pattern formation and change for a science of human behavior. In L. R. Bergman, R. B. Cairns, L.-G. Nilsson, and L. Nystedt, *Developmental Science and the Holistic Approach*. Mahwah, NJ: Erlbaum.

Leonardo da Vinci (1970). *The Notebooks of Leonardo da Vinci*. J. P. Richter (ed.). New York: Dover.

Nelson, S. (2004). Hebb and anti-Hebb meet in the brainstem. *Nature Neuroscience* 7, 687–688.

Paton, Bernadette (1995). *New-Word Lexicography and the OED*. Oxford: Oxford University Press.

Plerou, V., Gopikrishnan, P., and Stanley, H. E. (2003). Econophysics: Two-phase behavior of financial markets. *Nature* 421, 130–131.

Prigogine, I. (1984/1988). Wizard of time. In *The Omni Interviews*, Pamela Weintraub (ed.). New York: Ticknor and Fields, Omni Press.

Sheets-Johnstone, M. (2002). Preserving integrity against colonization. Paper given at Conference "Perils and Promises of Interdisciplinary Research," University of Copenhagen, 6 December.

Smith, L. B., and Thelen, E. (2003). Development as a dynamic system. *Trends in Cognitive Sciences* 7, 343–348.

Sporns, O., and Edelman, G. E. (1993). Solving Bernstein's problem: A proposal for the development of coordinated movements by selection. *Child Development* 64, 960–981.

Thelen, E., and Smith, L. B. (1994). *A Dynamic Systems Approach to the Development of Cognition and Action*. Cambridge, MA: MIT Press.

Vallacher, R. R., and Novak, A. (1997). The emergence of dynamical social psychology. *Psychological Inquiry* 8, 73–99. See also Commentaries.

Velupillai, K. (2003). Economics and the complexity vision: Chimerical partners or Elysian adventurers? Discussion Paper No. 7, Dipartimento di Economica, Università degli Studi di Trento.

INDEX

~ character. *See* Squiggle (~) character

Accommodation~assimilation, 210
Action~reaction, 64
Adaptation, 168, 170
Adenine, 2, 29
Adenine~thymine, 2, 29
Affordance, 47
Agency. *See also* Awareness; Goal-directedness; Intentionality
 biological, 9, 11
 concepts of, 104
 origins of, 104
 Santa Fe Institute approach, 104
 and self-organizing coordination tendencies, 11
 as a steering factor, 199
Agency~self-organization, 218
Agent-based modeling, 104
Akil, H., 78
Akil, M., 99
Alice in Wonderland metaphor, 16
Amazeen, N., 89
Ambiguity in leadership, 52
Anabolism~catabolism, 48–49
Anaxagoras, 13
Anesthesia, 119
Antiphase, 146–147, 165–174
Antiphase~in-phase, 146
Aquinas, T., 36, 54, 59–60, 184, 249, 252
Archetypal categories, 44, 56
Aristotle, 4, 10, 36, 54, 58–59, 184, 249, 252

Attraction, 162, 172. *See also* Attractor; Repeller
Attraction~repulsion, 162, 218
Attraction sans Attracteurs principle (ASA), 218, 226
Attractor. *See also* Fixed point
 in coordination dynamics, 10
 different types, 129
 in dynamical systems, 128–130
 fixed point, 12, 126–128, 132–133, 136–137, 162–174
 "ghosts" or "remnants" of, 175
 and multistability, 166
Attractor~repeller, 128, 163
Autorelaxation models in economics, 232
Awareness. *See also* Agency; Goal-directedness; Intentionality
 and conscious~unconscious, 44
 of contraries, 17
 and coordination dynamics, 9
 of self, 11

Babloyantz, A., 88
Bacon, F., 16, 26
Balasubramaniam, R., 89
Bardy, B., 89
Başar, E., 89
Basins of attraction, 229
Bateson, E., 89
Beauvoir, S. de, 37
Beckett, S., 77, 98
Beek, P., 89, 165
Beer, R., 89

Behaviorism, 204
Being~becoming, 7
Berkeley, G., 27
Bernstein, N., 88
Between~within, 218
Bifurcation. *See also* Instability; Phase transition; Threshold
  and creation of information, 102
  as a decision-making mechanism, 221
  essentially nonlinear, 165
  and neuronal spiking, 135
  pitchfork, 137, 167, 218
  saddle node (tangent), 222
  as a selection mechanism, 170
  symmetry-breaking, 138, 166–168
Bifurcation~bistability
  of buying~selling, 233
  of CD law, 164–168
  essentially nonlinear property of coordination, 138
  example of neuronal firing, 135
  as the heart of multifunctionality, 133
  occurs at critical values, 170
Bifurcation diagrams, 125, 132, 135, 137
Bifurcation~path, 218
Big Bang~Big Crunch, 42–43
Binary categorization scheme, 45
Binary opposites, 17
Binding, 141, 147, 174. *See also* Brain~mind; Coupling
Biology, 29–30
Birth~death, 218, 232
Bistability, 10, 134, 230
Bistability~monostability, 219, 230–231
Bit, 3, 101
Bizzi, E., 89
Black holes, 28
Blake, W., 17, 18
Body~mind, 7
Bohr, N. *See also* Quantum mechanics; Wave~particle
  coat of arms, 12, 35, 61, 62
  and complementarity, 3, 35, 61–62, 71, 83, 103
  contraria sunt complementa, 7, 35
  and coordination dynamics, 249
  and CP~CD, 189
  deep intuition for TCN, 252
  and discrete~continuous, 36
  opinion of paradox, 82
  and quantum mechanics, 3, 35–37, 61–62, 82–83, 103
  and the quantum postulate, 62
  quoted, 6, 62, 82, 83, 185
  and wave~particle, 3, 36, 82–83, 252
Bootsma, R., 89
Both/and. *See also* Either/or; Reconciliation
  and Aquinas, 59–60
  and Aristotle, 58
  concepts and methods, 78
  and Descartes, 54–55
  dualistic both/and interpretation, 50–51
  interpretations, 49
  and Kant, 56–57
  mutually inclusive both/and, 57–61
  and Spinoza, 60–61
  and Switters, 17–18
  and TCN, 19
Bottom-up~top-down, 219
Boundary~domain, 211. *See also* Control parameters; Levels
Brahma, 21
Brain. *See also* Brain~behavior; Brain~mind; Coordination dynamics
  activity, 235
  areas, 202
  complementary, 241–242
  complementary nature of, 84
  coordination, 140
  integration~segregation, 11, 91, 142, 143, 148, 174–177
  metastable nature of, 10, 11, 102–103, 106, 148–149, 171–174
  multifunctional potential of, 94–95, 135, 243
  neuroplasticity of, 211
  normal mode of operation, 84
  oscillations, 145–151, 153–155
  phase transitions in, 199
  rhythms, 145
  theories of, 11, 116, 143, 178, 188
  understanding in terms of coordination dynamics, 1, 10, 11, 75, 119, 143–151
  waves, 145

INDEX

Brain~behavior, 9
Brain~mind, 6, 9. *See also* Brain; Complementary pair~coordination dynamics
  and coexistent tendencies, 75
  contrary theories of, 143, 188
  creates functional information, 180
  how it works, 142–149
  how they work together, 149–151
  and integration~segregation, 144
  metastable, 148
  "talking" to each other, 149–150
Brainstorming complementary pairs, 238, 239, 242, 244
Bressler, S., 89
Brief History of Time, A (Hawking), 28
Buchanan, J., 89
Budbill, D., 255
Buddha, 6, 8, 36, 185, 252
Bunz, H., 167
Business cycle models, 232
Buying~selling, 15, 231–234
Buzsáki, G., 89
Byblow, W., 89

Ca1~ca2
  "ca1-ism" and "ca2-ism," 50
  and the complementary pair dictionary, 246–247
  and either/or~both/and interpretation, 50–51
  and interpretation of complementary pairs, 49, 50, 246
  and order of complementary aspects, 41, 246–247
Campbell, J., 202
Capra, F., 20–21, 37–38
  dynamical metaphor, 74, 177
  quoted, 21, 74
Carello, C., 89
Carson, R., 89
Cartesian. *See* Descartes, R.; Dualism
Categorical opposites, 7
Causality, 70, 115, 190–191
CD. *See* Coordination dynamics
CD law. *See* Coordination law
CD of CP. *See* Coordination dynamics of complementary pairs

Center for Complex Systems and Brain Sciences, 208
Center manifold theory, 113
CEVA. *See* Coexisting Equally Valid Alternatives
Chaos~order, 129
Cheyne, D., 89
Chimero, A., 89
Chinese philosophy, 19–21, 37
Chomsky, N., 37
Chua, R., 89
Churchland, P., 86
Clark, A., 89
Clark, J., 89
Closed system, 42. *See also* Big Bang~Big Crunch; Systems
Coexistence, 243
Coexisting Equally Valid Alternatives (CEVA) principle, 219, 226
Coexisting Opponent Tendencies (COT) principle, 219, 226
Coexisting tendencies. *See also* Metastability
  arising in metastable regime of CD, 75, 104
  and multifunctionality, 95
Collecting complementary pairs
  by application of the CP of CD collection, 240
  by brainstorming, 238, 239, 242, 244
  via complementary pair dictionary, 226, 244–248
  by fields of interest, 238–240, 242–243
  via a general CP collection, 240–241
Collective variable, 113. *See also* Coordination variables; Order parameter; Relative phase
Collins, J., 158, 228, 230
Communication, human, 149–151
Complementarism, 243
Complementarity. *See also* Quantum mechanics
  and Bohr, 3, 37, 61
  definition of, 3, 35, 103
  of DNA double helix, 30
  and Eastern philosophy, 3, 39, 61
  and Einstein, 70–71
  of functional information, 205

Complementarity (cont.)
  and human brain, 142, 242
  as interpretation of quantum mechanics, 254
  more general form of, 3
  nature entails, 253
  Pattee's definition of, 33
  practicing, 52
  principles of, 242
  scale dependence of, 38
  of twenty-first century, 12
  and Velupillai, 234
  viewpoint of quantum mechanics, 72
Complementary
  brain, 241–242
  colors, 5
  definition of, 39
Complementary aspects. *See also* Complementary pairs; Squiggle (~) character
  benefit of reconciling, 252
  coexistence of, 253
  danger of ignoring, 71
  definition of, 7, 39, 41, 49, 74
  directionality of, 246–247
  dynamics of, 74
  every conceivable, 253
  general dynamical principle of, 10
  and Grossberg's CP collection, 242
  inextricability of, 245
  and multifunctional dynamics, 74
  natural polarization of, 53
  as tendencies and states, 73
  tied to nonlinearity, 75
"Complementary Brain, The" (Grossberg), 241–243
Complementary nature, the (TCN). *See also* Complementary pair~coordination dynamics
  and CD law, 163, 171
  and complementary pairs, 3, 39, 75
  comprehensive theory of, 74
  and coordination dynamics, 173
  and CP~CD, 238, 251
  definition of, 38–40, 251
  experienced as CPs, 251, 252, 253
  and human brain~mind, 144
  human nature inextricable from, 252–253
  inherent in human brains and human behavior, 84–85
  intrinsic features of, 243
  key questions of, 17
  known via complementary pairs, 63
  at molecular level, 244
  more comprehensive view of, 73
  no fundamental level of, 254
  and organism~environment, 47
  paradigm, 254
  pervasiveness of, 178
  philosophy~science of, 184, 250–251
  revealing essence of, 171
  scale independence of, 39
  scientific basis for, 170, 171
  scientific theory of, 39
  TCN website, 247, 249
  thinking, 243
  what it is, 1–2, 38–40, 251–255
Complementary pair collections
  and complementary pair dictionary, 226
  by fields of endeavor, 238–240, 242–243, 257–262
  and Grossberg, 242
Complementary pair~coordination dynamics (CP~CD). *See also* Complementary pairs of coordination dynamics; Coordination dynamics of complementary pairs; Paradigm; Philosophy~science
  applying, 191
  as candidate philosophy~science of TCN, 15, 188–189
  as comprehensive theory of CP, 15
  definition of, 13–16, 250
  discussion of, 254
  embraces simple~complex, 197
  as interpretational extension of CD, 255
  metastable regime of, 251
  mind-set, 193, 216
  more than CP of CD plus CD of CP, 189, 249–251
  more than metaphor, 249
  nonlinear dynamics of, 253
  obeys the complementary nature, 253
  obstacles to pursuing, 190
  paradigm, 240
  philosophy~science of, 191, 250

INDEX

possible outcomes of, 197
prospect of yielding new facts, 190
putting to work, 14, 192–193, 236
as way out of self-reference conundrums, 13
Complementary pair dictionary
  basic layout of, 248–249
  call for complementary pairs, 249
  concept of, 240–241
  definition, 244–251
  prototype, 263–286
  small example of, 247–248
  why complementary pairs are entered twice, 247–248
Complementary pairs (CP)
  ambiguity of, 73
  archetypal, 44
  birth of, 162–164
  of brain processes, 242
  brainstorming for, 238–239, 241–242
  and coordination dynamics, 14, 76
  coordination dynamics of, 193
  definition of, 7, 39, 249
  demonstrating "real," 81, 90, 140, 168, 231
  as dynamical, 8–9, 63, 72–75, 76
  dynamics of, 7–9
  in economics, 231–233
  examples of, 42–49, 71
  by fields, 238–240, 242–244, 257–262
  interpretations of, 12, 49–61, 63, 73, 80
  language of, 242
  more than metaphor, 249
  multifunctional dynamics of, 80–81
  nucleotides, 2–3, 29
  number of, 248
  possible archetypal forms, 248
  and relative coordination, 141
  scientific basis of, 2, 3, 12, 122
  scientific reconciliation of, 63–72
  and squiggle character (~), 7, 40–41, 62, 187, 239, 245–247, 249
  as subject~object of research, 62, 241–243
  syntax for, 7, 13, 40–41, 191
  taxonomy of, 217, 248
  translated into complementary brain designs, 242

ubiquity of, 5–6, 17–34, 62, 63, 193
ways~means of discovering, 236–249
as windows into TCN, 39
Complementary pairs of coordination dynamics (CP of CD). *See also* CP of CD collection
  definition, 14, 193, 251
  as didactic tool, 193
  examples of, 134
  immediate use of, 216
  learning paradigm, 14, 209–211
  role of, 138
  strategy, 193–217, 236, 249, 254
  as upgrade of coordination dynamics, 194
  used to advance science of learning, 203–209
Complexity vision, 233–234
Complex~simple, 195–197
Complex systems, 201
Confucianism, 20
Consciousness, 44, 61
Conscious~unconscious
  and awareness, 44
  and Jung, 43–45
Conservation laws, 69
Construction~reduction, 109–110, 202
Context
  in evaluation of dynamical concepts, 133
  and laws of coordination, 9
  and understanding levels, 201–202
Continuous~discontinuous, 82
Contraria sunt complementa, 7, 35, 68
Contraries
  as coexistent and inextricable, 2
  as complementary, 2, 6–7, 12, 35–38, 61, 62, 63
  different names for, 1, 7, 17, 39
  of human awareness, 1
  impact of, 6, 34
  interpretation of, 6–7, 72, 251
  reconciling, 2, 18
  ubiquity of, 1, 5–6, 33–34
Control parameters (**cp**). *See also* Collective variable; Coordination variables; Haken-Kelso-Bunz model; Levels; Order parameter

Control parameters (**cp**) (cont.)
  in CD law, 156–157, 160–161
  of coordination dynamics, 124–127, 134–137
  and coordination variables, 160
  definition of, 116
  determining relevant, 117, 139
  and general anesthesia, 119
  inducers as, 229–230
  and nonlinear dynamics, 124
  nonspecificity~specificity of, 117–118
  and pharmacology, 118, 119
  what qualifies as, 199
Control parameters~coordination variables, 116, 219
Convergence~divergence
  as complementary pair, 10
  in coordination dynamics, 219
  of dynamical trajectories, 12, 124
  and HKB model, 215
  of metastable regime of coordination law, 173
  in rhythmical~discrete movement, 212–215
Cooperation~competition
  as complementary pair, 2, 46, 240
  in coordination dynamics, 12, 120, 219
  in CP of CD learning paradigm, 211
  fluid mechanics example, 120
Coordination. *See also* Coordination law; Relative coordination
  absolute~relative, 141–142
  and coupling, 170
  deep problem of, 85–88, 195
  dynamic laws of, 87, 156–174
  and "homunculi," 86, 92
  of living things, 87
  new functional states arise in, 93
  related terms, 86
  science of, 110
  self-organizing, 92
  and von Holst, 141
  within and between levels, 90
Coordination dynamics (CD). *See also* Complementary pair~coordination dynamics; Complementary pairs of coordination dynamics; Coordination;

Coordination dynamics of complementary pairs
  advancing, 15, 193
  aims of, 91, 195
  and anesthesia, 119
  challenges of, 139
  comparison with quantum mechanics, 103
  and complementary pairs, 2, 11, 13, 14, 76, 81, 92–110, 111, 180
  conditions~assumptions of, 151–156
  contributors to, 88–89
  and creation of functional information, 91
  definition of, 8, 90–92
  in different fields and levels, 89, 227, 254
  and embodied cognition, 89
  exhibits coexisting tendencies, 76
  and fluctuations, 121
  and gene regulation, 229
  historical roots of, 88–89, 250
  of human brain, 3, 144, 146
  as informational, 91
  language of, 110, 227
  and learning, 208
  and levels of description, 118
  and life, 121–122
  main ideas of, 92–111
  matter independence of, 100
  metastable, 10, 11, 75, 103, 148, 174
  and multifunctionality, 81
  multilevel scheme of, 108–110, 114, 116
  nonlinear features of, 166
  paradigm of, 81, 156, 194
  principles of
    Attraction sans Attracteurs (ASA), 218, 226
    Coexisting Equally Valid Alternatives (CEVA), 219, 226
    Coexisting Opponent Tendencies (COT), 219, 226
    the In-Between (PIB), xiii, 171
    Relative Levels (PRL), 222
    Selection via Instability (SVI), 221, 224, 226
  quantitative consequences in, 12, 201, 203
  and tendencies, 46, 91
  visualizing, 131
  ways to observe, 125, 135

INDEX

Coordination dynamics of complementary pairs (CD of CP). *See also* Complementary pair~coordination dynamics; Complementary pairs of coordination dynamics
  assumptions of, 227
  definition of, 193, 225, 250
  as different way to apply CD, 15
  in economics, 231–234
  strategy, 15, 225–238, 250, 254
Coordination law (CD law)
  archetypal form of, 156
  broken symmetry version of, 166, 170
  conditions~assumptions of, 151–156
  description of, 156–174
  general equation of, 156
  metastable regime of, 171–174
  prerequisites of, 139–140
  support for, 156
  visualizing, 158–174
Coordination patterns. *See also* Dynamic patterns
  in-phase, 166
  and levels, 114
  reciprocal causality of, 115
Coordination variable~control parameter
  and abrupt changes, 200
  complementary nature of, 199
  in coordination dynamics, 219
  identifying relevant, 199, 200
Coordination variables (**cv**). *See also* Collective variable; Order parameter; Relative phase
  in CD law, 127, 131, 135–136, 156–157
  of complex systems, 232
  definition of, 113
  determining relevant, 123, 199
  in economics, 232
  and pattern dynamics, 122, 124–125
  payoff from discovering, 199
Copenhagen interpretation of quantum mechanics, 3, 35, 83, 178, 252. *See also* Light; Wave~particle
Correspondence principle, 148
COT. *See* Coexisting Opponent Tendencies
Coupling. *See also* Coordination law; Oscillation~rhythm
  and coordination, 162
  mechanisms of, 155
  between neural assemblies, 144–145
  and neural oscillations, 145–146
  nonlinear, 156
  parameter in CD law, 159, 161–171, 173
  strength, 157, 166–170
Coupling~components, 220
Coupling~uncoupling, 149, 155–156
CP. *See* Complementary pairs
CP~CD. *See* Complementary pair~coordination dynamics
CP of CD. *See* Complementary pairs of coordination dynamics
CP of CD collection (base set)
  current, 217–225
  defined, 215–217
  as example of CP by fields, 238–240
  general, 241–242
  recasting CD as, 236–237
  as tool, 217, 249
Creation~annihilation, 146, 220
Creation~destruction
  of functional information, 252
  in Hinduism, 21
Crick, F., 30, 37, 141
Critical fluctuations, 223
Critical slowing down, 201, 223
cv-dot, 130–131, 135, 137, 156–157
Cycle~rest, 154, 232
Cycles, 153–154
Cytosine, 3

Daffertshofer, A., 88
Dalai Lama, 16
Dancing Wu Li Masters, The (Zukav), 21, 37–38
Darwin, C., 28, 30, 36. *See also* Evolutionary biology
De Broglie, L., 183
De Guzman, G., 88
$\delta\omega$, 161, 162–164, 166, 169, 171. *See also* Heterogeneity
  in CD law, 162–164, 166, 169, 171
  definition, 161
Denmark, 35

Descartes, R. *See also* Dualism; Either/or
  Cartesian partition, 54, 72
  and *Cogito ergo sum*, 54
  and coordination dynamics, 251
  dualism, 8, 51, 54–55
  hypothesis over observation, 26
  and Newton, 65
  quoted, 54
  "radical doubt" of, 54
  as a rationalist, 27
  and *res cogitans/res extensa*, 34, 54
  and Scholasticism, 54
  and universal mechanism metaphor, 55
Deterministic~stochastic, 157, 220
Deus ex machina, 92
Dialectic, 22–24, 58
Dialectical materialism, 26
Dichotomies, 7, 17, 94, 193
Dichotomizing, 81
  East and West, 23
  and human cognition, 5
Differentiation~integration, 65
Ding, M., 88
Dirac, P., 36, 102, 217
Discovery~implication, 71–72
"Discovery science," 79
Discrete~continuous, 65, 220
Disraeli, B., 209
Distribution of demand, 233
Ditzinger, T., 88
DNA
  and absolute~relative coordination, 141
  complementary base pairings of, 2–3
  double helix of, 3, 29–30, 141
  and organism~environment, 47
DNA~protein, 47
Double-aspect theory, 58
Double helix, 3, 29, 30, 141
Doyle, R., 203, 207
Dualism
  of Descartes, 51, 54–55
  of Kant, 51, 57
  mind/matter, 54–55
  and mutually exclusive either/or, 53
Duals, 17. *See also* Complementary pairs; Contraries
Dwell~escape, 220

Dwell times, 177, 223
Dynamical
  behavior, visualizing, 131
  instability, 102
  landscape, 164
  laws, 123, 199
  metaphors, 72–75
  science, 72
Dynamical systems. *See also* Coordination dynamics
  and coordination variables, 199
  description of, 123–124
  elementary concepts of, 122–139
  informationally coupled, xii, 91, 97
  parameter space of, 137
  self-organizing, 8
  varieties of, 124
Dynamic instability, 10, 97, 121, 167, 200, 208
Dynamic patterns. *See also* Coordination dynamics; Pattern dynamics; Self-organization
  in economy, 233
  and functional equivalence, 94–95
  informationally based, 97
  intrinsic nonlinearity of, 97
  and multifunctionality, 94–95
Dynamic Patterns (Kelso)
  emergent properties, 80
  "metastable mind," 148
  organism~environment, 47
  "stream of consciousness," 146
  supporting literature, 195
Dynamic patterns~pattern dynamics, 203, 220
Dynamics. *See also* Intrinsic dynamics
  economic, 233
  of genetic toggle switch, 229–230
  intrinsic, 105, 108, 204, 206
  multifunctional, 63

$E = mc^2$, 69, 196
East and west, 23
Eastern philosophy
  and complementarity, 3
  and Heraclitus, 57
  and quantum mechanics, 20

Economics
  of buying~selling, 231–234
  demand, 233, 234, 235
  dynamics, 232
  supply~demand, 226
Economy
  factors affecting, 232
  self-organizing, 234
Econophysics, 233
Edelman, G. M., 35, 89, 141, 155, 174, 205
Education. *See* Learning
Einstein, A.
  and complementarity, 70–71
  and energy~matter, 2, 36, 69
  general theory of relativity, 70
  and gravity~radiation, 69–70
  and hypothesis~deduction, 67
  and photoelectric effect, 68
  quoted, 67, 69, 70
  special theory of relativity, 69
  and time~space, 2, 36, 68–69, 252
  and wave~particle, 68
Either/or. *See also* Both/and; Dichotomizing
  and interpretation of CP, 75
  mind-set, 6, 7, 18–19, 24, 33, 44, 163, 191
  mutually exclusive, 49, 50, 51
  overcoming, 78–79
  in science, 75, 78–79, 190
  why it seems so fundamental, 1
Either/or~both/and
  and Bohr, 82
  and classical scientific mind-set, 72–73
  and definition of complementary pair, 49
  interpretation of complementary pairs, 51
  paradigm shift, 78
  reconciliation of, 49, 51
  science of, 177
  and Switters, 18
Eldredge, N., 28
Electricity~magnetism, 66–67, 252
Elementary particles, 83
Eliot, T. S., 254
Elsasser, W., 152
Emergentism, 80, 220
Emergentism~reductionism, 202

Emerson, R. W., 8, 45
Emit~absorb, 68
Empiricism, 27, 56
Energy-mass relation, 69
Energy~matter
  as complementary pair, 252
  and Einstein, 68–69, 252
Enhancement of fluctuations, 201, 233
Entropy, 42
Epigenetics
  and organism~environment, 47
  and Waddington, 30
Equation of CD law, 156
Equations of motion, 199
Ermentrout, B., 88
Evans, N., 203
Evolutionary biology, 27, 28–29, 30. *See also* DNA; Genes; Gradualism~saltationism; Molecular biology
Excitation~inhibition, 145
Experience~induction, 67

$F = ma$, 64, 196
Faith~reason, 37, 59–60
Feldman, A., 89
Feynman, R., 37, 180, 185, 196
Fields of endeavor
  boundaries between, 4, 189
  collecting CP by, 238–240, 257–262
Financial market, 232–233
Finger movement patterns, 161, 236
"Fingers do the walking" idea, 235–236
Fink, P., 89
First~third-person phenomenology, 235–237
Fitzgerald, F. S., 1, 189
Fixed point. *See also* Attractor; Bifurcation; Potential landscapes
  attractors, 12, 126–128, 132–133, 136–137, 162–174
  in CD law, 162–174
  and coordination dynamics, 12
  disappearance of, 171
  in pattern dynamics, 136–137
  and "potential landscapes," 126
  skateboarder analogy of, 126–128
  stable and unstable, 127

Fluctuation~determination, 132–133
Fluctuations
  and CD law equation (F), 157
  in coordination dynamics, 121
  and instability, 96
  noisy, 156, 158
  "probe" system stability, 133
  and recruitment~annihilation, 97
Fluctuations~states, 221
Foo, P., 89
Force~motion, 64
Foreground~background, 36, 226
Fowler, C., 89
Frank, T., 88
Frederiksborg castle, 35, 61
Freeman, W., 89
Friedrich, R., 88
Friston, K., 89
Fuchs, A., 88, 125, 135, 158
Fuller, R., 180
Functional equivalence, 94–95
Functional information
  as aspect of coordination dynamics, 198
  conveyed by different means, 151
  as a coupling mechanism, 155
  and CP~CD, 251
  definition of, 9, 98
  example of, in molecular biology, 99
  flows, 97
  importance of context, 99
  and language, 98
  modifies tendencies, 106–108
  and natural selection, 98
  origin of, 100
  stabilizes~destabilizes coordination, 106–107
Functional information~self-organization, 120, 221

Gaia hypothesis, 42–46
Gait transition, 228, 235–236
Galilei, G., 36, 65
Gandhi, M., 22, 36, 185–186, 252
Gell-Mann, M., 37, 195, 196
Gene regulation. *See also* Genes; Genetic; Molecular biology

  and coordination dynamics, 229
  and promotor~repressor, 15, 228–231
  regulatory networks, 97
Genes
  bias toward primary DNA sequences, 47
  and epigenetic interactions, 30
  networks, constructing, 228
  theory and paradigm of, 31
  transcription, 228
  turning on and off, 228
  and Watson and Crick, 30
Genetic. *See* Epigenetics; Gene regulation; Molecular biology
  determinism, 31, 71
  epigenetic factors, 47
  "instructions," 198
  material, 29–30
  switches, 198
  toggle switch, 228–231
Genotype~phenotype, 207
Geometrical~dynamical, 230
Geometry~dynamics, 65
Gibran, K., 38
Gibson, J., 36, 47
Glashow, S., 184
Glycolytic cycle, 153–154
Goal-directedness, 197–199. *See also* Agency; Intentionality
Gödel, Escher, Bach (Hofstadter), 253
Gogh, V. van, 5
Golden mean, 58
Goldman, E., 33
Golubitzky, M., 88
Gonzalez, E., 89
Goodman, D., 89
Goodwin, B., 32
Gould, S. J., 28, 37, 56
Gradual~abrupt change, 211
Gradualism, 28
Gradualism~saltationism, 37
Gravity~radiation, 69–70
Gray, C., 89
Greene, H., 102
Grillner, S., 89
Grossberg, S., 241–243
Guanine, 3
Guanine~cytosine, 29

Haken, H.
  contribution to coordination dynamics, 88
  and HKB model, 168
  and information processing, 175
  and order parameter~control parameter, 37
  quoted, 113
  and "slaving principle" (center manifold theorem), 113
  and synergetics, 106, 113
  and "universal nature," 158
Haken-Kelso-Bunz (HKB) model
  assumptions of, 167
  behaviors accounted for by, 167
  and convergence~divergence, 215
  and discrete behaviors, 213
Hamburger, V., 88
Haselager, P., 89
Hawking, S., 27–28. *See also* Big Bang~Big Crunch
Heaney, S., 212, 214–215
Hegel, G. W. F., 23, 24–25, 26, 36, 80
Heisenberg, W., 35, 36, 81–82
Heraclitus, 36, 38, 57
  quoted, 19, 57, 74
Heterogeneity, 152–153, 156. *See also* δω
Heterogeneity~homogeneity, 152–153
Hinduism, 21
Hippocampal formation, 145
Hitler, A., 52
Hodges, A., 33
Hofstadter, D., 253
Holistic theory, 11
Holmes, P., 88
Holroyd, T., 89
Holst, E. von, 88, 139–141, 177
Holt, K., 89
Homogeneous~heterogeneous, 156, 221
Homunculi, 86, 92
Human
  computer, 31
  existence, 253
  memory, 205
  mind, 31
  nature, 13, 52, 253
Human beings, 115, 252–253
Human brains. *See also* Brain; Brain~mind
  and coordination dynamics, 3
  and metastable regime, 174
  normal mode of operation, 84
  and Turing, 32
  two main theories of, 11
Hume, D., 27, 67
Hylomorphism. *See* Double-aspect theory
Hypotheses non fingo, 26
Hysteresis, 167, 244. *See also* Multistability; Nonlinearity

Iberall, A.
  contribution to coordination dynamics, 89
  and driving~dissipation, 120
  quoted, 118, 152
  theory of homeokinetics, 106
Ideational phase transition, 235–236
In-Between, Principle of the (PIB), xiii, 171
Individual~collective
  as complementary pair, 111
  and construction~reduction, 109
  as contraries, 2, 45–46
  and coordination dynamics, 112–113
  in CP of CD collection, 221
  as example of CP directionality, 246–247
  integration~segregation of, 176
Individual coordinating elements
  autonomy and collectivity of, 12
  in CD law, 160–168, 172–174
  collective behavior of, 112
  coupling of, 109, 176, 212
  defining dynamics of, 109
  and heterogeneity~homogeneity, 152–153, 159
  and integration~segregation, 177
  intrinsic differences between, 157
  maximum possible phase difference, 170
  measuring coupling strength of, 12
  in metastable regime, 103
  and problem of coordination, 87
  self-organization of, 93
  in traffic jams, 114–115
Individual inference~group inference, 210
Inducers as control parameters, 229–230
I-ness, 106. *See also* Agency; Awareness; Self; Self~other

Information. *See also* Coupling; Functional information; Metastability; Metastable; Quantum mechanics
  age, 34
  complementary types of, 243
  compressor of, 175
  creation of, 102–103, 174
  creation~destruction of, 147, 174–176
  environmental, 245
  functional, 97–101
  mechanisms of information processing, 174–176
  multiple meanings of, 97–98
Informational flooding, 142–143
Information~intrinsic dynamics, 221
Information sciences, 98
Innate patterns of behavior, 205
In-phase, 145, 146, 166–173, 212
In-phase~antiphase, 146–147
Instability. *See also* Cooperation~competition; Coordination dynamics; Dynamic instability; Fluctuations; Learning
  dynamic, 97, 121, 167
  and pattern dynamics, 96, 130–132
Integration~segregation
  in the brain, 11, 148
  as complementary pair, 46, 222
  as contraries, 2
  explaining, 140
  and individual coordinating elements, 177
  and James, 176
  and metastable coordination dynamics, 11, 91
  and statistical measures of the brain, 174
Intention, 107
Intentionality, 197–199. *See also* Agency; Goal-directedness; Planning~execution
Internet, 123, 239, 240, 246
Intrinsic dynamics. *See also* Information; Innate patterns of behavior; Learning
  adaptations of, 205
  benefits of knowing, 108
  definition of, 105
  and development of I-ness, 105
  and learning, 204
  methods to identify, 206
  modification by functional information, 204
Intrinsic dynamics~functional information, 204
Intrinsic dynamics~new information, 210
Isms, 55. *See also* Either/or

James, W.
  and consciousness, 176
  and integration~segregation, 177
  neutral monism, 60–61
  perchings and flights, 175–176
Jantzen, K., 89, 208
Jeka, J., 89
Jesus, 22, 36, 185, 252
Jirsa, V., 88, 158, 213–215
Jirsa-Kelso excitator theory, 213–215
Jorgensen, E., 95
Juarro, A., 89
Jung, C. G.
  and archetypal complementary pairs, 44
  and complementary pairs, 44
  and conscious~unconscious, 36, 148
  quoted, 43, 148

Kant, I.
  and agency, 199
  Critique of Pure Reason, 56
  and dialectic, 23
  and "formative power," 197
  and Hume, 56
  interpretation of complementary pairs, 45, 80
  phenomena/noumena dualism, 57
  pointing at coordination dynamics, 249
  quoted, 56, 197
  rationalism~empiricism, 27, 36, 56–57, 252
  and reconciliation of CP, 184–185
Katchalsky, A., 88
Katz, B., 37
Kauffman, S., 86, 198
Kay, B., 89
Keeping Together in Time (McNeill), 212

INDEX

Keijzer, F., 89
Kelso, J. A. S. *See also* Coordination dynamics; Haken-Kelso-Bunz model; Jirsa-Kelso excitator theory
  and broken symmetry version of CD, xiii
  and CD law, 158
  Dynamic Patterns, 47, 80
  and experimental ideas, 216–217
  "fingers do the walking" story, 235–236
  functional imaging work of, 208
  and "Principle of the In-Between," xiii
  quoted, 235
Kepler, J., 64
Keynes, J., 37
Kiemel, T., 88
Killeen, P., 12
Knowledge acquisition, 54, 78
Kopell, N., 88
Kugler, P., 89
Kuhn, T. S., 78, 79
Kuramoto, Y., 212
Kurtz, J., 88

Lagarde, J., 89
Language of living systems, 98, 110
Lao Tzu, 36, 249
Latash, M., 89
Laughlin, R., 80
Laws. *See also* Coordination law
  dynamical, 123
  of coordination, 8–9, 100
  of pattern generation, 95
  of planetary motion, 64
  of universal gravitation, 64, 70
Leadership, 52
Learning. *See also* Complementary pairs of coordination dynamics: learning paradigm; Intrinsic dynamics
  and CP of CD strategy, 203–209
  curve, 206
  and education policy, 209
  and intrinsic dynamics, 204
  landscape, 208
  and learners, 203–204
  paradigm, 209–211
  science of, 203, 205

Learning~memory, 222
Lee, T., 89, 206
Leibniz, G. W., 27
Leonardo da Vinci, 98, 209
Levels
  of description, 201
  every conceivable, 253–254
  micro~macro, 109, 114, 116, 201–203
  nature is complementary on all, 39
  and order~disorder, 43
  relating, 108–109, 152
  small number of interacting, 118
Life, 198
Ligand~receptor, 119
Light. *See also* Wave~particle
  behavior of, 3
  detectors, 82
  and Einstein, 68
  and Maxwell, 67
  and Newton, 65
  photons, 82
  wave or particle debate of, 82
Light/darkness dualism, 51
Linear~nonlinear, 166, 222, 232
Local~global, 143–145, 211, 222
Localization versus integration theories, 11, 143
Locke, J., 27
Lorenz, K., 205
Lovelace, A., 36
Lovelock, J., 42, 46
LTP~LTD, 211, 244

Macrocosm~microcosm, 253
Macro~micro, 109, 114, 116, 198, 201–203, 222
Mandelbrot, B., 158
Mandell, A., 89
Manicheans, 21–22, 51
Marx, K., 25–26, 36, 80, 254
Marxism, 20
Materialism, 6, 26
Mathematics
  of nonlinear dynamics, 124, 199
  unreasonable effectiveness of, 138
Matrix, The, analogy, 16
Matrix mechanics, 82

Maxwell, J. C.
  and electricity~magnetism, 2, 36, 63, 66–67, 252
  electromagnetic theory, 66–67
  equations, 63, 67
  quoted, 66
Mayer, G., 88
McCrone, J., xv, 7
McNeill, W., 212
Meade, M., 36
Measurement, 83
Mechanism, 55, 155
Mechanist mind-set, 55
Meijer, O., 89
Metabolism, 48
Metaphor
  mechanism, 55
  root, 58
  transcending, 39, 63, 77, 249
Metaphorical language, 2, 12
Metaphysical exigency, 54, 187
Metastability. *See also* Binding; Brain; Brain~mind; Complementary pair~coordination dynamics; Functional information
  and agency, 106
  coexisting tendencies of, 12
  dwell~escape, 220
  measures of, 220
  no fixed points in, 10–11
  not a state, 103
Metastability~information creation, 102–103, 222
Metastable
  brain, 180
  brain~mind, 148
  coordination dynamics, 11, 148
  quantum mechanics, 102–103
  regime of CD, 10, 12, 171–176
  tendencies, 188, 232
Meter, 83
Michaels, C., 89
Middle ground, 4, 8
Mind~body, 40, 250. *See also* Mind~matter
Mind/body problem, 51, 55, 60, 250
Mind~matter, 54–55, 61. *See also* Mind~body

Minkowski, H., 68, 196
Models, 123
Molecular biology, 29–31, 230, 238
Monism, 60–61
Monod, J., 37
Monostability, 134, 170, 230
Morowitz, H., 153–155, 224
Morphogenesis, 32, 33, 198
Morphogens, 32
Multifunctional. *See also* Bistability; Multistability
  dynamics, 11
  living things as, 170
  metastable coordination dynamics, 75
  oscillations, 145
  possibilities, 4
Multifunctional dynamics. *See also* Coordination dynamics; Metastability
  and the complementary nature, 11
  of complementary pairs, 63
  elucidating, 81
  of promotor~repressor toggle switch, 232
  understanding scientifically, 80–81
Multifunctionality, 95, 166, 245
Multifunctionality~functional equivalence, 222
Multiple learning rules, 243
Multistability, 12, 166
Multistability~metastability, 222
Multiverse, 253
Munhall, K., 89
Murray, J., 237–238
Mutually exclusive either/or, 50–54
Mutually inclusive both/and, 51, 57–61

Nano robots, 230
Natural selection, 30
Nature, 3, 8, 39, 80
Necker cube figures, 42, 73
Nelson, S., 244
Neorealists, 61
Neural circuitry
  and CD of the brain, 144–145
  reconfiguring itself, 95
  stability~plasticity of, 243
Neurons
  action potentials of, 135

INDEX

as basic cellular elements of information processing, 215
bistability of, 135–136
excitatory and inhibitory, 145
resting potentials of, 135
synaptic coupling of, 144, 145
Neuropharmacology, 118–119
Neurotransmitters, 118, 119, 209
Neutral monism, 60–61
Newell, K., 89
Newton, I.
 and agency, 199
 and Einstein, 67
 geometry~dynamics, 65
 Hypothesis of Light, 65
 and Kepler's laws, 64
 Principia Mathematica, 64
 quoted, 64, 72, 197
 and "self-motion," 178
 and terrestrial~celestial mechanics, 36, 64–65
 three laws of motion, 64, 70
 and Zeno's paradox, 65
Newtonian mechanics, 178
Nicolis, G., 88
Nietzsche, F., 45
Nonequilibrium market phase, 233
Non-Euclidean geometries, 196
Nonlinear
 condition~assumption of CD law, 156
 coordination dynamics, 231
 coupling, 156
 dynamics, 199
Nonlinearity, 75
Novak, A., 205
Novelty~experience, 210
Nucleotide base pairing, 2, 29
Nullclines, 230

Of Molecules and Men (Crick), 30
Open system, 30
Opposing tendencies, 20–21, 243–244. *See also* Coexisting Opponent Tendencies
Opposites
 dynamic unity of, 74
 as interdependent, 21

tension of, 121
Opposition, principle of, 23
Order~disorder, 42–43
Order of the Elephant, 35
Order parameter, 94, 113, 175, 232. *See also* Collective variable; Coordination variables; Relative phase
Organism~environment, 46–47, 223
Organisms, 32
Origin of Species (Darwin), 28
Oscillation~rhythm, 145, 153–155
Ostry, D., 89
Oullier, O., 89, 150
Oxford English Dictionary, 245
Oxygen~carbon dioxide, 49

Pairs, 2. *See also* Complementary pairs; Contraries
Pais, A., 67, 68, 69, 70
Paradigm
 analytical/reductionist, 55, 79
 CP learning, 209–211
 of coordination dynamics, 157
 shift, 78, 79, 84
 synthetic/holistic, 79
 TCN, 254
Paradox, 65, 82, 84, 90, 193
Parameters, 125, 198. *See also* Control parameters; δω
Parameter space, 137–138
Particle~wave, 27
Part~whole, 223
Pascal, B., 51
Pattee, H.
 and complementarity, 33
 contribution to CD, 89
 on coordination, 85
 quoted, 195
 and symbol~dynamics, 37, 224
Pattern dynamics. *See also* Coordination dynamics; Dynamic patterns; Self-organization
 definition, 8, 122
 and dynamic patterns, 94–97
 as equations of motion, 199
 foundations of CD, 122–139
 informationally based, 97

Pattern dynamics (cont.)
  main ideas of coordination dynamics, 94–97
  nonlinearity of, 97, 122
  observing, 125, 135
Pattern dynamics~dynamic patterns, 94
Pattern generators, 95
Pauli, W., 35, 217
Pauling, L., 155
Peper, L., 89
Perception~action, 46–47, 223
Persistence~change, 7, 76, 223
Persymphense, 93
Pharmacology, 119
Phase. *See also* Relative phase; Windshield wiper analogy
  definition of, 160
  plane trajectories, 130–131, 135–136, 159–174
  portraits, 125, 131, 138, 230
  relation, 146, 151, 162
  space, 170, 212
  synchrony, 177
  wrapping, 149, 172–173
Phase-locking
  and absolute coordination, 141
  binding, 141, 145, 147, 148
  in the brain, 141
  and dynamic linkages, 157
  oscillations, 145
  overemphasizing, 147
Phase-locking~phase-scattering, 147
Phase transition. *See also* Bifurcation
  in the brain, 199
  and creation of information, 175
  in the financial market, 233, 234
  ideational, 235
  nonequilibrium, 233
  self-organizing, 235
  of water, 117
Phenomenology
  first-person~third-person, 234–236
  mathematical, 134, 177
Phenomenon/noumenon, 51, 56, 57
Phenotype, 95
Φ, 159–160, 162–167, 169, 171, 173. *See also* Relative phase

phi-dot, 159
Philosophy of complementary pairs. *See also* Complementary pair~coordination dynamics; Philosophy~science; Reconciliation
  and attractor~repeller, 163
  and coordination dynamics, 13, 14
  and CP~CD, 13
  and CP of CD, 193
  and either/or thinking, 80
  four basic interpretations of complementary pairs, 63
  grounding in science, 10, 11, 13–14, 63, 72, 76–81, 87, 139, 172, 176, 188
  and multifunctional dynamics, 80–81
  and stability~instability, 163
  tenets of, 62–63, 191
  use of, 242
Philosophy~science
  of the complementary nature, 191, 231, 250, 254
  and CP~CD, 188, 192
  explaining itself, 251
  history of, 242
Photoelectric effect, 68
Photon, 68, 82
Piaget, J., 36, 208
Planck, M., 68
Planning~execution, 223
Plato
  and being~becoming, 24, 37
  and dialectic, 24
  and either/or, 24
  form~ideal, 36
  form versus change, 8, 24
  and one~many, 23, 185
  quoted, 23
  and Socrates, 23, 196
  and truth, 196
Pleiotropy, 95
Poincaré, H., 75
Polarization, natural process of, 186
Polarization~reconciliation, 1–5, 75, 177, 183–186
Polarized states, 10
Popper, K., 5
Position~momentum, 27

INDEX

Post-genomic era, 231
Potential landscapes, 125–126, 130, 135–136. *See also* Fixed point; Multistability; Skateboarder analogy
Preexisting capabilities, 206. *See also* Intrinsic dynamics
Preferences~exploration, 223
Prigogine, I., 37, 88, 106, 184, 187
Probability wave function of quantum mechanics, 83
Promoter~repressor, 15, 227–231, 237
Protein~DNA, 231
Punctuated equilibrium, 28–29
Pynchon, T., 228

Qualitative changes, 12, 166, 200, 201
Qualitative~quantitative, 9, 12, 200–201, 223
Quantum mechanics
 and Bohr, 3, 35–37, 62, 71, 82–83, 103
 complementarity viewpoint of, 72–73
 complexity of, 196
 Copenhagen interpretation of, 3, 35, 83, 178, 252
 dual nature of, 83
 and Eastern philosophy, 20–21, 35, 37–38
 and Einstein, 68
 and Heisenberg, 81–82
 and information creation, 101
 measurement process of, 175
 and measuring probabilities, 83
 and metastability, 102–104
 probability wave function of, 83
 and reductionism, 79
 and relativity, 28, 84
 and Schrödinger, 81–82

Rand, R., 88
Rapp, P., 88
Rate of change, 65, 130–131. *See also* **cv-dot**
Rationalism~empiricism, 27, 36, 56–57, 252
Rayleigh number, 120. *See also* Cooperation~competition
Reaction~anticipation, 223
Reaction~diffusion, 32, 37
Reciprocal
 causality, 115, 191
 connections, 144
 interactions, 150, 155
Reconciliation
 challenges of, 185–186
 of conflicting views, 4
 of CP~CD, 188
 of diametrically opposed opposites, 1
 effectiveness of, 185, 237
 of either/or~both/and, 78
 of energy~matter, 2, 36, 69
 of faith~reason, 36, 59–60, 252
 historical success of, 71
 of India and Pakistan, 186
 mastering, 72
 meaning of, 63
 of philosophy~science, 183–187
 of polarization~reconciliation, 186
 practical applications of, 185
 of rationalism~empiricism, 27, 36, 56–57, 252
 scientific, 63, 72
 of time~space, 2, 36, 68–69, 252
Recruitment~annihilation, 96, 97, 223
Reduction~construction, 224
Reductionism, 72, 79, 80, 202, 220
Reductionism~emergentism, 202, 220
Reductionism~holism, 79
Redundancy, 96
Relative coordination, 141, 148–149
Relative Levels, Principle of (PRL), 222
Relative phase, 159–160, 162–167, 169, 171, 173. *See also* Coordination variables; Φ
Relativity, 69–70, 84
Relativity~quantum mechanics, 27
Relaxation times, 133. *See also* Critical slowing down; Quantitative consequences in coordination dynamics
Repeller, 128, 137, 163. *See also* Attractor
Repulsion, 172. *See also* Attraction; Attractor; Repeller
Rhythmical~discrete CD, 212–215
Rhythmical finger movements, 161, 235–236
Rhythms, 212. *See also* Coordination dynamics; Oscillation~rhythm; Phase-locking; Relative coordination

RNA, 228
Robbins, T., 17–18
Roberton, M., 89
Rockwell, T., 89
Rosen, R., 89
Russell, B., 61

Saltzman, E., 89
Sartre, J.-P., 37
Savelsbergh, G., 89
Scale. *See* Levels
Scenario multiverse, 253
Schmidt, R. A., 89, 206
Scholasticism, 54
Scholz, J., 89
Schöner, G., 88
Schrödinger, E., 35, 81–82
Science. *See also* Coordination dynamics; Coordination dynamics of complementary pairs; Philosophy~science
 of the complementary nature, 80
 of complementary pairs, 12, 80, 180
 of coordination dynamics, 196
 definition of, 183
 of the in-between, 12
 objective of, 77
 paradigm shift in, 78
 and "reductionist prerogative," 72
 stumbling blocks of, 80
Science~religion, 36
Scientific method, 27, 78
Scientific revolution, 26–27
Selection via Instability (SVI) principle, 221, 224, 226
Self, 11, 44, 253. *See also* Agency; I-ness; Self~other
Self-assembly
 of pattern generators, 95
 of patterns and structures, 32
Self-organization. *See also* Coordination dynamics; Dynamic patterns; Functional information; Metastability; Pattern dynamics
 and coordination dynamics, 92, 93, 111–122, 198
 definition of, 11, 112
 directed, 106–108

examples of, 32, 112
functional information~self-organization, 197–199, 221
hexagon patterns in, 242
Self-organizing
 coordination dynamics, 191
 coordination tendencies, 11
 dynamical systems, 8
 processes, 245
 systems, 32, 253
Self~other, 1, 33, 253
Self-reference, 253
Selverston, A., 95
Separatrix, 230
Shakespeare, W., 36
Shamanism, 19–20
Shaw, R., 89
Sheets-Johnstone, M.
 and concepts of agency, 105
 contribution to CD, 89
 and coordination dynamics, 89
 and "fingers do the walking" experiment, 235–236
 first~third-person phenomenology, 234–236
 and ideational phase transitions, 235
 quoted, 9, 234, 235, 236
Sherrington, C., 36
 contribution to CD, 88
Simple~complex, 91, 195–197, 224
Simplicity, 195–197
Singer, W., 89
Skandas, 8
Skateboarder analogy, 126–128
Smith, L., 89, 205
Society for Neuroscience, 78
Socrates, 22–23, 36, 185, 196
Source~sink, 154, 224
Space~time, 83, 195, 224
Spatial~temporal reorganization, 211
Sperry, R., 88
Spinoza, B., 27, 60–61
Splitting world into pairs, 81, 179. *See also* Dichotomies; Dichotomizing
Sporns, O., 174, 205
Squiggle character (~). *See also* Complementary aspects; Complementary pair~

coordination dynamics; Complementary pairs; Philosophy of complementary pairs
  and avoiding either/or thinking, 40
  as a category of word association, 41
  definition of, xiv, 7, 40, 62, 245
  directionality, 41, 246–247
  meaning of, xiv, 187, 246–247, 249
  "not a bridge," 41, 187
Stability. *See also* Attractor; Bistability; Fixed point; Fluctuations; Instability; Metastability; Multistability
  different kinds of, 138
  in economics, 231–232
  measures of, 131, 220
  of system, 121
Stability~instability
  as complementary pair, 2, 9, 163
  and cooperation~competition, 121
  of coordination dynamics, 96, 224
  in CP learning paradigm, 211
  of dynamical systems, 130–131
  in economics, 232
Stabilization~destabilization, 224
Stable~unstable, 224
Stanley, G., 232
States. *See also* Tendencies
  as attractors of dynamical system, 10
  brain, 12
  mental, 12, 31, 142
  no fixed, 171
  polarized, 10
  vernacular usage of, 142
States~tendencies, 12, 142, 172, 225
Steinberg, F., 208
Sternad, D., 89
Stewart, I., 88
Strange loops, 253
Strogatz, S., 88
Strohman, R., 31, 156, 237
Structure~function, 80, 86–88, 100, 224. *See also* Levels; Self-organization
Structure of Scientific Revolutions, The (Kuhn), 79
Strutt, J. (Lord Rayleigh), 120
Subject~object, 13
Sugden, D., 89
Summers, J., 89
Supply~demand, 40, 234
Sustainable development, 46
SVI. *See* Selection via Instability
Swift, G., 85
Swinnen, S., 89
Switching, 10, 229–231, 237. *See also* Phase transition
Symbol~dynamic, 224
Symmetry~broken symmetry, xiii, 10, 225, 243
Symmetry~dynamics, 225
Symmetry~symmetry breaking, 136–137, 168. *See also* Coordination dynamics; Coordination law
Synapses, 144, 243. *See also* LTP~LTD
Synaptic connectivity, 209
Synchronization, 145–147, 150–151
Synonym~antonym, 248
Synonymic redundancy, 248
Syntax of complementary pairs, 13, 40–41, 62–63, 191–192
Systems. *See also* Dynamical systems
  closed, 42
  complex, 114
  of interest, 15
  open, 30

Taoism, 20, 37. *See also* Chinese philosophy
Tao of Physics, The (Capra), 20, 37–38, 74
Tass, P., 89
TCN. *See* Complementary nature, the
Temprado, J., 89
Tendencies. *See also* Contraries; Cooperation~competition; Individual~collective; Integration~segregation; Local~global; Metastability; Phase-locking~phase-scattering
  coexisting, xiii, 10, 104, 147, 172, 173, 222, 225, 243
  in coordination dynamics, 10, 46
  in metastable regime of coordination law, 172
  opposing tendencies, 148
  and states, in CD law, 142, 172, 173
Teuber, H. L., 88
Thalamus, 145

Thelen, E., 89, 205
Theoretical~empirical, 92
Theory~experiment of coordination dynamics, 90
Theory~practice, 13
Thesis~antithesis, 24–25, 36
Theta rhythm, 145
Thinking, 176
Thompson, E., 89
Thorner, J., 243–244
Threshold, 164, 218
Thymine, 2, 29
Tilde. *See* Squiggle character (~)
Time~space
  as complementary pair, 27, 252
  and Einstein, 68–69, 70
  and Minkowski, 69
  mutability of, 70
Tinbergen, T., 205
Tirtha, B., 21
Togetherness~apartness, 45, 172, 225
Tononi, G., 89, 174
Top-down~bottom-up, 202
Traffic jam metaphor, 114–116
Trajectories, 124–126, 130, 135, 138
Transition, 167. *See also* Bifurcation; Phase transition; Selection via Instability; Switching
Transmitter~receptor, 36
"Trapping and wrapping," 173
Treffner, P., 89
Tuller, B., 89
Tunga people, 19
Turing, A., 31–33, 37, 198
Turvey, M., 89, 165
"Two brain~minds talking" experiment, 149–151
Two-valued identifications, 52

Ubiquity
  of complementary pairs, 193
  of contraries, 1, 5–6, 17–34
  of coordination dynamics, 88–91
Ulrich, B., 89
Unified theory, 27
Universe, 42–43

Unstable. *See* Fixed point; Fluctuations; Instability; Stability

Vallacher, R., 205
Van Gelder, P., 89
Van Orden, G., 89
Variables. *See also* Collective variable; Coordination variables; Order parameter
  definition of, 122–123
  key, 8, 94
  pattern, 15
Velupillai, K., 231–234

Waddington, C., 30, 88
Wallace, S., 89
Walter, C., 89
Warren, W., 89
Watson, J., 29–31, 32
Watson, J. B., 204. *See also* Behaviorism
Wave, 82
Wave~particle
  and Bohr, 36, 252
  as complementary pair, 40, 252
  debate, 82–84
  and Einstein, 68
Ways~means, 244
Web~weaver, 253
Weinberg, H., 89
Weiss, P., 88
Weizsacker, C. F. von, 101
Western science, 8
Wheeler, J.
  and Big Bang, 42
  and complementarity, 72
  and "it from bit," 101
  and measurement, 83
  and metastability, 103
  quoted, 42, 83, 101, 192
Whitall, J., 89
Whitehead, A., 55, 61
Whole~part, 90–91
Wigner, E., 138
Wilson, E., 56, 196–197
Windshield wiper analogy, 213, 214
Wiring diagram, 95
Within~between, 90–91, 111, 218, 225

INDEX

Wolpert, L., 32
Woolf, V., 142

Yates, G., 89
Yates-Iberall conjecture, 224
Yin~yang
  on Bohr's shield, 35, 37–38, 62
  definition of, 20–21
  and Heraclitus, 57
  and Lao Tzu, 36
  level independence of, 38
  as a metaphorical complementary pair, 177

Zanone, P., 89, 208
Zeno's paradox, 65
Zoroastrianism, 21–22. *See also* Manicheans
Zukav, G., 21, 37–38